*Edited by
Satoshi Horikoshi and
Nick Serpone*

**Microwaves in
Nanoparticle Synthesis**

Related Titles

Loupy, A., de la Hoz, A. (eds.)

Microwaves in Organic Synthesis

Third, Completely Revised and Enlarged Edition

2013
ISBN: 978-3-527-33116-1

Quinten, M.

Optical Properties of Nanoparticle Systems

Mie and Beyond

2011
ISBN: 978-3-527-41043-9

Gruttadauria, M., Giacalone, F. (eds.)

Catalytic Methods in Asymmetric Synthesis

Advanced Materials, Techniques, and Applications

2011
ISBN: 978-0-470-64136-1

Gubin, S. P. (ed.)

Magnetic Nanoparticles

2009
ISBN: 978-3-527-40790-3

Amabilino, D. B. (ed.)

Chirality at the Nanoscale

Nanoparticles, Surfaces, Materials and more

2009
ISBN: 978-3-527-32013-4

Edited by Satoshi Horikoshi and Nick Serpone

Microwaves in Nanoparticle Synthesis

Fundamentals and Applications

WILEY-VCH

WILEY-VCH Verlag GmbH & Co. KGaA

The Editors

Prof. Satoshi Horikoshi
Sophia University
Dep. of Materials and
Life Sciences
7-1 Kioicho, Chiyodaku
Tokyo 102-8554
Japan

Prof. Nick Serpone
Visiting Professor
Universita di Pavia
Dipartimento di Chimica
Gruppo Fotochimico
Via Taramelli 10
Pavia 27100
Italy

The cover picture shows Scanning Electron Micrographs of γ-MnO2 synthesized for 2 h, in lower and higher magnification. Taken from Chapter 10 of this book, Figure 18, with permission.

All books published by **Wiley-VCH** are carefully produced. Nevertheless, authors, editors, and publisher do not warrant the information contained in these books, including this book, to be free of errors. Readers are advised to keep in mind that statements, data, illustrations, procedural details or other items may inadvertently be inaccurate.

Library of Congress Card No.: applied for

British Library Cataloguing-in-Publication Data
A catalogue record for this book is available from the British Library.

Bibliographic information published by the Deutsche Nationalbibliothek
The Deutsche Nationalbibliothek lists this publication in the Deutsche Nationalbibliografie; detailed bibliographic data are available on the Internet at <http://dnb.d-nb.de>.

© 2013 Wiley-VCH Verlag GmbH & Co. KGaA, Boschstr. 12, 69469 Weinheim, Germany

All rights reserved (including those of translation into other languages). No part of this book may be reproduced in any form – by photoprinting, microfilm, or any other means – nor transmitted or translated into a machine language without written permission from the publishers. Registered names, trademarks, etc. used in this book, even when not specifically marked as such, are not to be considered unprotected by law.

Print ISBN: 978-3-527-33197-0
ePDF ISBN: 978-3-527-64815-3
ePub ISBN: 978-3-527-64814-6
mobi ISBN: 978-3-527-64813-9
oBook ISBN: 978-3-527-64812-2

Cover Design Simone Benjamin, McLeese Lake, Canada

Typesetting Toppan Best-set Premedia Limited, Hong Kong

Printing and Binding Markono Print Media Pte Ltd, Singapore

Printed on acid-free paper

Contents

Preface *XI*
List of Contributors *XIII*

1 **Introduction to Nanoparticles** *1*
 Satoshi Horikoshi and Nick Serpone
1.1 General Introduction to Nanoparticles *1*
1.2 Methods of Nanoparticle Synthesis *8*
1.3 Surface Plasmon Resonance and Coloring *10*
1.4 Control of Size, Shape, and Structure *12*
1.4.1 Size Control of Nanoparticles *12*
1.4.2 Shape Control of Nanoparticles *15*
1.4.3 Structure Control of Nanoparticles *17*
1.5 Reducing Agent in Nanoparticle Synthesis *18*
1.6 Applications of Metallic Nanoparticles *19*
1.6.1 Application of Nanoparticles in Paints *20*
1.6.2 Application in Chemical Catalysis *20*
1.6.3 Application of Nanoparticles in Micro-wiring *22*
1.6.4 Application of Nanoparticles in Medical Treatments *22*
 References *23*

2 **General Features of Microwave Chemistry** *25*
 Satoshi Horikoshi and Nick Serpone
2.1 Microwave Heating *25*
2.2 Some Applications of Microwave Heating *26*
2.3 Microwave Chemistry *29*
2.3.1 Microwaves in Organic Syntheses *29*
2.3.2 Microwaves in Polymer Syntheses *30*
2.3.3 Microwaves in Inorganic Syntheses *31*
2.3.4 Microwave Extraction *32*
2.3.5 Microwave Discharge Electrodeless Lamps *32*
2.4 Microwave Chemical Reaction Equipment *33*
 References *36*

3	**Considerations of Microwave Heating** 39
	Satoshi Horikoshi and Nick Serpone
3.1	General Considerations of Microwave Heating 39
3.1.1	Electromagnetic Waves and a Dielectric Material 39
3.1.2	Heating a Substance by the Microwaves' Alternating Electric Field 40
3.1.3	Heating a Dielectric by the Microwaves' Alternating Magnetic Field 45
3.1.4	Penetration Depth of Microwaves in a Dielectric Material 45
3.1.5	Frequency Effects in Chemical Reactions 46
3.2	Peculiar Microwave Heating 47
3.2.1	Special Temperature Distribution 47
3.2.2	Superheating 49
3.2.3	Selective Heating in Chemical Reactions 50
3.3	Relevant Points of Effective Microwave Heating 52
	References 53

4	**Combined Energy Sources in the Synthesis of Nanomaterials** 55
	Luisa Boffa, Silvia Tagliapietra, and Giancarlo Cravotto
4.1	Introduction 55
4.2	Simultaneous Ultrasound/Microwave Treatments 58
4.3	Sequential Ultrasound and Microwaves 63
4.3.1	Sequential Steps of the Same Reaction 63
4.3.2	Sequential Reactions 69
4.4	Conclusions 72
	References 72

5	**Nanoparticle Synthesis through Microwave Heating** 75
	Satoshi Horikoshi and Nick Serpone
5.1	Introduction 75
5.2	Microwave Frequency Effects 76
5.2.1	Synthesis of Ag Nanoparticles through the Efficient Use of 5.8-GHz Microwaves 77
5.2.2	Metal Nanoparticle Synthesis through the Use of 915-MHz Microwaves 79
5.3	Nanoparticle Synthesis under a Microwave Magnetic Field 81
5.4	Synthesis of Metal Nanoparticles by a Greener Microwave Hydrothermal Method 84
5.5	Nanoparticle Synthesis with Microwaves under Cooling Conditions 85
5.6	Positive Aspects of Microwaves' Thermal Distribution in Nanoparticle Synthesis 87
5.7	Microwave-Assisted Nanoparticle Synthesis in Continuous Flow Apparatuses 90
5.7.1	Microwave Desktop System of Nanoparticle Synthesis in a Continuous Flow Reactor 91
5.7.2	Synthesis of Metal Nanoparticles with a Hybrid Microreactor/Microwave System 92

5.7.3	Other Examples of Continuous Microwave Nanoparticle Synthesis Equipment *94*	
5.7.4	Microwave Calcination Equipment for the Fabrication of Nanometallic Inks *95*	
5.7.5	Synthesis of Metal Nanoparticle Using Microwave Liquid Plasma *96*	
5.7.6	Compendium of Microwave-Assisted Nanoparticle Syntheses *96*	
	References *103*	

6 Microwave-Assisted Solution Synthesis of Nanomaterials *107*
Xianluo Hu and Jimmy C. Yu
6.1 Introduction *107*
6.2 Synthesis of ZnO Nanocrystals *110*
6.2.1 Synthesis of Colloidal ZnO Nanocrystals Clusters *111*
6.2.2 Controlled Growth of Basic and Complex ZnO Nanostructures *113*
6.2.3 Synthesis of ZnO Nanoparticles in Benzyl Alcohol *113*
6.3 Synthesis of α-Fe$_2$O$_3$ Nanostructures *114*
6.3.1 α-Fe$_2$O$_3$ Hollow Spheres *115*
6.3.2 Monodisperse α-Fe$_2$O$_3$ Nanocrystals with Continuous Aspect-Ratio Tuning and Precise Shape Control *116*
6.3.3 Self-Assembled Hierarchical α-Fe$_2$O$_3$ Nanoarchitectures *118*
6.4 Element-Based Nanostructures and Nanocomposite *118*
6.4.1 Silver Nanostructures *118*
6.4.2 Te Nanostructures *122*
6.4.3 Selenium/Carbon Colloids *123*
6.5 Chalcogenide Nanostructures *125*
6.5.1 Cadmium Chalcogenides *125*
6.5.2 Lead Chalcogenides *129*
6.5.3 Zinc Chalcogenides *131*
6.6 Graphene *132*
6.7 Summary *135*
References *135*

7 Precisely Controlled Synthesis of Metal Nanoparticles under Microwave Irradiation *145*
Zhi Chen, Dai Mochizuki, and Yuji Wada
7.1 Introduction *145*
7.1.1 General Introduction – Green Chemistry *145*
7.1.2 Microwave Chemistry for the Preparation of Metal Nanoparticles *147*
7.2 Precise Control of Single Component under Microwave Irradiation *152*
7.2.1 Spheres *152*
7.2.1.1 Au Nanoparticles *152*
7.2.1.2 Ag Nanoparticles *154*
7.2.1.3 Pt Nanoparticles *156*

7.2.1.4 Pd, Ru, and Rh Nanoparticles *157*
7.2.1.5 Other Transition Metals *158*
7.2.2 Nanorods and Nanowires *160*
7.2.2.1 Ag Nanorods and Nanowires *160*
7.2.2.2 Au, Pt, Ni Nanorods and Nanowires *161*
7.2.3 Other Morphologies *162*
7.2.3.1 Au *162*
7.2.3.2 Ag *163*
7.2.3.3 Pt, Pd, Ni, and Co *163*
7.3 Precise Control of Multicomponent Structures under Microwave Irradiation *164*
7.3.1 Multicomponent Nanoparticles *164*
7.3.1.1 Core–Shell Structures *164*
7.3.1.2 Alloys *168*
7.3.2 Metal Nanoparticles on Supports *170*
7.3.2.1 Metal Oxide Supports *170*
7.3.2.2 Carbon Material Supports *171*
7.3.2.3 Other Supports *176*
7.4 An Example of Mass Production Oriented to Application *178*
7.5 Conclusion *180*
References *180*

8 Microwave-Assisted Nonaqueous Routes to Metal Oxide Nanoparticles and Nanostructures *185*
Markus Niederberger

8.1 Introduction *185*
8.2 Nonaqueous Sol–Gel Chemistry *186*
8.3 Polyol Route *189*
8.4 Benzyl Alcohol Route *191*
8.5 Other Mono-Alcohols *197*
8.6 Ionic Liquids *198*
8.7 Nonaqueous Microwave Chemistry beyond Metal Oxides *199*
8.8 Summary and Outlook *201*
References *202*

9 Input of Microwaves for Nanocrystal Synthesis and Surface Functionalization Focus on Iron Oxide Nanoparticles *207*
Irena Milosevic, Erwann Guenin, Yoann Lalatonne, Farah Benyettou, Caroline de Montferrand, Frederic Geinguenaud, and Laurence Motte

9.1 Introduction *207*
9.2 Biomedical Applications of Iron Oxide Nanoparticles *208*
9.3 Nanoparticle Synthesis *211*
9.3.1 Synthesis in Aqueous Solution *211*
9.3.1.1 Coprecipitation Method *211*
9.3.1.2 Forced Hydrolysis *211*

9.3.1.3 Hydrothermal Method *212*
9.3.1.4 Aqueous Sol–Gel Method *212*
9.3.1.5 Direct Micelles Microemulsion Method *212*
9.3.2 Synthesis in Non-Aqueous Solvent *213*
9.3.2.1 Reverse Micelle Microemulsion Method *213*
9.3.2.2 Non-Aqueous Sol–Gel Method *213*
9.3.2.3 Polyol Synthesis *213*
9.3.2.4 Thermal Decomposition *214*
9.4 Nanoparticle Surface Functionalization *214*
9.4.1 Hydrophobic Nanocrystals *215*
9.4.1.1 Ligand Exchange *215*
9.4.1.2 Surface Chemical Modification *216*
9.4.1.3 Tails Interdigitation *216*
9.4.1.4 Silica or Polymer Shell *217*
9.4.2 Water Soluble Nanocrystals *217*
9.4.2.1 Direct Surface Functionalization *217*
9.4.2.2 Two-Step Surface Functionalization *219*
9.5 Microwave-Assisted Chemistry *222*
9.5.1 Microwave-Assisted Synthesis of Nanoparticles *223*
9.5.1.1 Microwave-Assisted Hydrothermal Method *223*
9.5.1.2 Microwave-Assisted Solvothermal Method *227*
9.5.2 Microwave-Assisted Functionalization of Nanoparticles *229*
9.5.2.1 Gold Nanoparticle Microwave Functionalization *230*
9.5.2.2 Iron Oxide Nanoparticle Microwave Functionalization *231*
9.5.2.3 Microwave-Assisted Silica Encapsulation of Iron Oxide Nanoparticles *234*
9.5.2.4 Europium Oxide Nanoparticle Microwave Functionalization *235*
9.6 Conclusions *236*
References *236*

10 Microwave-Assisted Continuous Synthesis of Inorganic Nanomaterials *247*
Naftali N. Opembe, Hui Huang, and Steven L. Suib
10.1 Introduction and Overview *247*
10.2 Microwave-Assisted Continuous Synthesis of Inorganic Nanomaterials *249*
10.3 Types of Microwave Apparatus Used in Continuous Synthesis *250*
10.4 Microwave Continuous Synthesis of Molecular Sieve Materials *253*
10.5 Microwave Continuous Synthesis of Metal Oxides and Mixed Metal Oxide Materials *259*
10.6 Microwave Continuous Synthesis of Metallic Nanomaterials *267*
10.7 Conclusions and Outlook *268*
References *269*

11	**Microwave Plasma Synthesis of Nanoparticles: From Theoretical Background and Experimental Realization to Nanoparticles with Special Properties** *271*
	Dorothée Vinga Szabó
11.1	Introduction *271*
11.2	Using Microwave Plasmas for Nanoparticle Synthesis *272*
11.2.1	General Comments on Plasmas *272*
11.2.2	Considerations in a Microwave Plasma *274*
11.2.3	Particle Formation *277*
11.2.4	Characterization of Nanoparticles *278*
11.3	Experimental Realization of the Microwave Plasma Synthesis *279*
11.3.1	Custom-Made Applicators *279*
11.3.2	Coated Nanoparticles and Particle Collection *280*
11.4	Influence of Experimental Parameters *282*
11.4.1	Precursor Selection *284*
11.4.2	Influence of Precursor Concentration *287*
11.4.3	Interdependence of Microwave Power, Pressure, Temperature, and Gas Velocity *288*
11.4.4	Influence of Residence Time in the Plasma on Particle Size *292*
11.4.5	Summary of Experimental Parameters *292*
11.5	Nanoparticle Properties and Application *294*
11.5.1	Ferrimagnetic Nanoparticles *294*
11.5.2	Gas-Sensing Nanoparticles *297*
11.5.3	Nanoparticles for Anodes in Li-Ion Batteries *299*
11.6	Summary *300*
	References *301*
12	**Oxidation, Purification and Functionalization of Carbon Nanotubes under Microwave Irradiation** *311*
	Davide Garella and Giancarlo Cravotto
12.1	Introduction *311*
12.2	Oxidation and Purification *313*
12.3	Functionalization *316*
12.4	Conclusion *321*
	References *321*

Index *325*

Preface

The special optical characteristics imparted by metallic nanoparticles have been used in producing colored glass ever since the 4th century AD, even though the craftsmen were unable to see the nanoparticles and thus explain the true character of metallic colloids. The first scientific evaluation of a colloid (gold) was done by Michael Faraday in 1857; he remarked that colloidal gold sols have properties different from bulk gold (Chapter 1, Table 1.2). The history of nanomaterials dates back to 1959, when Richard P. Feynman, a physicist at Cal Tech, forecasted the advent of nanomaterials. In one of his classes he stated that "there is plenty of room at the bottom" and suggested that scaling down to the nano-level and starting from the bottom-up was the key to future technologies and advances. The remarkable progress in characterizing nanoparticles and unravelling novel physical and chemical properties of nanoparticles has opened the possibility of new materials. Simple preparation methods using various techniques to produce high-quality nanoparticles are now available (Chapter 1, Figure 1.4), one of which is the use of microwave heating that has attracted considerable attention worldwide. Several books have been written mostly on microwave-assisted organic syntheses in the past decade, yet none have dealt specifically with microwaves and inorganic materials except perhaps in the use of microwave radiation in the sintering of ceramics. The latter notwithstanding, research in nanoparticle syntheses with microwaves has seen a remarkable growth in the last several years.

The main purpose of this book is to give an overview of nanoparticle synthesis using the microwave method, with the first chapter providing an introduction to nanoparticles followed by two other chapters that explain some of the fundamentals of microwave heating (Chapters 2 and 3). In the remaining chapters several specialists in the field describe some of the specifics and variations in nanoparticle synthesis. As the data available in the literature were enormous, we had to make the difficult choice of including only the most relevant and up-to-date literature; we apologize to the reader if we missed to include other worthwhile contributions. Prominent in the book are abundant chemical information and some beautiful TEM data that define the structural features of nanoparticles. We are thankful to all the contributors who have answered the call, and also to the Wiley-VCH editorial staff for their thorough and professional assistance. The data presented would not have been possible without the fruitful collaboration of many university and industrial researchers, and not least without the cooperation of students whose names appear in many of the earlier publications. We are indeed very grateful for their effort.

We hope this book becomes a starting point for researchers in other fields to become interested in pursuing microwave chemistry, in general, and microwave-assisted nanoparticle syntheses, in particular.

January 2013
Satoshi Horikoshi
Nick Serpone

List of Contributors

Farah Benyettou
UMR 7244 CNRS, University of Paris 13
CSPBAT Laboratory
74 rue Marcel Cachin, 93017 Bobigny
France

Luisa Boffa
Università di Torino
Dipartimento di Scienza e Tecnologia del Farmaco
via P. Giuria 9
10125 Torino
Italy

Zhi Chen
Tokyo Institute of Technology
Department of Applied Chemistry
Graduate School of Science and Engineering
2-12-1 Ookayama
Meguro, Tokyo 152-8552
Japan

Giancarlo Cravotto
Università di Torino
Dipartimento di Scienza e Tecnologia del Farmaco
via P. Giuria 9
10125 Torino
Italy

Caroline de Montferrand
UMR 7244 CNRS, University of Paris 13
CSPBAT Laboratory
74 rue Marcel Cachin, 93017 Bobigny
France

Davide Garella
Università di Torino
Dipartimento di Scienza e Tecnologia del Farmaco
via P. Giuria 9
10125 Torino
Italy

Frederic Geinguenaud
UMR 7244 CNRS, University of Paris 13
CSPBAT Laboratory
74 rue Marcel Cachin, 93017 Bobigny
France

Erwann Guenin
UMR 7244 CNRS, University of Paris 13
CSPBAT Laboratory
74 rue Marcel Cachin, 93017 Bobigny
France

List of Contributors

Satoshi Horikoshi
Sophia University
Faculty of Science and Technology
Department of Materials and Life Sciences
7-1 Kioicho
Chiyodaku, Tokyo 102-8554
Japan

Xianluo Hu
Huazhong University of Science and Technology
College of Materials Science and Engineering
State Key Laboratory of Material Processing and Die & Mould Technology
Wuhan 430074
China

Hui Huang
University of Connecticut
Department of Chemistry
55 North Eagleville Road
Storrs, CT 06269-3060
USA

Yoann Lalatonne
UMR 7244 CNRS, University of Paris 13
CSPBAT Laboratory
74 rue Marcel Cachin, 93017 Bobigny
France

Irena Milosevic
UMR 7244 CNRS, University of Paris 13
CSPBAT Laboratory
74 rue Marcel Cachin, 93017 Bobigny
France

Dai Mochizuki
Tokyo Institute of Technology
Department of Applied Chemistry
Graduate School of Science and Engineering
2-12-1 Ookayama
Meguro, Tokyo 152-8552
Japan

Laurence Motte
UMR 7244 CNRS, University of Paris 13
CSPBAT Laboratory
74 rue Marcel Cachin, 93017 Bobigny
France

Markus Niederberger
ETH Zürich
Laboratory for Multifunctional Materials
Department of Materials
Wolfgang-Pauli-Strasse 10
8093 Zürich
Switzerland

Naftali N. Opembe
University of Connecticut
Department of Chemistry
55 North Eagleville Road
Storrs, CT 06269-3060
USA

Nick Serpone
Universita di Pavia
Dipartimento di Chimica
Gruppo Fotochimico
Via Taramelli 10
Pavia 27100
Italy

Steven L. Suib
University of Connecticut
Department of Chemistry
55 North Eagleville Road
Storrs, CT 06269-3060
USA

Silvia Tagliapietra
Università di Torino
Dipartimento di Scienza e Tecnologia
del Farmaco
via P. Giuria 9
10125 Torino
Italy

Dorothée Vinga Szabó
Karlsruhe Institute of Technology
Institute for Applied
Materials – Materials Process
Technology
Hermann-von-Helmholtz-Platz 1
76344 Eggenstein-Leopoldshafen
Germany

Yuji Wada
Tokyo Institute of Technology
Department of Applied Chemistry
Graduate School of Science and
Engineering
2-12-1 Ookayama
Meguro, Tokyo 152-8552
Japan

Jimmy C. Yu
The Chinese University of Hong Kong
Department of Chemistry
Shatin, New Territories, Hong Kong
China

1
Introduction to Nanoparticles
Satoshi Horikoshi and Nick Serpone

1.1
General Introduction to Nanoparticles

Nanotechnology is the science that deals with matter at the scale of 1 billionth of a meter (i.e., 10^{-9} m = 1 nm), and is also the study of manipulating matter at the atomic and molecular scale. A nanoparticle is the most fundamental component in the fabrication of a nanostructure, and is far smaller than the world of everyday objects that are described by Newton's laws of motion, but bigger than an atom or a simple molecule that are governed by quantum mechanics. The United States instituted the National Nanotechnology Initiative (NNI) back in 2000, which was soon followed (2001) by a plethora of projects in nanotechnology in nearly most of the U.S. Departments and Agencies [1]. About 20 Research Centers were subsequently funded by the National Science Foundation (NSF), an agency responsible solely to the President of the United States and whose mandate is to fund the best of fundamental science and technology projects. NSF was the lead U.S. agency to carry forward the NNI. The word "nanotechnology" soon caught the attention of various media (TV networks, the internet, etc.) and the imagination and fascination of the community at large.

In general, the size of a nanoparticle spans the range between 1 and 100 nm. Metallic nanoparticles have different physical and chemical properties from bulk metals (e.g., lower melting points, higher specific surface areas, specific optical properties, mechanical strengths, and specific magnetizations), properties that might prove attractive in various industrial applications. However, how a nanoparticle is viewed and is defined depends very much on the specific application. In this regard, Table 1.1 summarizes the definition of nanoparticles and nanomaterials by various organizations.

Of particular importance, the optical property is one of the fundamental attractions and a characteristic of a nanoparticle. For example, a 20-nm gold nanoparticle has a characteristic wine red color. A silver nanoparticle is yellowish gray. Platinum and palladium nanoparticles are black. Not surprisingly, the optical characteristics of nanoparticles have been used from time immemorial in sculptures and

Microwaves in Nanoparticle Synthesis, First Edition. Edited by Satoshi Horikoshi and Nick Serpone.
© 2013 Wiley-VCH Verlag GmbH & Co. KGaA. Published 2013 by Wiley-VCH Verlag GmbH & Co. KGaA.

Table 1.1 Definitions of nanoparticles and nanomaterials by various organizations: International Organization for Standardization (ISO), American Society of Testing and Materials (ASTM), National Institute of Occupational Safety and Health (NIOSH), Scientific Committee on Consumer Products (SCCP), British Standards Institution (BSI), and Bundesanstalt für Arbeitsschutz und Arbeitsmedizin (BAuA).

	Nanoparticle	Nanomaterial
ISO	A particle spanning 1–100 nm (diameter)	–
ASTM	An ultrafine particle whose length in 2 or 3 places is 1–100 nm	–
NIOSH	A particle with diameter between 1 and 100 nm, or a fiber spanning the range 1–100 nm.	–
SCCP	At least one side is in the nanoscale range.	Material for which at least one side or internal structure is in the nanoscale
BSI	All the fields or diameters are in the nanoscale range.	Material for which at least one side or internal structure is in the nanoscale
BAuA	All the fields or diameters are in the nanoscale range.	Material consisting of a nanostructure or a nanosubstance

Figure 1.1 Photographs of the famous Lycurgus cup which displays a different color depending on whether it is illuminated externally (a) or internally (b). For details, consult the website of the British Museum [2].

paintings even before the 4th century AD. The most famous example is the Lycurgus cup (fourth century AD) illustrated in Figure 1.1.

This extraordinary cup is the only complete historic example of a very special type of glass, known as dichroic glass, that changes color when held up to the light. The opaque green cup turns to a glowing translucent red when light is shone through it internally (i.e., light is incident on the cup at 90° to the viewing direction). Analysis of the glass revealed that it contains a very small quantity of tiny

(~70 nm) metal crystals of Ag and Au in an approximate molar ratio of 14 : 1, which give it these unusual optical properties. It is the presence of these nanocrystals that gives the Lycurgus Cup its special color display. The reader can marvel at the cup now in the British Museum [2].

Until the Middle Ages, the reputation of soluble gold was based mostly on its fabulous curative powers of various diseases, for example, heart and venereal diseases, dysentery, epilepsy, and tumors; it was also used in the diagnosis of syphilis. The history of the nanoparticle from ancient times to the Middle Ages has been summarized by Daniel and Astruc [3]. The first book on colloidal gold was published in 1618 by the philosopher and medical doctor Francisci Antonii. This book includes considerable information on the formation of colloidal gold sols and their medical uses, including successful practical cases. The book noted that soluble gold appeared around the fifth or fourth century B.C. in Egypt and China. On the other hand, industrial manufacturing of stained glass with colloidal particles was established by Kunckel in the seventeenth century (1676). He also published a book whose Chapter 7 was concerned with "drinkable gold that contains metallic gold in a neutral, slightly pink solution that exerts curative properties for several diseases" [4]. He concluded that gold must be present in aqueous gold solutions to a degree of contamination such that it is not visible to the human eye. A colorant in glasses, that is, the "Purple of Cassius", was a colloid resulting from the presence of gold particles and tin dioxide and was highly popular in the seventeenth century. A complete treatise on colloidal gold was published in 1718 by Helcher [5]. In the treatise, this philosopher and doctor stated that the use of boiled starch in its drinkable gold preparation noticeably enhanced its stability. These ideas were common in the eighteenth century, as indicated in a French chemical dictionary dated 1769 [6], under the heading "*or potable*" it was said that drinkable gold contained gold in its elementary form, albeit under extreme sub-division suspended in a liquid. In 1794, Fuhlame reported in a book that she had dyed silk with colloidal gold [7]. In 1818, Jeremias Benjamin Richters suggested an explanation for the differences in color shown by various preparations of drinkable pink or purple gold solutions in that these solutions contained gold in the finest degree of subdivision, whereas yellow solutions were found when the fine particles had aggregated. In 1857, in a well-known publication, Michael Faraday [8] reported the formation of deep red solutions of colloidal gold by reduction of an aqueous solution of chloroaurate ($AuCl_4^-$) by phosphorus in CS_2 (a two-phase system). He also investigated the optical properties of thin films prepared from dried colloidal solutions and observed reversible color changes of the films upon mechanical compression (from bluish-purple to green). Since that pioneering work, thousands of scientific papers have been published on the synthesis, modification, properties, and assembly of metal nanoparticles, using a wide variety of solvents and other substrates.

Nanotechnology is easily evident in various old churches. A well-known application of early nanotechnology is the ruby red color that was used for stained glass windows during the Middle Ages. Beautiful examples of these applications can be found in glass windows of many Gothic European cathedrals, among which the

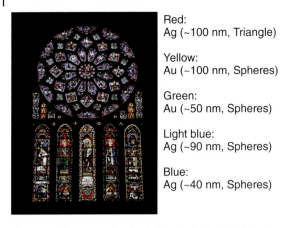

Figure 1.2 Rosace nord stained glass in the Cathédrale Notre-Dame de Chartres (France), color changes depend on the size and shape of gold and silver nanoparticle. From ref. [9].

León Cathedral (Spain) represents one of these unique masterpieces, located on the medieval French pilgrimage path to Santiago de Compostela (Spain); its impressive 2000 m^2 colored windows offer a unique view that certainly warrants a visit. Of course, the medieval artisans were unaware that they were using nanotechnology. They just knew that a particular process produced a beautiful effect. For example, the stained glass of a wonderful rose can be seen at the world heritage Cathédrale Notre-Dame de Chartres in France. The stained glass made in medieval times is displayed in Figure 1.2. Later chemistry clarified the reasons behind the generation of the color. These vivid colors were controlled by the size and the form (or shape) of the nanoparticles of gold and silver. The relation between particles and their associated colors has been discussed recently by Jin and coworkers [9]. In an article of 22 February 2005, the New York Times [10] summarized the relationship between the color of stained glass and the size/shape of the nanoparticles (see Figure 1.3). After several decades, the ingredients present in the stained glass (colored glass) of various churches were clarified subsequent to the development of analytical instruments. People without professional expertise in nanotechnology are also increasingly contributing to the technology.

An outline of the historical background in connection with nanoparticles (nanotechnology) is summarized in Table 1.2. The current technology that deals with nanoparticles, or simply nanotechnology, began from the special optical phenomenon and the establishment of a theory to describe the various physical phenomena that were followed subsequent to the development of analytical instruments. This continues as we speak, with various nanostructures being proposed and discovered, and their applications described.

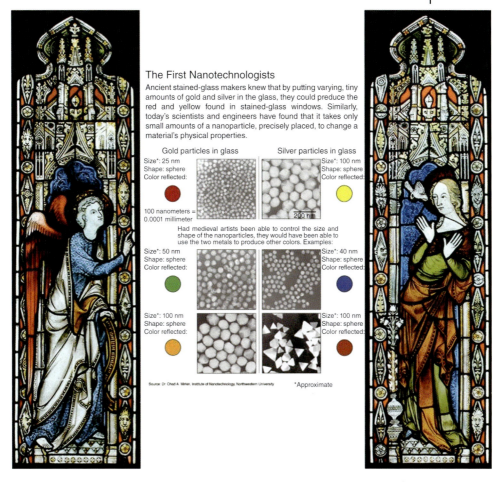

Figure 1.3 Comparison of the effect of size and shape of nanoparticles on the coloring of stained glass (Stained Glass Museum, Great Britain); see ref. [10].

Industrial production of nanomaterials saw its origins in the twentieth century. For example, nanoparticles of carbon black (tire soot) have been used in the fabrication of rubber tires of automobiles from the beginning of the twentieth century. Pigments such as SiO_2 and TiO_2 have been prepared by a high-temperature combustion method. Since the 1970s, the innovative development of nanoparticles is due to a combination of theory and experiments in the fields of physics, chemistry, materials science, and biosciences. Specific phenomena (chemical properties and physical properties), other than the optical property of a nanoparticle, have led to new possibilities in various fields. Applications of nanoparticles in various fields require an inexpensive and simple process of synthesizing high quality shaped-nanoparticles. In this regard, recent years have witnessed significant research being done in the use of microwave radiation in nanoparticle syntheses.

Table 1.2 Chronological table of nanotechnology.

Year	Remarks	Country/people
1200–1300 BC	Discovery of soluble gold	Egypt and China
290–325 AD	Lycurgus cup	Alexandria or Rome
1618	First book on colloidal gold	F. Antonii
1676	Book published on drinkable gold that contains metallic gold in neutral media	J. von Löwenstern-Kunckel (Germany)
1718	Publication of a complete treatise on colloidal gold	Hans Heinrich Helcher
1857	Synthesis of colloidal gold	M. Faraday (The Royal Institution of Great Britain)
1902	Surface plasmon resonance (SPR)	R. W. Wood (Johns Hopkins University, USA)
1908	Scattering and absorption of electromagnetic fields by a nanosphere	G. Mie (University of Göttingen, Germany)
1931	Transmission electron microscope (TEM)	M. Knoll and E. Ruska (Technical University of Berlin, Germany)
1937	Scanning electron microscope (SEM)	M. von Ardenne (Forschungslaboratorium für Elektronenphysik, Germany)
1959	Feynman's Lecture on "There's Plenty of Room at the Bottom"	R. P. Feynman (California Institute of Technology, Pasadena, CA, USA)
1960	Microelectromechanical systems (MEMS)	I. Igarashi (Toyota Central R&D Labs, Japan)
1960	Successful oscillation of a laser	T. H. Maiman (Hughes Research Laboratories, USA)
1962	The Kubo effect	R. Kubo (University of Tokyo, Japan)
1965	Moore's Law	G. Moore (Fairchild Semiconductor Inc., USA)
1969	The Honda–Fujishima effect	A. Fujishima and K. Honda (University of Tokyo, Japan)
1972	Amorphous heterostructure photodiode created with bottom-up process	E. Maruyama (Hitachi Co. Ltd., Japan)
1974	Concept of nanotechnology proposed	N. Taniguchi (Tokyo University of Science, Japan)

Table 1.2 (*Continued*)

Year	Remarks	Country/people
1976	Carbon nanofiber	M. Endo (Shinshu University, Japan)
1976	Amorphous silicon solar cells	D. E. Carlson and C. R. Wronski (RCA, USA)
1980	Quantum hall effect (Nobel Prize)	K. von Klitzing (University of Würzburg, Germany)
1982	Scanning tunneling microscope (STM) (Nobel Prize)	G. Binnig and H. Rohrer (IBM Zurich Research Lab., Switzerland)
1986	Atomic force microscope (AFM)	G. Binnig (IBM Zurich Research Lab., Switzerland)
1986	Three-dimensional space manipulation of atoms demonstrated (Nobel Prize)	S. Chu (Bell Lab., USA)
1987	Gold nanoparticle catalysis	M. Haruta (Industrial Research Institute of Osaka, Japan)
1990	Atoms controlled with scanning tunneling microscope (STM)	D. M. Eigler (IBM, USA)
1991	Carbon nanotubes discovered	S. Iijima (NEC Co., Japan)
1992	Japan's National Project on Ultimate Manipulation of Atoms and Molecules begins	
1995	Nano-imprinting	S. Y. Chou (University of Minnesota, USA)
1996	Nano sheets	T. Sasaki (National Institute for Research in Inorganic Materials, Japan)
2000	National Nanotechnology Initiative (NNI), USA	
2003	21st Century Nanotechnology Research and Development Act, USA	
2005	Nanosciences and Nanotechnologies: An action plan, Europe	

Synthesis of nanoparticles using microwave heating has been on the increase in recent years. Fabrication of high quality nanoparticles can be achieved by simple operations compared with the more conventional nanoparticle synthetic methods. This chapter describes the various features and provides examples of the use of microwaves in nanoparticle synthesis. Moreover, in order to clarify the feature (or features) of the microwave method, a general description of the nanoparticle is also included.

1.2
Methods of Nanoparticle Synthesis

Various preparation techniques for nanoparticles (nanomaterials) are summarized in Figure 1.4. Two approaches have been known in the preparation of ultrafine particles from ancient times. The first is the breakdown (top-down) method by which an external force is applied to a solid that leads to its break-up into smaller particles. The second is the build-up (bottom-up) method that produces nanoparticles starting from atoms of gas or liquids based on atomic transformations or molecular condensations.

The top-down method is the method of breaking up a solid substance; it can be sub-divided into dry and wet grinding. A characteristic of particles in grain refining processes is that their surface energy increases, which causes the aggregation of particles to increase also. In the dry grinding method the solid substance is ground as a result of a shock, a compression, or by friction, using such popular methods as a jet mill, a hammer mill, a shearing mill, a roller mill, a shock shearing mill, a ball mill, and a tumbling mill. Since condensation of small particles also takes place simultaneously with pulverization, it is difficult to obtain particle sizes of less than 3 µm by grain refining. On the other hand, wet grinding of a solid substrate is carried out using a tumbling ball mill, or a vibratory ball mill, a planetary ball mill, a centrifugal fluid mill, an agitating beads mill, a flow conduit beads mill,

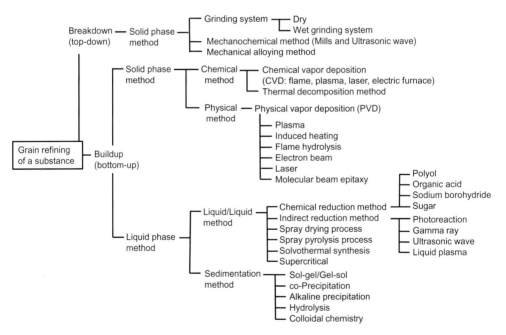

Figure 1.4 Typical synthetic methods for nanoparticles for the top-down and bottom-up approaches.

an annular gap beads mill, or a wet jet mill. Compared with the dry method, the wet process is suitable for preventing the condensation of the nanoparticles so formed, and thus it is possible to obtain highly dispersed nanoparticles. Other than the above, the mechanochemical method and the mechanical alloying method are also known top-down methods.

The bottom-up approach is roughly divided into gaseous phase methods and liquid phase methods. For the former, the chemical vapor deposition method (CVD) involves a chemical reaction, whereas the physical vapor deposition method (PVD) uses cooling of the evaporated material. Although the gaseous phase methods minimize the occurrence of organic impurities in the particles compared to the liquid phase methods, they necessitate the use of complicated vacuum equipment whose disadvantages are the high costs involved and low productivity. The CVD procedure can produce ultrafine particles of less than 1 μm by the chemical reaction occurring in the gaseous phase. The manufacture of nanoparticles of 10 to 100 nm is possible by careful control of the reaction. Performing the high temperature chemical reaction in the CVD method requires heat sources such as a chemical flame, a plasma process, a laser, or an electric furnace. In the PVD method, the solid material or liquid material is evaporated and the resulting vapor is then cooled rapidly, yielding the desired nanoparticles. To achieve evaporation of the materials one can use an arc discharge method. The simple thermal decomposition method has been particularly fruitful in the production of metal oxide or other types of particles and has been used extensively as a preferred synthetic method in the industrial world.

For many years, liquid phase methods have been the major preparation methods of nanoparticles; they can be sub-divided into liquid/liquid methods, and sedimentation methods. Chemical reduction of metal ions is a typical example of a liquid/liquid method, whose principal advantage is the facile fabrication of particles of various shapes, such as nanorods, nanowires, nanoprisms, nanoplates, and hollow nanoparticles. With the chemical reduction method it is possible to fine-tune the form (shape) and size of the nanoparticles by changing the reducing agent, the dispersing agent, the reaction time and the temperature. The chemical reduction method carries out chemical reduction of the metal ions to their 0 oxidation states (i.e., $M^{n+} \rightarrow M^0$); the process uses non-complicated equipment or instruments, and can yield large quantities of nanoparticles at a low cost in a short time. Of particular interest in this regard is the use of microwave radiation as the heat source that can produce high quality nanoparticles in a short time period. Besides the chemical reduction method which adds a reducing agent (direct reduction method), other reduction methods are known, such as photoreduction using gamma rays, ultrasonic waves, and liquid plasma which can be used to prepare nanoparticles. These methods that do not use a chemical reducing substance have the attractive feature that no extraneous impurities are added to the nanoparticles. Other than these methods, spray drying, spray pyrolysis, solvothermal synthesis, and the supercritical method are also known.

The general technique in the sedimentation method is a sol–gel process, which has been used extensively for the fabrication of metal oxide nanoparticles. This procedure transforms a solution of a metal alkoxide into a sol by hydrolysis, followed by polycondensation to a gel. Several books are available that provide details of the sol–gel process (see e.g., [11]). The wet process (liquid phase method) guarantees a high dispersivity of nanoparticles compared to the dry method. However, if the resulting nanoparticles are dried, aggregation of the particles soon follows. In this case, re-dispersion can be carried out according to the process used in the solid phase method.

Although various techniques have been summarized in Figure 1.4, there are some features to consider that are common to all the methods. That is, the synthesis of nanoparticles requires the use of a device or process that fulfills the following conditions:

- control of particle size, size distribution, shape, crystal structure and composition distribution
- improvement of the purity of nanoparticles (lower impurities)
- control of aggregation
- stabilization of physical properties, structures and reactants
- higher reproducibility
- higher mass production, scale-up and lower costs

1.3
Surface Plasmon Resonance and Coloring

The physical phenomenon of surface plasmon resonance (SPR) was reported long ago by Wood who could detect sub-monomolecular coverage [12]. Not only did Wood discover the plasmon resonance phenomenon, but also found that it changed with the composition of the liquid in touch with the metal surface. Although he speculated on how the light, grating and the metal interacted with each other, a clear rationalization of the phenomenon was not provided. He observed a pattern of "anomalous" dark and light bands in the refracted light when he shone polarized light on a mirror with a diffraction grating on its surface. The first theoretical treatment of these anomalies was put forward by Rayleigh in 1907 [13]. Rayleigh's "dynamical theory of the grating" was based on an expansion of the scattered electromagnetic field in terms of outgoing waves only. With this assumption, he found that the scattered field was singular at wavelengths for which one of the spectral orders emerged from the grating at the grazing angle. He then observed that these wavelengths, which have come to be called the Rayleigh wavelengths, λ_R, correspond to the Wood anomalies. Further refinements were made by Fano [14], but a complete explanation of the phenomenon was not possible until 1968 when Otto [15], and in the same year Kretschmann and Raether [16], reported the excitation of the surface plasmon band. Surface plasmon resonance has also been similarly researched in solid state physics in recent years in application studies, especially in such applied research as biosensing, solar cells, and super high-density recording. Details on surface plasmon resonance from the point of view of solid state physics have been given by Schasfoort and Tudos [17].

In this section we present a simple outline of the relation between surface plasmon resonance and the color of nanoparticles. In solid state physics, the plasmon represents the collective oscillation of a free charge in a metal, and may be considered as a kind of plasma wave. The positive electrical charge in the metal is fixed and the free electron is free to move around it. An applied external electric field, as from a light source, causes the free electrons at the surface of the metal to vibrate collectively, giving rise to surface plasmons.

Since electrons are also particles with an electric charge, when they vibrate they also generate an electric field, and when the electric field from the vibration of free electrons and the applied external electric field (e.g., electromagnetic waves) resonate the resulting phenomenon is referred to as a surface plasmon resonance that takes place at the surface of the metal. However, if light irradiates a solution that contains dispersed metal nanoparticles smaller than the wavelength of light, then depending on the electric field of light, the deviation produces a free electron at the surface of the metal. As a result, the weak or thick portions of the electric field appear on the nanoparticle surface (Figure 1.5) and can be considered as a kind of polarization. Such localized plasmon resonance is called localized surface plasmon resonance (LSPR).

The LSPR is typically concentrated in a very narrow region on the surface of a nanoparticle. The electric field distribution on the nanoparticle surface caused by LSPR can be visualized using an electromagnetic field analysis software with a finite element method (Comsol Multiphysics 4.2a). The LSPR distribution on the surface of a 20-nm (diameter) Au nanoparticle is shown in Figure 1.6a. When visible light at 520 nm, which corresponds to the maximum position of the LSPR band in the Au nanoparticle, is used to irradiate the nanoparticle, the electric field generated is concentrated on the right and left side of the Au nanoparticle, perpendicular to the incident light direction. On the other hand, the electric field generated in two adjacent Au nanoparticles is concentrated within the gap between the two particles. Figure 1.6b illustrates two particles with a 4-nm gap, while Figure

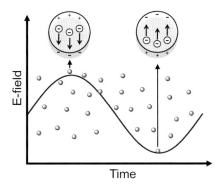

Figure 1.5 Mechanism of a localized surface plasmon resonance.

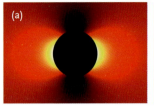 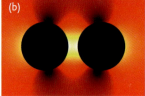 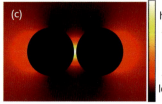

Figure 1.6 Images of the electric field distribution on the Au nanoparticle surface (20 nm diameter) under visible light irradiation visualized with Comsol Multiphysics 4.2a: (a) single Au nanoparticle, (b) two Au nanoparticles with 4 nm gap, and (c) two Au nanoparticles with 1 nm gap.

1.6c displays two particles separated by a gap of 1 nm. As the gap becomes narrower and narrower, the density of the generated electric field gets bigger and bigger.

The wavelength corresponding to the LSPR depends on the kind of metal, the shape of the metal nanoparticle, and the extent of aggregation of the metallic nanoparticles. Moreover, the surface plasma vibration also changes with the dielectric constant and the quality of the carrier fluid. The plasma oscillations in the metal occur mainly in the ultraviolet (UV) region. However, in the case of Au, Ag, and Cu, the plasma shifts nearer to the visible light domain with the band due to electrons in the s atomic orbitals. For example, the wavelength of the surface plasmon resonance band maximum of a spherical Au nanoparticle is 520–550 nm. If a colloidal Au nanoparticle solution is now irradiated with visible light at these wavelengths (520–550 nm), the visible light corresponding to the green color is absorbed and the particles now display a red purple color, which is the complementary color to green. In a colloidal Ag nanoparticle solution which has a plasmon resonance band maximum near 400 nm, the blue color of the visible light is absorbed and the Ag particles now take on a yellow color, the complementary color to blue.

1.4
Control of Size, Shape, and Structure

1.4.1
Size Control of Nanoparticles

The physical and chemical properties of nanomaterials depend not only on their composition but also on the particle size [18] and shape [19]. *Accordingly, a high quality synthesis protocol must first of all provide control over particle size and shape.* For example, if the diameter of an Au nanosphere is made to increase, the surface plasmon resonance will be gradually shifted from 530 nm to the longer wavelength side (see Figure 1.7) [20]. Thus, if nanoparticles differ in size, their optical characteristics will also change significantly.

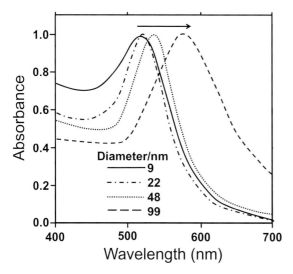

Figure 1.7 Visible-light spectra of Au nanospheres with various particle sizes. From ref. [20]. Copyright 2006 by the American Chemical Society.

In optical applications of nanoparticles, simplification of the size distribution of the particles becomes a very important factor. *Therefore, it is important to fabricate nanoparticles with a single target size in mind.* Generally, in order to prepare monodispersed nanoparticles, it is imperative that the nanoparticles grow very slowly after the rapid generation of the seed particles [21]. If the size of the nanoparticles decreases (i.e., increase in specific surface area), then the increase in the surface energy of such nanoparticles will facilitate their aggregation. Consequently, after their growth to the desired optimal size, it will be necessary to stabilize the particulate surface by addition of a dispersing agent. The historical use of dispersing agents in nanoparticle syntheses is not new; for example, Ag colloids protected by citrate were reported by Lea way back in 1889 [22]. However, where the concentration of nanoparticles is unusually high, the decentralized stabilization will fall, because the protective action of the organic substrate (citrate) is no longer strong enough to prevent aggregation. Thus, several studies of dispersing agents that maintain a high dispersivity of the nanoparticles, and also at various concentrations, have been reported. According to the hard and soft acids and bases (HSAB) rule [23], Ag^+, Au^+, Pd^{2+}, Pt^{2+} are classified as soft acids in the Lewis sense (SA), and substrates possessing the thiol (R–SH) and the phospine (P–R_3) functional groups, classified as soft bases, have proven to be suitable dispersing agents [24]. Early research that examined organic thiol molecules as possible dispersing agents was reported by Brust and coworkers [25]. As shown in Figure 1.8, if 1-dodecanethiol is used as the dispersing agent in Au nanoparticle synthesis the 1-dodecanethiol molecule can form a monomolecular layer on the Au nanoparticle surface, and firmly stabilize the dispersed Au nanoparticles. This particular paper has been the third most cited article in the journal Chemical Communications ever since 1965

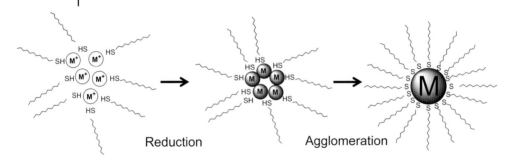

Figure 1.8 Mechanism for the capping of metal nanoparticles with 1-dodecanethiol. From ref. [26]. Copyright 1994 by the Royal Society of Chemistry.

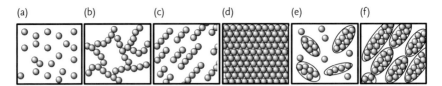

Figure 1.9 Typical group diagram of a nanoparticle; (a) Random structure; (b) fractal structure (c) structural alignment; (d) close-packed structure; (e) ordered structure (dispersion); (f) ordered structure (dense). From ref. [27].

[26], and so can easily be said to have had a significant impact in the chemistry field. Moreover, an increase in the size of the nanoparticles can be achieved by the change in alkyl chain length from 1-dodecanethiol to alkyl chains of octane, decane and hexadecane, as well as with a decrease in steric hindrance. For this reason, the 1-dodecanethiol dispersing agent is also used to control the particle size. Development of polymers as the dispersing agents has also been studied. In this case, the protection capability is determined by the affinity of the nanoparticle surface and by the molecular weight of the polymer.

It is important to realize that the physical properties of a nanoparticle can change with the aggregation ratio, even though the colloidal solution may contain nanoparticles of identical size [27]. The images and the characteristics of the state of aggregation of nanoparticles are depicted in Figure 1.9. In the dispersed random structure shown in Figure 1.9a, the dynamical physical properties and the optical properties are significant. On the other hand, the electronic properties are displayed by the fractal structure shown in Figure 1.9b, and ion and electronic transport properties appear in the structure orientation of Figure 1.9c. The optical properties appear in the close-packed structure of the nanoparticles (Figure 1.9d), whereas the discrete structures or otherwise orderly structures of Figure 1.9e and f display dynamical physical properties, magnetism, optical, and electronic properties.

Methods to separate out particles of a given target size from a colloidal solution which contains nanoparticles of various sizes are known. They are (i) separation by precipitation, (ii) centrifugal separation, (iii) gel filtration column, and (iv) gel electrophoresis. As a feature of each screening method, the precipitation separation is suitable for a large distribution of colloid nanoparticles in the solution. The centrifugal separation and the gel filtration column are well suited for solutions of colloidal nanoparticles with a narrow size distribution. Gel electrophoresis is a suitable method to separate nanoparticles taking advantage of the difference in charge density of the particles, and is suitable for separating particles with a small cluster size. In fact, a combination of these various methods might prove advantageous. However, a problem with sorting the various sized nanoparticles using these methods is that only a fraction of the nanoparticles of a given size may be collected, and then only in small quantities. The digestive ripening method and high temperature melting technique have been proposed to resolve this problem [28].

1.4.2
Shape Control of Nanoparticles

The shape of nanoparticles is an important factor that determines the nature of the surface plasmon resonance band just as the size of the nanoparticles did (see Figure 1.7). Absorption spectra in the visible spectral region of various Au rod-shaped nanoparticles (i.e., nanorods) with changes in the aspect ratio (length of long side and short side) are shown in Figure 1.10 [29]. The diameters of the Au nanorods espousing a pillar form and used in this experiment ranged from 5 to 20 nm and the lengths from 20 to 150 nm. It is worth noting that the change in

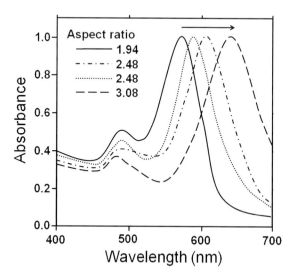

Figure 1.10 Visible-light spectra of Au rod-shaped nanoparticles with various aspect ratios (long side to short side in the rod-shaped nanoparticle). From ref. [29]. Copyright 2006 by the American Chemical Society.

the ratio of a nanorod is related to the size ratio of a crystal face. An increase in the size ratio (aspect ratio) shifts the maximal absorption band to longer wavelengths. Therefore, the physical composition of the nanorods can easily change their spectroscopic features, such that various studies have been required to understand these characteristics.

The preparation of Au nanorods using surfactants has been reported by Yu and coworkers [29]. Gold nanorods were synthesized using an Au anode under ultrasonic irradiation with a template consisting of the cationic surfactant hexadecyltrimethylammonium bromide (CTAB). Au exfoliates as a cluster from the electrode and is molded into the shape of a rod through the interaction with the CTAB micelle (at concentrations above the cmc). In the growth mechanism of nanorods, the CTAB dispersing agent is selectively adsorbed onto the {100} and {110} crystal faces of the Au nanoparticles. For this reason, the {111} crystal face grows and a rod-like metal nanoparticle is generated as a result. The use of CTAB as the dispersing agent subsequently quickly led to reports on nanoparticle research [30, 31]. Nanoparticles of various forms and shapes have been prepared using the adsorption characteristics of a dispersing agent. Chen and coworkers reported an unusual composition of branched Au nanoparticles (see Figure 1.11) using high concentrations of CTAB [32]. Evidently, the molecular association of a surfactant as a dispersing agent determines the various shape features of metallic nanoparticles.

The physical aspects of Au nanorods prepared using a hard template, such as mesoporous alumina, are similar to those when using a soft template like CTAB. In an early report that made use of a hard template, the Au nanorods were

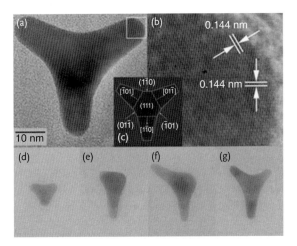

Figure 1.11 (a) TEM image of a regular tripod nanocrystal and (b) high-resolution image of the pod end as marked by a white frame in panel (a); (c) diagram showing the crystal planes and pod directions. The lower row of panels exhibits the particles developed at various stages: (d) embryo of a triangular shape, (e) monopod, (f) V-shaped bipod, and (g) Y-shaped tripod. From ref. [32]. Copyright 2003 by the American Chemical Society.

synthesized in the inner fine pores of mesoporous alumina [33]. In the initial stage, the nanosize porous alumina electrode is produced and the metal is then electrochemically deposited sequentially in the fine pores, which provide a firm mold; the diameter of the short axis of the Au nanorod which grows inside the fine pores is regulated by the size of the pores. Subsequently, the alumina mold is dissolved and removed; the so-formed Au nanorods are then taken out of the template.

An interesting feature of the above method is the fabrication of nanorods of multiple layers of different metals such as Au–Ag–Au. Therefore, a nanoparticle with various features can be synthesized. As for the plural layer-type nanorod, applied research can lead to a nanosize system that might be considered a nano-sized bar code [34].

1.4.3
Structure Control of Nanoparticles

Nanoparticles that are composed of two or more metals differ in their catalytic, magnetic, and optical characteristics from nanoparticles that consist of a single metal. Such nanoparticles can be sub-divided into three kinds of structures: (i) the alloy structure that exists randomly in a crystal (Figure 1.12a); (ii) the core–shell structure in which the metal at the center differs from the peripheral metal (Figure 1.12b); and (iii) the twinned hemisphere structure wherein two sorts of hemispheres are joined. The latter heterojunction structure facilitates phase separation (Figure 1.12c). Nanostructures consisting of complex metal nanoparticles tend to hide the various new features.

The core–shell structure is comparatively easy to fabricate in complex metal nanoparticles with effective functional control, which has led to several studies and reports in the literature. For instance, although the color of an Au nanoparticle liquid dispersion is purplish red (the purple of Cassius) and that of an Ag nanoparticle liquid dispersion appears yellow, whenever Au forms the core and Ag the shell the structure then takes an orange color. Moreover, if a structured matter

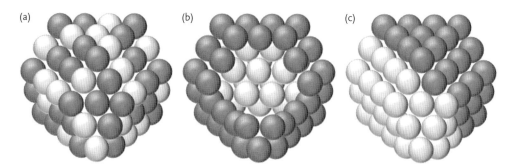

Figure 1.12 Schematic images of bimetal nanoparticles: alloy structure (a), core–shell structure (b), and heterojunction structure (c) of complex metal nanoparticles.

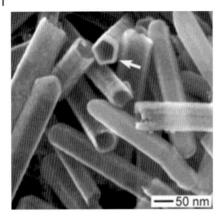

Figure 1.13 SEM image of gold nanotubes that had been broken through sonication to show their cross-sections. The gold nanotubes were prepared by reacting silver nanowires with an aqueous HAuCl$_4$ solution. From ref. [37]. Copyright 2003 by the American Chemical Society.

has magnetic properties, such as magnetite nanoparticles, then the magnetic metal particles could be used to form the structure's core, such that the structure will now be embodied with both magnetic and optical characteristics.

Synthetic methods of preparing core–shell nanoparticles are roughly divided into two categories: (i) involving a simultaneous reduction reaction and (ii) involving a sequential one-electron reduction reaction. As an example of the simultaneous reduction reaction, consider the core being made up of Pt nanoparticles and the shell composed of Pd nanoparticles [35]. A unique method that uses differences in the oxidation potentials of Ag and Au has also been reported [36]. Here, a silver nanoparticle is added to an HAuCl$_4$ solution, following which an oxidation–reduction reaction takes place (Eq. 1.1) wherein gold is deposited on the surface of a Ag nanoparticle yielding the core–shell structure.

$$3Ag(s) + HAuCl_4\,(aq) \rightarrow Au(s) + 3AgCl\,(aq) + HCl\,(aq) \tag{1.1}$$

The development of this method has led to the fabrication of Au nanotubes by first making the pentagonal prismatic Ag nanowires to use as the template (Figure 1.13) [37].

1.5
Reducing Agent in Nanoparticle Synthesis

Gold hydrosols have been synthesized through reduction of a gold chloride solution under an atmosphere of phosphorus; the method succeeded in controlling the size of Ag nanoparticles [8]. Decades later, Turkevich and coworkers examined the mechanism of metal salt reduction with citric acid [38]. The latter acid has been used widely as a reducing agent to fabricate various metal nanoparticles. The reduction of hexachloroplatinic acid with sodium borohydride has also been investigated, together with hydroxylamine hydrochloride, dimethylaminoborane, sodium citrate, hydrazine hydrate, sodium formate, boranetrimethylamine complex, sodium borohydride and formaldehyde as the reducing agents; the various features of the resulting nanoparticles were examined [39].

More recently, the polyol method using ethylene glycol has become quite popular. Ethylene glycol can be both the reaction solvent and the reducing agent in the synthesis of nanoparticles. What made ethylene glycol attractive was its polar nature, which is useful in dissolving the metal salt and can also play the role of a dispersing agent, and since it has a high boiling point (198 °C) it is suitable for the preparation of the base metal. On the other hand, the disadvantage of this polyol is the high boiling point making removal of the solvent difficult.

1.6
Applications of Metallic Nanoparticles

The various characteristics of different nanoparticles relative to bulk metals are summarized below.

Optical function: The surface absorption plasmon of Au and Ag can express various colors by changing the size of the particle, the form or shape of the particle, and the rate of condensation. A new *paint* that has the durability of an inorganic pigment and the vivid color of an organic substrate can be made. Nanoparticles smaller than the wavelength of light can be used to make high-penetration conductivity materials (there is little absorption, dispersion, and reflection).

Catalyst function: Reaction efficiencies can be enhanced since the specific surface area of such nanoparticles is large compared with existing particles; to the extent that the surface terrace is regular at the atomic level, a hyperactive catalyst with high selectivity can be made: for example, Au nanoparticles.

Thermal function: When the particle diameter is small (less than 10 nm), the melting point is also lower than a bulk metal. Electronic wiring can be made with nanoparticles that have a low boiling point, for example, a polymer.

Electrical function: Since superconductivity transition temperature rises so that particle diameter is small (less than 1 nm), it can be used to make high-temperature superconductivity material.

Mechanical function: Since the mechanical characteristics improve, mechanical strength can be sharply raised by mixing the nanoparticles with metals or ceramics.

Magnetic function: The attractive force of a magnetic metal increases on reduction of the particle diameter, such that soft-magnetic materials can be made in the form of an alloy of nanoparticles. Moreover, a permanent magnet can be made if the nanoparticles are smaller than the magnetic domain made to magnetize.

1.6.1
Application of Nanoparticles in Paints

One of the most interesting aspects of metal nanoparticles is that their optical properties depend strongly upon the particle size and shape. Bulk Au looks yellowish in reflected light, but thin Au films look blue in transmission. This characteristic blue color steadily changes to orange, through several tones of purple and red, as the particle size is reduced down to ~3 nm. The nanoparticles attracted attention as color materials and the possibility of their use has been examined in various fields. Figure 1.14 illustrates the photograph of a car to which was applied a "clear colored coating" containing gold nanoparticles on a base coating containing red pearl mica [40]. Spraying with the clear colored coating containing the nanoparticles increased the depth of the red background even more, and since the car is in the shade there is almost no diffuse reflection. The red color becomes a feature of paints containing nanoparticles. Paints that contain nanoparticles cannot be removed as easily as can classical paint. However, because of high costs, paints with nanoparticles are used only in limited applications.

Metal nanoparticles have also been used in enamel color paints in pottery. Conventional enamel color has used paints with mixed transition metals in the pulverization (glass frit) of glass. If, instead of transition metal paints, Au nanoparticles were used, then high quality red paint could be made with high transparency. Research into iron oxide nanoparticles in paints has also been carried out.

1.6.2
Application in Chemical Catalysis

Ni, Pd, Ag, and Pt have been used as typical metal catalysts in chemical reactions. However, the dissociative adsorption of hydrogen or oxygen molecules cannot be

Figure 1.14 Photograph of a car to which was applied a clear coating containing gold nanoparticles on a red colored base coating. From ref. [40].

carried out on an Au smooth surface and at a temperature of less than 200 °C [41]. Therefore, such a gold material is inactive as a catalyst in hydrogenation and oxidation reactions. However, when Au nanoparticles are used, they work effectively as catalysts, as discovered by Haruta [42]. The ratio of the corner to the edge of an Au nanoparticle of several nanometers in size becomes large compared with Au particles of larger size. Thus, both the adsorption and the catalytic characteristics of the Au surface increase. In an icosahedron consisting of 2054 Au atoms, the percentage of Au atoms exposed on the surface is 15% in Au nanoparticles of 4.9 nm size (external diameters), whereas the surface exposure of an Au atom reaches 52% in 2.7 nm sized nanoparticles (309 Au atoms). If the number of Au atoms in a nanoparticle was to become even smaller, the number of atoms that constitutes the whole, as well as the electrons, will become limited since the electronic structure of the nanoparticle would then become discontinuous [43]. This state takes on a cluster structure and a quantum size effect shows up in the physical properties.

Generally, metals that form clusters of such a small size tend to be unstable to the atmosphere. However, Au clusters are stable and so Au can be used as a catalyst. The catalytic action is rapid as the size of the Au nanoparticle is small in catalyzed oxidation reactions. For example, in order to oxidize CO with a Pt catalyst, a temperature of not less than 100 °C is needed. However, with an Au nanoparticle as the catalyst, reaction can occur even at temperatures below 0 °C (see Figure 1.15).

These effects are the result of changes in the so-called surface plasmon resonance [44], which is observed at the frequency at which conduction electrons oscillate in response to the alternating electric field of the incident electromagnetic radiation. However, only metals with free electrons (essentially Au, Ag, Cu, and the alkali metals) possess plasmon resonances in the visible spectral region, which give rise to such intense colors for these metals. Elongated nanoparticles (ellipsoids and nanorods) display two distinct plasmon bands related to transverse and longitudinal electron oscillations.

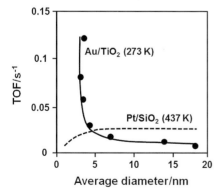

Figure 1.15 Particle dependence of Au catalyst and Pt catalyst in the CO oxidation reaction (TOF: turnover frequency).

1.6.3
Application of Nanoparticles in Micro-wiring

Metal nanoparticle paste is used for circuit pattern formation of a printed wired board in the electronic industry [45]. The melting point of metal nanoparticles decreases relative to bulk metals, so that circuit formation impossible on polymer base material is attainable using a conventional electric conduction paste. Furthermore, whenever particles at the nanoscale are used, the wiring width is thin to a nano level. Formation of nanoparticle wiring can use an ink-jet method, a method that is both inexpensive and requires shorter times than vacuum evaporation and photolithographic methods that are typically used. Generally, Au is used to make the metal nanoparticle paste. However, it is expensive, so that substitution of Cu nanoparticles has been proposed. Cu nanoparticles tend to be oxidized so that the process requires the presence of anti-oxidants.

1.6.4
Application of Nanoparticles in Medical Treatments

Just as the surface plasmon resonance is seen in a metal nanoparticle, an increase in the quantity of nanoparticles raises the scattering intensity. Taking advantage of this feature, the application to specific molecule recognition in a living body tissue is expected (see e.g., refs. [46, 47]). For example, by covering the cancer cell surface it becomes possible to distinguish a healthy cell from a cancer cell by the presence of antibodies joined to the Au nanoparticle. Although the Au nanoparticle junction with the antibody is nicely distributed in the healthy cell (Figure 1.16a), when a cancer cell exists the antibodies are concentrated mostly at the Au nanoparticle (Figure 1.16b). The imaging at various wavelengths is performed by a change in the shape of the nanoparticle [48]. Moreover, if a protein and a functional molecule were joined to the Au nanoparticle, it could also be used for imaging cells other than cancer cells. In addition, to the extent that Au nanorods

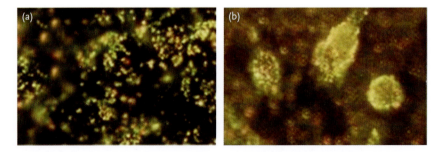

Figure 1.16 Molecular-specific imaging of cancer cells using Au nanoparticle/anti-EGFR conjugates. (a) Dispersed Au nanoparticle in healthy cells, (b) Concentrated Au nanoparticles about a cancer cell. From ref. [48]. Copyright 2008 by the American Chemical Society.

also display a plasmon resonance in the near-infrared domain, and that such Au nanorods congregate about the circumference of an abnormal cell, it becomes possible to treat cancer using a near-infrared laser.

References

1 For details about NNI see the website: www.nano.gov.
2 For details about the British Museum see the website: http://www.britishmuseum.org/explore/highlight_image.aspx?image=k741.jpg&retpage=20945.
3 Daniel, M.-C., and Astruc, D. (2004) *Chem. Rev.*, **104**, 293–346.
4 Kunckel, J. (1676) Ars Vitraria Experimentals oder Vollkommene Glasmacherkunst, Frankfurt, Germany.
5 Helcher, H.H. (1718) *Aurum Potabile Oder Gold Tinstur*, J. Herbord Klossen, Breslau and Leipzig, Germany.
6 (1769) *Dictionaire de Chymie*, Lacombe, Paris.
7 Fulhame, M. (1794) *An Essay on Combustion with a View to a New Art of Dying and Painting*, J. Cooper, London, UK.
8 Faraday, M. (1857) *Philos. Trans. R. Soc. London*, **147**, 145–181.
9 Jin, R., Cao, Y., Mirkin, C.A., Kelly, K.L., Schatz, G.C., and Zheng, J.G. (2001) *Science*, **294**, 1901–1903.
10 For the details about The New York Times article of February 22, 2005 see: http://www.nytimes.com/imagepages/2005/02/21/science/20050222_NANO1_GRAPHIC.html.
11 See e.g. (a) Brinker, C.J. and Scherer, G.W. (1989) *Sol-Gel Science: The Physics and Chemistry of Sol-Gel Processing*, Elsevier Science, Amsterdam; (b) Hench, L.L., and West, J.K. (1990) *Chem. Rev.*, **90**, 33–72.
12 (a) Wood, R.W. (1902) *Proc. Phys. Soc. London*, **18**, 269–275; (b) Wood, R.W. (1902) *Philos. Mag.*, **4**, 396–402; Wood, R.W. (1912) *Philys. Mag.*, **23**, 310–317.
13 Rayleigh, L. (1907) *Proc. R. Soc. London Ser. A*, **79**, 399–416.
14 Fano, U. (1941) *J. Opt. Soc. Am.*, **31**, 213–222.
15 Otto, A. (1968) *Z. Phys.*, **216**, 398–410.
16 Kretschmann, E., and Raether, H. (1968) *Z. Naturforsch. Teil A*, **23**, 2135–2136.
17 see e.g. Schasfoort, R.B.M. and Tudos, A.J. (2008) *Handbook of Surface Plasmon Resonance*, The Royal Society of Chemistry Publishing, Cambridge, UK.
18 Henglein, A. (1989) *Chem. Rev.*, **89**, 1861–1873.
19 Burda, C., Chen, X., Narayanan, R., and El-Sayed, M.A. (2005) *Chem. Rev.*, **105**, 1025–1102.
20 Liz-Marzán, L.M. (2006) *Langmuir*, **22**, 32–41.
21 Sugimoto, T. (2000) *Fine Particles: Synthesis, Characterization, and Mechanisms of Growth*, Surfactant Sci. Series Vol. 92, Marcel Dekker Inc., New York.
22 Lea, M.C. (1889) *Am. J. Sci.*, **37**, 476–491.
23 Pearson, R.G. (1963) *J. Am. Chem. Soc.*, **85**, 3533–3539.
24 Prasad, B.L.V., Stoeva, S.I., Sorensen, C.M., and Klabunde, K.J. (2003) *Chem. Mater.*, **15**, 935–942.
25 Brust, M., Walker, M., Bethell, D., Schiffrin, D.J., and Whyman, R. (1994) *Chem. Commun.*, 801–802.
26 See http://www.rsc.org/Publishing/Journals/cc/News/Top40MostCitedArticles.asp.
27 Yamaguchi, Y. (2008) *Kagakukougaku*, **72**, 344–348.
28 Stoeva, S., Klabunde, K.J., Sorensen, C.M., and Dragieva, I. (2002) *J. Am. Chem. Soc.*, **124**, 2305–2311.
29 Yu, Y.-Y., Chang, S.-S., Lee, C.-L., and Wang, C.R.C. (1997) *J. Phys. Chem. B*, **101**, 6661–6664.
30 (a) Jana, N.R., Gearheart, L., and Murphy, C.J. (2001) *J. Phys. Chem. B*, **105**, 4065–4067; (b) Jana, N.R., Gearheart, L.,

and Murphy, C.J. (2001) *Chem. Mater.*, **13**, 2313–2322.
31 Nikoobakht, B., and El-Sayed, M.A. (2001) *Langmuir*, **17**, 6368–6374.
32 Chen, S., Wang, Z.L., Ballato, J., Foulger, S.H., and Carroll, D.L. (2003) *J. Am. Chem. Soc.*, **125**, 16186–16187.
33 van der Zande, B.M.I., Böhmer, M.R., Fokkink, L.G.J., and Schöneberger, C. (1997) *J. Phys. Chem. B*, **101**, 852–854.
34 Nicewarner-Peña, S.R., Freeman, G.P., Reiss, B.D., He, L., Peña, D.J., Walton, I.D., Cromer, R., Keating, C.D., and Natan, M.J. (2001) *Science*, **294**, 137–141.
35 Toshima, N., Yonezawa, T., and Kushihashi, K. (1993) *J. Chem. Soc. Faraday Trans.*, **89**, 2537–2543.
36 Sun, Y., Mayers, B., Herricks, T., and Xia, Y. (2003) *Nano Lett.*, **3**, 955–960.
37 Sun, Y., and Xia, Y. (2002) *Science*, **298**, 2176–2179.
38 Turkevich, J., Stevenson, P.C., and Hillier, J. (1951) *Discuss. Faraday Soc.*, **11**, 55–75.
39 Rheenen, P.R.V., McKelvy, M.J., and Glaunsinger, W.S. (1987) *J. Solid State Chem.*, **67**, 151–169.
40 Iwakoshi, A. (2008) *Techno-cosmos*, **21**, 32–38.
41 Saliba, N., Parker, D.H., and Koel, B.E. (1998) *Surf. Sci.*, **410**, 270–282.
42 Haruta, M. (2003) *Chem. Rec.*, **3**, 75–87.
43 Haruta, M. (2008) *J. Vac. Soc. Jpn.*, **51**, 721–726.
44 Kreibig, U., and Vollmer, M. (1995) *Optical Properties of Metal Clusters*, Springer-Verlag, Berlin.
45 Kawazome, M., Kim, K., Hatamura, M., and Suganuma, K. (2006/2007) *Funsai*, **50**, 27–31.
46 Bendayan, M. (1989) *Colloidal Gold, Volume 1: Principles, Methods, and Applications* (ed. M.A. Hayat), Academic press, Inc., London, pp. 34–88 (Chapter 3).
47 Wang, L., Li, J., Song, S., Li, D., and Fan, C. (2009) *J. Phys. D Appl. Phys.*, **42**, 203001–203012.
48 Jain, P.K., Huang, X., El-Sayed, I.H., and El-Sayed, M.A. (2008) *Acc. Chem. Res.*, **41**, 1578–1586.

2
General Features of Microwave Chemistry

Satoshi Horikoshi and Nick Serpone

2.1
Microwave Heating

Microwave radiation is electromagnetic radiation spanning the frequency range 30 GHz to 300 MHz (i.e., from a wavelength of 1 m to 1 cm). It is used widely in communications and in heating processes, especially foodstuff. Historically, the powerful interaction of microwaves with materials was discovered in 1946 from the melting of chocolate, a process attributed to microwave heating, whereas the first commercial microwave oven was developed in 1952 at the Raytheon Company [1]. This discovery is frequently given as an example of serendipity. Before the discovery of microwaves, high frequency induction heating was commonly used. The patent of dielectric heating by means of high frequency induction was issued in 1933 [2].

Applications of microwave radiation can be classified into communication and heating. Communications typically use an adjusted wave in terms of frequency, phase, and amplitude in order to carry information. On the other hand, the microwave output power and efficient irradiation apparatus are important factors when microwave radiation is used for heating purposes. Therefore, the composition and fabrication of microwave devices are highly different and fusion of various technologies is called for. Microwaves have also been suggested as a possible vehicle to transmit electrical energy. For instance, a satellite consisting of huge solar cells and in a geostationary orbit has been proposed by Shinohara [3]; the solar power satellite (SPS) concept is illustrated in Figure 2.1. To supply the electricity generated to the surface of the Earth would require the fusion of high power handling techniques (in the field of heating) and highly precise microwave transmission techniques as are available in the telecommunication field.

Microwave chemistry has been a subject that has generated much excitement of late. Prompt and high-quality syntheses can be achieved using the microwave technique. Some of the fundamental research in microwave radiation has been developed by industry. The next stage will involve the improvement of efficient chemical syntheses as might be required for possible industrial applications in the

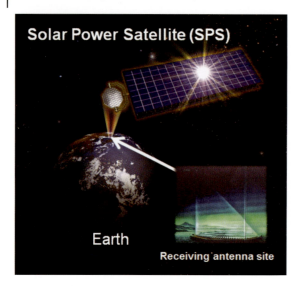

Figure 2.1 The satellite with solar cell in space and microwave power transmission from space to the Earth (Offer from Research Institute for Sustainable Humanosphere, Kyoto Univ.).

preparation of significant quantities of high-value chemicals. For this to occur, however, will necessitate a combination of microwave chemistry with highly coherent microwaves, as established in the communication field. In this regard, such coherence of electromagnetic waves is typically found in lasers. The amplification by interference of a highly coherent wave promises to enhance the efficiency of microwave chemistry accompanied with high energy saving.

2.2
Some Applications of Microwave Heating

To the extent that microwaves are electromagnetic waves they move at the speed of light. Since a sample is directly heated by microwaves, the time of heat conduction is shortened considerably relative to conventional heating and this regardless of however complex the reactor design might be, since reactor design has no influence on reaction times. Accordingly, microwaves have become an important heat source in a wide range of applications superseding the more traditional thermal methods. Figure 2.2 summarizes some of the applications where microwaves have become a common source of thermal energy, for example, in such fields as attenuation of environmental pollution, medicine, printing, paints, foodstuff, fabrication of thin films, agriculture, and drying of wood, among others. Not least microwaves have also been used for other purposes, for example, as a plasma source in the treatment of semiconductor surfaces, or in the fabrication of industrial diamonds. Moreover, electron spin resonance (ESR) spectroscopy used in chemistry makes

Environmental
- Sintering of asbestos
- Decomposition of dioxin and PCB
- Oil recovery from plastics
- Process treatment of waste oil
- Processing of medical wastes
- Processing of radioactive contaminated substances
- Degradation of chlorofluorocarbons
- Decomposition of VOCs
- Exhaust gas treatment
- Incineration of garbage
- Electrodeless lamps
- Biomass
- Oxidization of soot
- Enhancement of activity of photocatalysts
- Recycling of scrap tires
- Solidification of waste plastics

Medical uses
- Hyperthermia
- Sterilization
- Muscular warming
- Cutting of blood vessels

Usage for foodstuffs
- Sterilization
- Processing
- Defrosting
- High-speed cooking
- Reduced-pressure drying
- Activation of enzyme and yeast

Ink and paint
- Drying of printing ink
- Drying of paints

Films and paper
- Dryness of book cover
- Selective heating of pastes
- Sintering of electronic wiring
- Film curating

Agriculture
- Sterilization of soil
- Drying of wood
- Extermination of noxious insects in wood
- Extraction and degradation of contaminants

Usage in Wood treatments
- Adhesion processing
- Bending
- Drying of wood

Microwave chemistry

Organic chemistry
- Rapid syntheses
- Solid-phase syntheses
- Non-solvent and non-catalytic processes
- Organometallic complexes
- Combinatorial chemistry

Analytical chemistry
- Carbonization processing
- Acid and alkali treatment
- High-speed concentration, extraction and degradation

Biochemistry
- Enzyme reactions
- PCR
- Heating brain of rat
- DNA dyeing

Polymer
- Vulcanization and firing of rubber
- Higher selectivity in polymer syntheses
- Heat processing of plastics
- Size control of polymers

Catalytic chemistry
- Heterogeneous catalysis (reduction reactions, etc...)
- Homogeneous catalysis (Suzuki coupling, & others)
- Syntheses of catalysts and catalyst supports (e.g., zeolites)

Inorganic materials and metal chemistry
- Functional material composition
- Composition of quality nanomaterials
- Hydrothermal syntheses
- Particulate coatings
- Control of crystallinity
- Syntheses of nitrides
- Drying of refractory products
- Calcination of ceramics & formation of interfacial junctions
- Microwave-assisted iron manufacturing
- Metallic powder metallurgy
- Carbonization
- High-speed treatments of lightweight fire-resistant building materials
- Heat treatments of specially glasses
- Syntheses of artificial zeolites
- Drying of extruded molding ceramics
- Syntheses of artificial bones

Photochemistry
- Electrodeless lamps
- UV hardening of transparent coatings
- Photochemical syntheses

Figure 2.2 Fields of applications of microwave radiation.

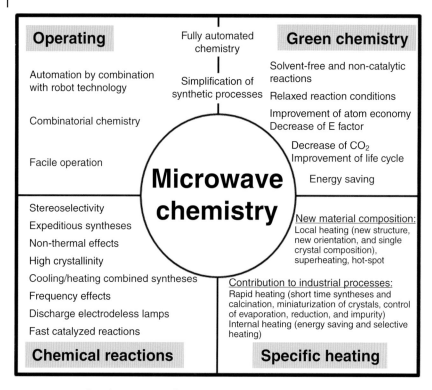

Figure 2.3 Benefits of microwave radiation in microwave chemistry.

use of 9 GHz microwaves. The microwave radiation used in the technology for military communications also found a domestic niche in the widely used microwave ovens. The technology has developed into various other fields, one of which that is relevant here is microwave chemistry, an area that has witnessed unparalleled growth in the last decade. The applications of microwaves in the chemical field are also summarized in Figure 2.2, the major ones being organic chemistry, analytical chemistry, biochemistry, polymer chemistry, catalysis, photochemistry, and inorganic chemistry of materials.

Microwave chemistry can be subdivided into four different areas: operating, green chemistry, chemical reactions, and specific heating (see Figure 2.3). The various features of microwave chemistry are by no means limited to those illustrated. The following section examines microwave chemistry, albeit non-exhaustively.

Researchers at the Johnson Space Center (Houston, TX, USA) have investigated the potential for using high-frequency microwave energy for the selective killing of bacteria while producing minimal damage to surrounding healthy human tissue [4]. The technology examined could be used to treat bacterial infections with minimal use of antibiotics as well as for decontamination of selective subsystems onboard space transport vehicles, the Space Station, and future lunar outposts. Potential Earth-based applications have also included the treatment of infections, pain associated with compression fractures or broken vertebrae, and tumors. Although no current methods are available for killing bacteria while minimizing damage to healthy tissue, it is theoretically possible since bacteria have unique properties of conductivity as a function of frequency [4]. The overall purpose of their studies was to determine whether specific microwave frequencies could effectively destroy bacteria at temperatures lower than frequencies that could damage human tissue. Results indicated that the Ka-band microwave frequency could effectively kill *Burkholderia cepacia*, common bacteria, with 20-GHz microwaves (but not <20 GHz), while producing minimal damage to healthy human cells. The authors concluded that the fabricated device could be useful for the prevention, treatment, and cure of medical infections in astronauts engaged in space exploration [4].

2.3
Microwave Chemistry

2.3.1
Microwaves in Organic Syntheses

Two pioneering papers that made use of microwaves in organic synthesis were published by Gedye and Giguere nearly three decades ago in 1986. Gedye and coworkers [5] carried out organic syntheses using a Teflon reactor and a commercial/domestic microwave oven; this resulted in a remarkable decrease in reaction time compared to conventional heating methods. Giguere and coworkers [6] pointed out that reactor, solvent and temperature management were important factors in microwave-assisted organic synthesis. Research into such application of microwaves increased, but slowly and gradually. By contrast, the number of publications in microwave chemistry has increased dramatically since 2000 [7], due in large part to the availability of commercial microwave devices intended for organic synthesis that have led many chemists to convert from the microwave oven to these newly fabricated devices (see below). Safety, reproducibility, control of microwave output power, temperature and pressure were some of the important and attractive features of such devices.

Currently, microwave organic synthesis systems have become fully automated by combining with robot technology. Such systems are playing an active part in fields such as combinatorial chemistry. This has also led research of microwave chemistry from the academic laboratory level to the industrial level where microwaves are being used as a heating tool in organic syntheses. One may now ask where is the charm of microwaves in organic synthesis?

The biggest attraction of microwaves in this field is the dramatic enhancement of reaction rates. An example is the Suzuki–Miyaura cross coupling reaction which when performed with microwaves occurs in a reaction time of about 50 s, whereas when using the traditional heating methods the reaction times are several hours [8]. Similar findings have been reported in various other organic reactions. Several synthetic examples of solvent-free and non-catalytic reactions have also been reported. Not surprisingly then, the use of microwaves in chemical syntheses is particularly attractive in the field of green chemistry. Many excellent books are now available that describe some of the details of microwave-assisted organic syntheses [9].

2.3.2
Microwaves in Polymer Syntheses

In the field of polymer synthesis, drying and polymerization of an epoxy resin was reported as an example of the use of microwaves at the end of the 1960s [10]. Resins used as denture bases are known examples. However, there are but a scant number of papers in the literature on the synthesis of macromolecules in comparison with the number of reports on organic syntheses.

Vulcanization of rubber has a long history as an industrial process which made use of microwaves as they became available [11]. Most recently, a patent application [12] described a vulcanization process of rubber tires with application of a premolded profile in an autoclave with the use of microwaves, wherein the rubber tires are vulcanized through microwave heating, resulting in significant reduction of vulcanization time when compared to processes of conventional vulcanization. This allowed for greater productivity, and the elimination of some steps of the chemical transformation of rubber. In a technical note, Parodi [13] describes the advantages of microwave vulcanization over other existing technologies with the principal advantages of the microwave technology summarized in Table 2.1.

Polymerization of amino acids in a short time by microwave heating was reported in 1990 by Yanagawa and coworkers [14]. They described the efficient reaction of amino acid amides to produce polypeptides in response to microwave heating during repeated hydration–dehydration cycles. The polypeptides so formed from a mixture of glycinamide, alaninamide, valinamide, and aspartic acid α-amide had molecular weights ranging from 1000 to 4000 Da. Except for glycine, the amino acids were incorporated into the polypeptides in proportion to the starting concentrations. The polypeptides had some definite secondary structure (e.g., an α-helix and β-sheets) in aqueous media. The reaction provided not only a convenient method for abiotic peptide formation but also a convenient method for the chemical synthesis of peptides. In another study by the same laboratory, Ito et al. [15] reported that polypeptides synthesized from a mixture of amino acid amides by microwave heating during repeated hydration–dehydration cycles, showed hydrolase- and oxidoreductase-like catalytic activities. In this regard, the resulting polypeptides catalyzed the reduction of the ferricyanide ion $[Fe(CN)^{3-}]$ with NADH. The polypeptides seemed to have a strong affinity for adenine nucleotides, because the reaction was inhibited by adenine derivatives such as NAD^+ and AppA. The authors also entertained the notion for the emergence of primitive protein enzymes.

Table 2.1 Microwave Vulcanization *versus* Other Rubber Processing Methods [13].

Microwave vulcanization	Other processing methods
Much faster and deeper heating of rubber items	Relative to all vulcanization methods (*advantage increasingly strong with the rising thickness of rubber articles to be processed*)
Much lower specific energy consumption (*Watts/kg of rubber processed*)	Relative to all vulcanization methods
Superior working safety	Over molten nitrate salt-baths, molten metal baths, and overheated steam under pressure
Absence of any environmental impact risks and ecological issues	As compared to molten salt-baths
Higher manufacturing throughput rates	Relative to all vulcanization methods
Minimized plant maintenance costs	As compared to molten salt-baths and vulcanization by overheated steam under pressure
Absence of rubber pigment or dye discoloration phenomena	Relative to vulcanization by overheated steam and molten salt-baths
Negligible process start-up times (*no time lags and minimal generation of rubber scraps to reach steady-state conditions*)	As compared to all vulcanization methods

The synthesis of polymers typically leads to an increase in the viscosity of the sample. Experiments carried out in a microwave oven frequently led to a thermal runaway of the sample. However, such problems have been mitigated by the expressed availability of dedicated devices for microwave chemical synthesis. Research examples of various polymerizations have been reported in recent years. For example, the synthesis of polylactic acid is a simple process when using microwaves and the reaction times are relatively short [16].

2.3.3
Microwaves in Inorganic Syntheses

The calcination and sintering of ceramics by microwaves were reported in 1972 [17]. Fundamental and application studies soon followed around 1980. The conventional heating methods for such tasks are heat conduction and radiation heating. With regard to microwave irradiation, the sample itself generates heat. Therefore, a sintered compact product is obtained at low temperature for a short time. Various kinds of ceramics have been treated by the microwave method [18]. In recent years, the calcination of special ceramics with a specific microwave method has also been reported [19].

2.3.4
Microwave Extraction

Microwave extraction technology is already in use in a variety of fields. Microwave extraction is roughly divided into extraction of an active ingredient, pre-treatment of an analytical sample, and extraction/removal of contaminants from polluted soils. As an example in the early stages of the extraction technology of an active ingredient, we need only mention the extraction of an essential oil from vegetable oil [20]. Moreover, extraction of useful ingredients from citrus fruits using microwaves requires only a few minutes relative to more conventional extraction methods that require no less than 18 days in some cases [21]. In the pre-treatment (decomposition) of a sample for use in atomic absorption spectrometry (AAS), in inductively coupled plasma atomic emission spectroscopy (ICP-AES), and in inductively coupled plasma-mass spectrometry (ICP-MS), microwaves have served well as standard analytical methods [22].

2.3.5
Microwave Discharge Electrodeless Lamps

Research on electrodeless and wireless lamps involving a microwave energy source is surprisingly over a century old. In the 1890s, Tesla [23] demonstrated wired and wireless transfer of power to electrodeless fluorescent and incandescent lamps in his lectures and articles. A research report in the late twentieth century examined the principles and the mechanism(s) of excitation of such lamps by microwave radiation [24]. However, both the size of the microwave generator as the energy source of the electrodeless lamp and the corresponding price tag were major issues in practical application. Recent years have witnessed increased practical applications of microwave discharge electrodeless lamps (MDELs) as a result of their high performance, their miniaturization and the lower costs of magnetrons and semiconductor microwave generators. Since the wavelength and the photon energy of the 2.45 GHz microwaves are only about 12.24 cm and 1×10^{-5} eV, respectively, the microwave energy is about five orders of magnitude smaller than the vibrational energy of typical molecules.

Consequently, if microwaves could replace ultraviolet radiation to drive chemical reactions, applications of microwaves would spread like wildfire. The advantages of MDELs over existing (Hg) lamps [25] are (i) the relatively long lifetime of electrodeless lamps, (ii) the lack of complications in lamp shape because the lamps are electrodeless, (iii) the absence of variations in light intensity, (iv) the shorter ignition time to lighting the MDEL device relative to a typical Hg electrode lamp, (v) the UV radiation can be supplied external to the reactor to avoid absorption of the microwave radiation by the reactor contents, (vi) the facility with which the lamps can be replaced, (vii) filling compounds (gases) that might be deleterious to metals and plastics can be used, (viii) the electrical energy can be transmitted from outside the reactor by a wireless process.

Figure 2.4 Coating of photo-curing resin using microwave discharged electrodeless lamp (courtesy of Japan fusion UV systems).

Different types of MDELs are used in various industrial processes depending on the type of applications, such as in paint hardening (drying) equipment [26], a technology developed in the United States in the early 1970s to cast a polymer rapidly by means of a UV hardening method starting from a monomer material. The process yielded a thin film of variable thickness compared to the more conventional drying process. Moreover, volatile organic compounds (VOCs) generated from the evaporation of organic solvents can be controlled by this system. The MDELs emit a stable light irradiance for long time periods so that the UV hardening method with an MDEL system fits nicely in an industrial process. At present, MDELs are widely used, among others, in the printing process, in paints (Figure 2.4), in coatings of a DVD surface, and in the plastic coating of optical fibers. They have been proposed as environmental treatment lamps [25]. Since they are electrodeless (i.e., wireless), they can be used directly in waste waters, such that contaminants present in waste waters can be decomposed with the VUV/UV-light emitted from these lamps; a particular type of MDEL is the high surface area system provided by putting bead-like MDELs into a reactor [27]. MDELs have also been proposed as a light source to activate a photocatalyst [25].

2.4
Microwave Chemical Reaction Equipment

Since the early use of microwave chemical reaction equipment led to some accidents, such as the explosive break-up of a reactor, the manufacturers of such equipment installed pressure- and temperature-sensing devices to obviate such accidents. Moreover, the malfunction of thermocouples caused by microwaves has now been remedied by introduction of fiber-optic thermometers. Multimode-type microwave chemical synthesis equipment became commercially available in the 1990s. This was followed by the availability of single-mode type microwave chemical synthesis equipment, albeit not till 2000. A year later, equipment that combined robot technology to automate the equipment also became available and was widely applied in combinatorial chemistry studies. A bench-scale microwave device for possible industrial applications was also introduced. At present, people interested in pursuing microwave chemistry can choose equipment according to the target reaction. Examples of microwave chemical synthesis and calcination equipment are shown in Figure 2.5.

The microwave chemical reaction equipment displayed in Figure 2.5 can be divided into two different kinds of cavity:

Multimode microwave apparatus: Reactants used in chemical syntheses typically have various dielectric losses. Therefore, the distribution of the electric field (E field) and magnetic field (H field) produced by the standing waves within the multimode cavity will be extremely complex on introducing the reactor into the cavity. In order to minimize this problem, commercial microwave chemical apertures use a mode stirrer and/or a turntable. In multimode devices the applicator is larger than the wavelength of the microwaves. If a mode diffuser

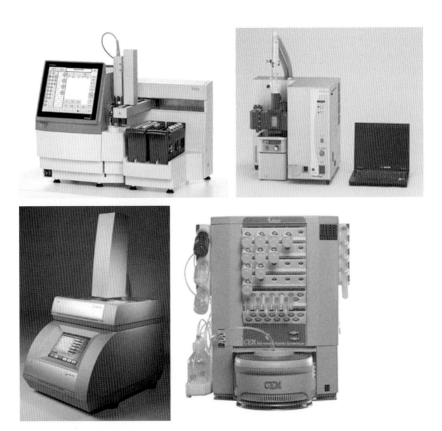

Figure 2.5 Various microwave chemical synthesis and microwave calcination equipments.

2.4 Microwave Chemical Reaction Equipment

and/or a turntable is not used, the sample will not be irradiated uniformly by the microwaves, which may cause temperature hot spots to occur.

Single-mode microwave apparatus: The use of single-mode or quasi-single-mode cavities allows one to define the precise positions within the cavities where the electric field and magnetic field strengths are at maximum densities. Moreover, the electric field strength in single-mode cavities is much higher than in multimode devices, the effective cavity power being three orders of magnitude

Figure 2.5 *Continued*

higher. However, distribution of the electric field and magnetic field in a single-mode cavity is very sensitive to the dielectric loss of a sample.

For equipment design, important points in the efficient heating of a solution at some given microwave frequency are the microwaves' penetration depth into the sample solution and the impedance matching to the sample solution. The electric power consumed becomes maximal in a general DC circuit when the impedance of the sample is equal to the impedance of the power supply. This phenomenon becomes a factor that determines the heating efficiency in microwave heating. Generally, the microwave single-mode system is constructed with a waveguide as the resonance cavity, a short plunger, a three-stub tuner (and/or an iris), a power monitor, and an isolator. Impedance matching can be ascertained by measuring the input power and the refracted power using a power monitor. When impedance does not match, it is displayed as reflected microwaves. In this case, impedance matching is adjusted with the components of the system (*E–H* tuner or three stub tuner). High resonance can be generated between the short plunger and the tuner (or iris). It is relevant to note that the impedance can change with changes in temperature. Moreover, the chemical composition of the sample changes during the synthesis. Therefore, it is difficult to maintain impedance matching in organic syntheses. A change of impedance can cause the electromagnetic waves of the microwave radiation to be reflected.

References

1 Spencer, P.L. (1950) U.S. Patent, 2,495,429 January 24.
2 McArthur, T. (1933) US Patent, 1,900,573, March 7.
3 Shinohara, N. (2011) *IEEE Microw. Mag.*, **12**, S64–S73.
4 Arndt, D., Byerly, D., Ngo, P., Phan, C., Dusl, J., Sognier, M., and Ott, M. (2005) Microwave Radiation–Therapeutic Application for Cure of Subcutaneous Bacterial Infections, Space Life Sciences, Johnson Space Center, Houston, TX, USA, see http://research.jsc.nasa.gov/PDF/SLiSci-7.pdf (accessed 1 August 2012).
5 Gedye, R., Smith, F., Westaway, K., Ali, H., Baldisera, L., Laberge, L., and Rousell, J. (1986) *Tetrahedron Lett.*, **27**, 279–282.
6 Giguere, R.J., Bray, T.L., Duncan, S.M., and Majetich, G. (1986) *Tetrahedron Lett.*, **27**, 4945–4948.
7 Kappe, C.O., and Stadler, A. (2005) *Microwaves in Organic and Medicinal Chemistry* (eds R. Mannhold, H. Kubinyi, and G. Folkers), Wiley-VCH Verlag GmbH, Weinheim, Germany, Chapter 1, pp. 4–5.
8 Namboodiri, V.V., and Varma, R.S. (2001) *Green Chem.*, **3**, 146–148.
9 See e.g. de la Hoz, A., and Loupy, A. (eds) (2012), *Microwaves in Organic Synthesis*, 3rd edn, Wiley-VCH Verlag GmbH, Weinheim, Germany.
10 Nishi, M. (1968) *J. Osaka Dep. Univ.*, **2**, 23–40.
11 Meredith, R.J. (1998) *Engineers' Handbook of Industrial Microwave Heating (Power & Energy Series)*, The Institute of Electrical Engineers, Herts., UK.
12 Takahashi, J. (2010) Vulcanization Process of Rubber Tires with the Use of Microwaves, United States Patent Application Publication, USPatent 2010/0090373 A1, April 15; see also http://www.faqs.org/patents/

app/20100090373 (accessed 1 August 2012).
13 Parodi, F. (2001–2011) Microwave Chemicals for the Rubber Industry: Novel, Specialty Microwave Heating Susceptors for the Fast UHF Vulcanization of White & Colored Rubber Compounds; see http://www.fpchem.com/fap_6a2-en.html (accessed 1 August 2012).
14 Yanagawa, H., Kojima, K., Ito, M., and Handa, H. (1990) *J. Mol. Evol.*, **31**, 180–186.
15 Ito, M., Handa, N., and Yanagawa, H. (1990) *J. Mol. Evol.*, **31**, 187–194.
16 Nakamura, T., Nagahata, R., Kunii, K., Soga, H., Sugimoto, S., and Takeuchi, K. (2010) *Org. Process Res. Dev.*, **14**, 781–786.
17 Shimomura, K., Miyazaki, T., Taniguchi, N., and Tutiya, A. (1972) *Annu. Rep. RIKEN Jpn.*, **48**, 11–19.
18 See e.g. (a) *Materials Research Advisory Board* (1994) *Microwave Processing of Materials*, National Research Council, Publication NMAB-473, National Academy Press, Washington, DC; (b) Folz, D.C., Booske, J.H., Clark, D.E., and Gerling, J.F. (eds) (2003) Microwave and radio frequency applications. Proceedings from the Third World Congress on Microwave and RF Processing, American Ceramic Society, Westerville, OH.
19 Takizawa, H. (2012) *J. IEE. Jpn.*, **132**, 17–19.
20 Paré, J.R.J., Sigouin, M., and Lapointe, J. (1991) U.S. Patent No. 5,002,784, March 26.
21 Tsukayama, M., Ichikawa, R., Yamamoto, K., Sasaki, T., and Kawamura, Y. (2009) *Nippon Shokuhin Kagaku Kogaku Kaishi*, **56**, 359–362.
22 See e.g. http://www.epa.gov/osw/hazard/testmethods/sw846/pdfs/3015a.pdf (accessed 1 Aug 2012).
23 Tesla, N. (1891) Lecture Delivered before the American Institute of Electrical Engineers, Columbia College, NewYork, USA.
24 Inoue, A. (1997) *J. IEE Jpn.*, **117**, 155–158.
25 Horikoshi, S., Abe, M., and Serpone, N. (2009) *Photochem. Photobiol. Sci.*, **8**, 1087–1104.
26 Ashikaga, K., and Kawamura, K. (2006) *Ind. Coat.*, **201**, 40–44.
27 Horikoshi, S., Tsuchida, A., Sakai, H., Abe, M., and Serpone, N. (2011) *J. Photochem. Photobiol. A Chem.*, **222**, 97–104.

3
Considerations of Microwave Heating
Satoshi Horikoshi and Nick Serpone

3.1
General Considerations of Microwave Heating

3.1.1
Electromagnetic Waves and a Dielectric Material

Microwaves are electromagnetic waves that move at the speed of light, and are reflected when they are used to irradiate a metal surface. By contrast, when they irradiate a dielectric material, various phenomena occur according to the nature of the electromagnetic waves. The influence of electromagnetic waves on dielectric materials in various ranges of the electromagnetic spectrum is summarized in Figure 3.1, which also shows the various analytical instruments and their usefulness in the ranges indicated.

Atomic species: If a dielectric is placed into an alternating electric field of the electromagnetic waves, the positions of the atomic nucleus (protons) and of the electrons that constitute the dielectric tend to follow the course of the electric field. The spatial relationship of a proton and an electron in an atom becomes distorted when subjected to the electric field (referred to as electronic polarization). Since the electric dipole that reflects the electric deviation (a proton and an electron) inside an atom is very small, the resonance phenomenon will occur with light of short wavelengths, such as X-rays and ultraviolet rays.

Molecular species: The distortion of the electric charge by electromagnetic waves also occurs in a molecule with a dipole moment and in a salt. For example, in the structure of the NaCl salt, composed of Na^+ and Cl^- ions, distortion of the structure arises when subjected to an alternating electric field (referred to as ionic polarization). Compared with an atom, the dipole moment of a molecule is large, so that resonance occurs in the infrared domain where the wavelengths are longer. Moreover, changes in bond lengths in an organic molecule occur through stretching and bending vibrations. These phenomena occur in various

---

Microwaves in Nanoparticle Synthesis, First Edition. Edited by Satoshi Horikoshi and Nick Serpone.
© 2013 Wiley-VCH Verlag GmbH & Co. KGaA. Published 2013 by Wiley-VCH Verlag GmbH & Co. KGaA.

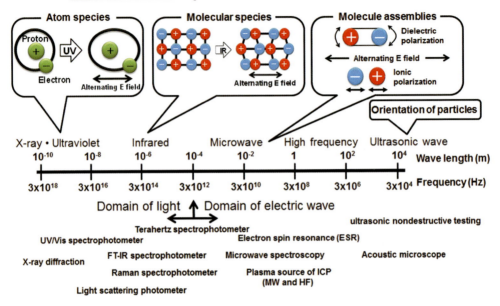

Figure 3.1 Resonance of dielectric to electromagnetic waves, and positioning of the analytical equipment.

absorption wavelength regions depending on the chemical structure and can be observed by means of an infrared absorption spectrophotometer.

Molecular assemblies: Orientation polarization of a dielectric possessing an electric dipole takes place on interaction with the electromagnetic waves in the microwave range. Moreover, as for the case of ions in solution, Joule heating takes place by space-charge polarization. For example, when microwave heating an electrolyte–water solution dielectric heating and Joule heating occur simultaneously compared with pure water, and thus exothermic efficiency becomes remarkably high. Accordingly, dielectric heating by orientation polarization of water and resistance heating by the Joule process are enhanced in electrolyte–water media.

3.1.2
Heating a Substance by the Microwaves' Alternating Electric Field

An optical fiber is a useful tool currently used in order to transmit light to its destination. On the other hand, in the domain of microwaves, the waveguide, consisting of a metal tube, is used to transmit the microwaves. The cross-sectional size of the waveguide tube (Figure 3.2) changes according to the frequency (wavelength) of the microwave radiation generated by the magnetron or by the

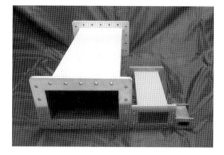

Figure 3.2 Actual photograph displaying the different sizes (and sectional areas) of the waveguides used for 915-MHz microwaves (left), 2.45-GHz microwaves (center) and 5.8-GHz microwaves (right) [1].

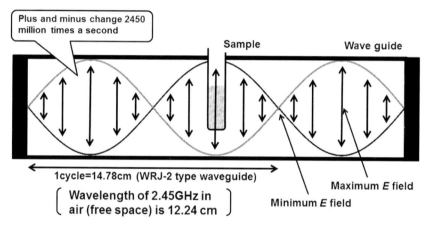

Figure 3.3 Distribution of the alternating electric field (sinusoidal synthesis wave) inside a resonant waveguide.

semiconductor generator. When metal plates (e.g., short plunger) are attached to the extremities of the waveguide tube, the waveguide then transmits the microwave radiation at multiples of 1/2 waves such that the microwaves are in a resonance state (Figure 3.3). For the alternating electric field distributed in the interior of the waveguide, the electric charges (plus and minus) interchange at the upper and lower ends of the waveguide. At the microwave frequency of 2.45 GHz, the changes of plus/minus in the waveguide occur 2450 million times in 1 s. Depending on the characteristics of the dielectric, the dielectric tends to modify the alternating electric field through structural changes.

As an example, when a water sample is placed in a reactor and subjected to an alternating electric field (Figure 3.3), the electric deviation (electric dipole) of the water molecule tends to follow the alternating electric field (Figure 3.4). However, since water consists of clusters of water molecules through hydrogen bonding, not all can follow the alternating electric field. The result is the dielectric heating of water by the irradiating microwaves.

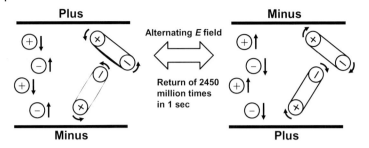

Figure 3.4 Action of the alternating electric field on a dielectric and on ions.

However, even though a molecule is polar, the heating efficiency changes according to the state of the molecular assemblies: solid, liquid, gas. The density of molecules in such states is an important factor. For example, although liquid water can be heated by microwaves, steam is not since the molecular density of the vapor (molecules move freely in air) is small, and consequently the alternating electric field of microwaves has no influence. On the other hand, ice is a structure in which water molecules are positioned in regular lattice sites of the ice and are strongly interacting with each other through the hydrogen bond. Therefore, the ice in which the water molecules solidifies strongly cannot resonate with the frequencies in the microwave range. Heating of ice needs infrared rays (Figure 3.1). Frozen foodstuffs can be defrosted in a microwave oven because the presence of salts in the foodstuffs enhances the heating process.

An important property of a dielectric material is the permittivity (ε) which is a measure of the ability of a material to be polarized by an electric field. The dielectric constant (ε') of a material is the ratio of its permittivity ε to the permittivity in vacuum ε_o, such that $\varepsilon' = \varepsilon/\varepsilon_o$. Since the dielectric constant represents the ratio of two similar quantities, it is dimensionless. It should also be noted that the dielectric constant is also a function of frequency in some materials, for example, polymers, primarily because polarization is affected by frequency. Another important property of a dielectric is the *dielectric loss factor* (ε'') which represents the loss of energy in a dielectric material through conduction, slow polarization currents, and other dissipative phenomena. In addition, the dielectric loss is the portion of the energy of an alternating electrical field in a dielectric medium that is converted into heat. As such it provides a good indication of the heating efficiency of the microwave radiation. As an example, water molecules are electrically neutral but their V-type structure (the H–O–H angle being 104.45°) imparts a dipole moment on the molecules. Accordingly, when water interacts with the microwaves' alternating electric field, the molecules' dipoles follow the electric field causing the molecules to rotate. The degree (magnitude) of the polarization at this time is reflected in the dielectric constant (ε'). Note that the heating efficiency is not reflected in the values of the dielectric constants. For instance, if a molecule is assumed to be a pendulum which swings with the microwaves, then however large the swings may be (i.e., however large the dielectric constant), no heat is generated by this process.

In microwave heating the dielectric loss (ε'') consists of the sum of two terms (Eq. 3.1): (i) dielectric heating and (ii) Joule heating – see ref. [2] for a detailed discussion of this equation.

$$\varepsilon'' = \frac{\varepsilon_S - \varepsilon_\infty}{1+\omega^2\tau^2}\omega\tau + \frac{\sigma}{\omega} \tag{3.1}$$

Taking the hydroxide ion as a typical ionic species with both ionic and dipolar characteristics, conductive loss effects can become larger than dipolar relaxation in solutions that contain large quantities of ionic species. Such losses tend to be slight at ambient temperatures in the case of solids, but can change substantially with increase in temperature. For the solid sample, a typical example is alumina (Al_2O_3) whose dielectric losses are negligibly small (~10^{-3}) at ambient temperature but can reach fusion levels in a matter of minutes in a microwave cavity. This effect originates from a strong increase in losses of electric conduction associated with the thermal activation of electrons as they migrate from the oxygen 2p valence band to the aluminum's 3s3p conduction band. In addition, losses of electric conduction in solids tend to be enhanced by defects in materials, which sharply reduce the energy needed to generate electrons and holes in the conduction and valence bands, respectively. In the case of carbon black powder, conduction losses tend to be rather high, which makes this material an attractive impurity additive to induce losses within solids with too small dielectric losses.

The loss angle (tan $\delta = \varepsilon''/\varepsilon'$) is a measure of reactance (resistance in a capacitor) of a molecule. The loss angle is the rate of heating against the ease of polarization. The loss angle also expresses the phase difference between the electric field and the polarization of the material. Dielectric loss factors and dielectric constants of some common polar solvents are shown in Figure 3.5a [1]. A higher dielectric loss and lower dielectric constant is connected with a higher loss angle. The dielectric constant of water at room temperature is 77.8; it is a remarkably high value compared with other solvents. On the other hand, the dielectric loss of water is relatively low at 9.6 (high polarization; few losses). Therefore, heating of water by microwaves (2.45 GHz) is not effective. The microwave frequency most suitable for heating water is about 17 GHz [3]. The dielectric constants and dielectric losses after heating (55 and 90 °C) of each sample and the relation of dielectric losses to dielectric constants are shown in Figure 3.5b and c. The ratio of dielectric loss to dielectric constant decreases upon heating. Therefore, the heating efficiency by microwaves falls with the rise in heat of a sample. For example, dimethyl sulfoxide (DMSO) has the highest dielectric loss at 25 °C. However, DMSO is replaced by ethylene glycol at 55 and 90 °C. On the other hand, the decrease ratio for ε'' of ethylene glycol by heating was lower than that for DMSO. Therefore, ethylene glycol has high heating efficiency compared with DMSO. Note that the dielectric parameters at room temperature may not be reflected in the actual heating.

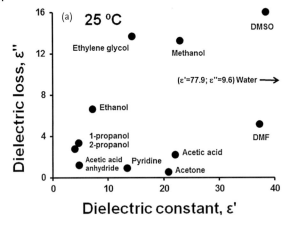
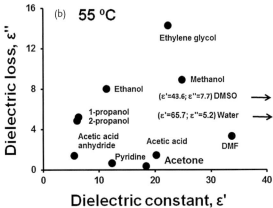
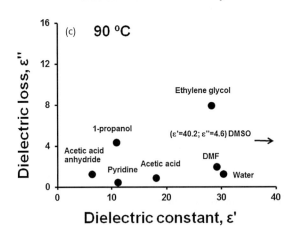

Figure 3.5 Dielectric losses versus dielectric constants of some common polar solvents at (a) 25 °C, (b) 55 °C, (c) 90 °C. Figure drawn from the data reported in ref. [4].

3.1.3
Heating a Dielectric by the Microwaves' Alternating Magnetic Field

Heating by microwaves often tends to be misunderstood as being only dielectric heating; there are other types of heating induced by microwaves, namely Joule heating (electric field) and magnetic loss heating (magnetic hysteresis heating and eddy current heating). Magnetic losses occur in the microwave region for metal oxides such as ferrites and other magnetic materials. Such losses are different from hysteresis or eddy current losses because they are induced by domain wall and electron-spin. For optimal absorption of the microwave energy, these materials are best positioned at locations of maximal magnetic field density. Transition metal oxides possess high magnetic losses and so can be used as added loss impurity additives to induce losses within solids for which the dielectric loss is too small. For instance, induction heating caused by the microwaves' magnetic field occurs in catalyzed reactions that involve solids, an example being the rapid induction heating of magnetite (Fe_3O_4); this excludes hematite (Fe_2O_3) a non-magnetic material [5].

3.1.4
Penetration Depth of Microwaves in a Dielectric Material

Even though the dielectric loss of a solvent may be high, the heating efficiency is sometimes low. This may be caused by the shallow penetration depth of the microwaves. For example, though solar light is absorbed by sea water it is nonetheless dark at greater depths. A similar phenomenon occurs with microwaves in that as the microwave energy is changed into heat it decreases. Therefore, if a large size reactor is used, the microwaves may not heat the center of the solution. The penetration depth D_p (in cgs units) is the depth at which microwaves pervade into the material and the power flux has fallen to 1/e (= 36.8%) of its surface value; That is, it denotes the depth at which the power density of the microwaves is reduced to 1/e of its initial value. D_p can be estimated by Eq. 3.2 [6],

$$D_p = \frac{\lambda}{4\pi} \left[\frac{2}{\varepsilon'\left(\sqrt{1+(\varepsilon''/\varepsilon')^2} - 1\right)} \right]^{1/2} \tag{3.2}$$

where λ is the wavelength of the radiation, $\lambda_{(2.45\,GHz)} = 12.24$ cm. Since the dielectric constant and the dielectric loss are components of this equation, we can expect that the penetration depth will change with increase in temperature. For example, since the penetration depth at 25°C is about 1.8 cm for water, even if we used a reactor 10 cm in diameter, the microwaves would not reach the center. However, at 50°C, the penetration depth is 3.1 cm and at 90°C, it is 5.4 cm so that now the microwaves do reach the center of the water sample. The penetration depths of common solvents at room temperature are listed in Figure 3.6 [1]. It turns out that the penetration depth of the microwaves into almost all nonpolar solvents is

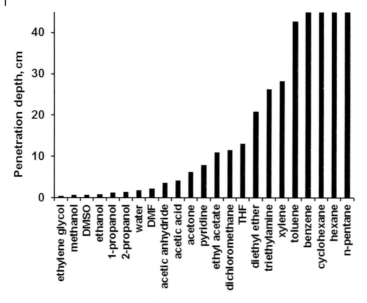

Figure 3.6 Penetration depths of the 2.45-GHz microwaves for some common solvents. Figure drawn from the data of ref. [1].

very deep compared with polar solvents. On the other hand, if ions are added to water, the penetration depth then becomes remarkably shallow. For example, when NaCl (0.25 M) is added to pure water, the penetration depth decreases to 0.5 cm at room temperature (for pure water it is 1.8 cm) [7]. Moreover, even if this electrolyte–water system were to be heated, the penetration depth would hardly change (0.4 cm at 90 °C). Therefore, in the case of solutions containing ions, since microwave heating does not occur, temperature hot spots will occur only at the surface of the reactor, and it will be necessary to carry out powerful mixing to control these. Hence, it is necessary to examine the heating efficiency and the penetration depth of microwaves into a substance to achieve optimal microwave heating.

3.1.5
Frequency Effects in Chemical Reactions

Microwave frequency effects were examined long ago in the heating of food. If the frequency changed, it was known that the heating efficiency and the penetration depth of the microwaves would also change. However, an examination of frequency effects in chemical reactions has too often been disregarded. Figure 3.7 illustrates the changes in the dielectric loss for ethanol and methanol as a function of frequency at 25 °C [8]. It is apparent that appreciable values of the dielectric loss exist over a wide frequency range. The optimal frequency differs for the two alcohols. Moreover, the frequency of the maximum dielectric loss shifts to the high

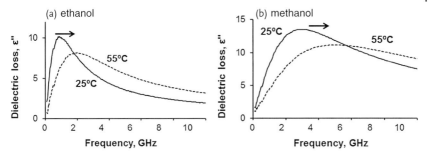

Figure 3.7 Changes of dielectric loss with frequency for (a) ethanol and (b) methanol (at 25 and 55 °C) [8].

frequency side on heating. Recently, novel microwave chemical apparatuses have been proposed for the 915-MHz and 5.8-GHz microwave frequencies [1]. By understanding the characteristics of each frequency, microwave chemistry can be made more effective.

In photochemistry, light of various wavelengths can be used to carry out a chemical reaction. However, with microwaves this is not the case. Specific microwave frequency bands have been reserved internationally for the use of radio frequency energies employed for industrial, scientific and medical (ISM) purposes other than communications [9]. Examples of applications of these bands include radio-frequency process heating, microwave heating source, and for medical diathermy machines. The powerful microwave emissions of these devices can create electromagnetic interference and disrupt radio communications that use the same frequencies, so these devices have been limited only to certain frequency bands. The ISM bands are defined by the ITU-R (International Telecommunication Union–radio communications sector). The globally used frequencies are 2.45, 5.8 and 24 GHz; other frequencies are established by each country. The global distribution map of microwave frequencies commonly used is shown in Figure 3.8.

3.2
Peculiar Microwave Heating

3.2.1
Special Temperature Distribution

Heat sources often used in chemistry are the oil bath, the electric heater, a gas flame, hot wind, and warm water. Typical heating occurs by convection, conduction, and by radiation from the portion near a heat source. On the other hand, only a dielectric material that can absorb microwave radiation can be heated by microwaves. For a microwave-induced chemical reaction, a reactor with low dielectric loss (Quartz, Teflon, polyetheretherketone, and some ceramics) and a sample

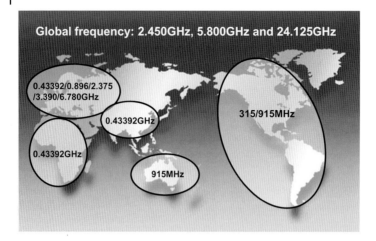

Figure 3.8 Global distribution map of frequency bands (GHz) used for ISM purposes [1].

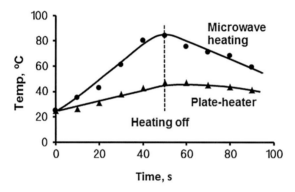

Figure 3.9 Heating rate and cooling rate of water by microwave heating and conventional heating.

with high dielectric loss are optimal combinations for the reaction to occur. While microwave heating of a substance through conduction heats the substance quickly, if microwave irradiation were to be stopped, the substance would cool fairly rapidly. Temperature changes of water heated with an electric plate-heater and with microwaves are shown in Figure 3.9. The water temperature rose to 84 °C under microwave irradiation for 50 s after which when the microwave irradiation was stopped for 40 s, the water temperature fell to 25 °C. For the plate-heater, the temperature of water fell by only 6 °C after the heating stopped. In heating by microwaves, since the water itself is heated, stopping microwave heating will cause the heat of the water to be lost to the container and to the surrounding atmosphere. If pulsed microwave irradiation were used, strict thermal control should become possible. On the other hand, stopping the heating from a heater makes control

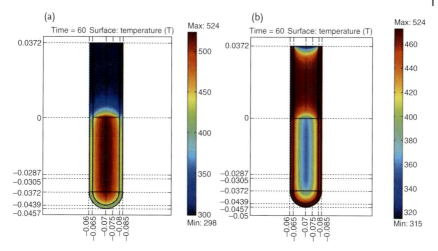

Figure 3.10 Temperature profile after 60 s as affected by microwave irradiation (a) compared to treatment in an oil-bath (b). Microwave irradiation raises the temperature of the whole reaction volume simultaneously, whereas in the oil heated tube, the reaction mixture in contact with the vessel wall is heated first. Temperature scale is given in K. '0' on the vertical scale indicates the position of the meniscus [10].

of temperature difficult since there is remaining heat in the heater and the container.

Conventional heating heats the reactions from the outside in, and the walls of the reaction vessel are generally the hottest part of the system, especially during the initial ramp to the desired temperature. Therefore, the temperature of water falls rapidly from the outside in (Figure 3.10). On the other hand, in heating with microwaves, the temperature of water falls from inside out. Indeed, when heated with microwaves, the heat of the water escapes into the open air through the reactor walls, such that the water temperature at the reactor walls is low compared with the temperature at the center of the reactor. The temperature distribution of the microwaves and conventional heating are then the reverse of each other. The image of such a temperature distribution was introduced by Schanche [10].

3.2.2
Superheating

The boiling point of the solvent used in a chemical reaction is one of several important factors in organic syntheses. Next to specific microwave effects is the phenomenon of macroscopic superheating. Even though the solution may have reached its boiling point on microwave heating, it may not boil and the solution is said to be in a metastable state. Solvents will boil only when they are in contact with their own vapor and, if this is not the case, they can be heated to above their normal (atmospheric) boiling point without the onset of boiling. Imperfections in glassware or on boiling chips have areas that cannot be wetted by the solvents, and thus create small solvent-vapor pockets, termed nucleation sites. Without nucleation sites, solvents are only in contact with their own vapor at the top of the vessel, and thus boiling (and hence release of heat) is limited to this relatively small interface. Since the most likely sites for nucleation in the absence of boiling chips are the pits and scratches on glassware walls, under microwave irradiation these sites are likely the coolest part of the system and so nucleation events are considered less likely to take place [11].

The superheating rates for some 23 common solvents when subjected to 5.8-GHz and 2.45-GHz microwave radiation are reported in Table 3.1 [12]. Except for ethylene glycol and pyridine, superheating occurred with the 2.45-GHz in all cases. With a 5.8-GHz single-mode microwave applicator, superheating occurred in almost all polar organic solvents (when equipment and reactor change, the value also tends to change). Moreover, DMF, acetic acid, 1-propanol, acetone, ethyl acetate, THF, and dichloromethane increase in temperature by 20 °C or more above the boiling point. In a nonpolar solvent, since heating hardly occurs with 2.45-GHz microwaves, unless it is under optimal heating conditions, superheating does not occur. However, if the frequency were changed to 5.8 GHz, superheating will then be easily observed in nonpolar solvents. On the other hand, bumping of low-boiling-point solvents occurs on superheating, a disadvantage of superheating.

3.2.3
Selective Heating in Chemical Reactions

The efficiency of microwave heating changes according to the dielectric loss of a substance. Organic syntheses that use this characteristic have been reported by Raner [13]. Two-phase water–chloroform systems (1:1 by volume; 100 ml) were heated, and typically after 40 s the temperatures of the aqueous and organic phases were 105 and 48 °C, respectively, due to differences in the dielectric properties of the solvents. A sizable differential could be maintained for several minutes before cooling. Comparable conditions would be difficult to obtain by traditional heating methods. Differential heating is particularly advantageous in carrying out Hofmann eliminations (Figure 3.11). A mixture of N-[2-(4′-ethoxybenzoyl)ethyl]-N,N,N-trimethylammonium iodide, water and chloroform was heated with stirring for 1 min at 110 °C (temperature of the aqueous phase). During the reaction, the product 4′-ethoxyphenylvinyl ketone was extracted and diluted into the poorly microwave-absorbing but cooler organic phase. The mixture was then subjected to rapid post-reaction cooling; the ketonic monomer was obtained in 97% yield. On the other hand, selective heating can occur in a heterogeneous catalyzed reaction.

The catalyst consisting of activated carbon as the support for a metallic catalyst has been used in an environmental clean-up and in organic synthesis. For example, if the activated carbon in the metal/support catalyst system were distributed in the nonpolar solvent, only the catalyst would be heated selectively by the microwaves (Figure 3.12). The temperature of the catalyst is remarkably high relative to a

Table 3.1 Boiling points and difference between superheating temperature and boiling point for some 23 common solvents under non-stirring conditions [12].

Solvent	Nature of solvent	Boiling point (°C)	Difference between superheating temp. and boiling point in °C (time in min to reach superheating temp.)	
			2.45 GHz	5.8 GHz
Ethylene glycol	Protic/Polar	198	— (39)	2 (39)
DMSO	Aprotic/Polar	189	2 (30)	13 (27)
DMF	Aprotic/Polar	153	10 (25)	24 (25)
Acetic anhydride	Aprotic/Polar	140	2 (25)	18 (15)
Pyridine	Aprotic/Polar	115	— (30)	7 (7)
Acetic acid	Protic/Polar	101	25 (20)	29 (9)
Water	Protic/Polar	100	2 (17)	— (30)
1-Propanol	Protic/Polar	97	20 (10)	20 (10)
2-Propanol	Protic/Polar	82	15 (6)	10 (6)
Ethanol	Protic/Polar	78	20 (5)	19 (8)
Methanol	Protic/Polar	65	24 (8)	15 (6)
Acetone	Aprotic/Polar	57	15 (5)	22 (5)
Xylene	Nonpolar	139	— (66)	1 (66)
Toluene	Nonpolar	111	— (30)	3 (30)
Triethylamine	Nonpolar	90	— (30)	8 (9)
Cyclohexane	Nonpolar	81	— (30)	5 (27)
Benzene	Nonpolar	80	— (30)	8 (28)
Ethyl acetate	Nonpolar	77	— (30)	26 (7.8)
Hexane	Nonpolar	69	— (30)	1 (28)
THF	Nonpolar	66	2 (20)	22 (5)
Dichloromethane	Nonpolar	40	— (30)	22 (2)
n-Pentane	Nonpolar	36	— (30)	11 (10)
Diethyl ether	Nonpolar	35	5 (14)	5 (5)

Sample volume: 30 ml, microwave input power: 30 W.

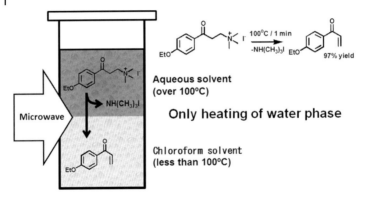

Figure 3.11 Hofmann elimination reaction taking place by selective heating [13].

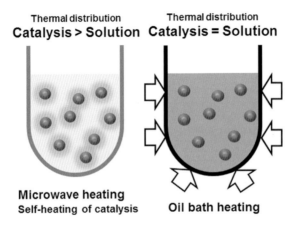

Figure 3.12 Selective heating in a heterogeneous catalyzed reaction.

nonpolar solvent. By contrast, the temperature of a catalyst is no higher than the temperature of the solvent under conventional heating conditions. When the solution has reached the reaction temperature, the temperature on the surface of the catalyst will be quite high [14]. On the other hand, if it is irradiated with microwaves of a superfluous input power, hot spots are formed on the catalyst surface [15], an event that will frequently deactivate a catalyst.

3.3
Relevant Points of Effective Microwave Heating

When performing chemical syntheses, some of the points that must be examined in order to achieve best performance by microwave heating are summarized in

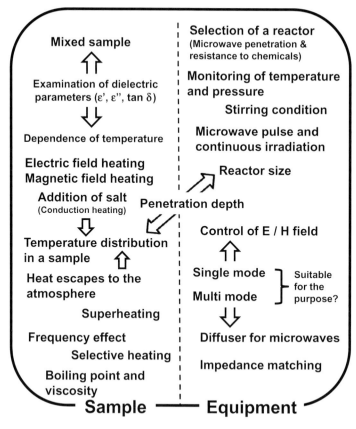

Figure 3.13 Relevant points in performing a microwave-assisted chemical reaction effectively.

Figure 3.13. Note that even if all these were not to occur, microwave chemistry could still take place.

References

1 Horikoshi, S., and Serpone, N. (2012) Microwave frequency effects in organic synthesis, in *Microwaves in Organic Synthesis*, 3rd edn (eds A. de la Hoz and A. Loupy), Wiley-VCH Verlag GmbH, Weinheim, Germany, pp. 377–423.

2 Stuerga, D., and Delmotte, M. (2006) *Microwaves in Organic Synthesis*, 2nd edn (ed. A. Loupy), Wiley-VCH Verlag GmbH, Weinheim, Germany, Ch. 1, p. 27.

3 Hasted, J.B. (1974) *Aqueous Dielectrics*, Chapman and Hall, London, UK.

4 Horikoshi, S., Matsuzaki, S., Mitani, T., and Serpone, N. (2012) *Radiat. Phys. Chem.*, **81**, 1885–1895.

5 Roy, R., Agrawal, D., Cheng, J., and Gedevanishvil, S. (1999) *Nature*, **399**, 668–670.

6 Metaxas, A.C., and Meredith, R.J. (1983) *Industrial Microwave Heating*, Peter Peregrinus, London, UK.

7 Horikoshi, S., Sumi, T., Abe, M., and Serpone, N. (2012) *J. Microwave Power Electromagnetic Energy*, **46**, 215–228.

8 Barthel, J., Bachhuber, K., Buchner, R., and Hetzenauer, H. (1990) *Chem. Phys. Lett.*, **165**, 369–373.

9 Technical report by A.C. Metaxas; see http://www.pueschner.com/downloads/MicrowaveHeating.pdf#search='40680ISMband3390' (accessed 27 July 2012).

10 Schanche, J.-S. (2003) *Mol. Divers.*, **7**, 293–300.

11 Baghurst, D.R., and Mingos, D.M.P. (1992) *Chem. Commun.*, 674–677.

12 Horikoshi, S., Iida, S., Kajitani, M., Sato, S., and Serpone, N. (2008) *Org. Process Res. Dev.*, **12**, 257–263.

13 Raner, K.D., Strauss, C.R., Trainor, R.W., and Thorn, J.S. (1995) *J. Org. Chem.*, **60**, 2456–2460.

14 Suttisawat, Y., Sakai, H., Abe, M., Rangsunvigit, P., and Horikoshi, S. (2012) *Int. J. Hydrogen Energy*, **37**, 3242–3250.

15 Horikoshi, S., Osawa, A., Abe, M., and Serpone, N. (2011) *J. Phys. Chem. C*, **115**, 23030–23035.

4
Combined Energy Sources in the Synthesis of Nanomaterials

Luisa Boffa, Silvia Tagliapietra, and Giancarlo Cravotto

This chapter deals with the combined use of ultrasound (US) and microwave irradiation (MW) for the preparation of nanomaterials. These two nonconventional energy sources play an important role in the synthesis of such materials, moreover, when used in combination they have been shown to possess specific additive or synergic effects.

In 1995, Maeda and Amemiya [1] first studied chemical effects under simultaneous US/MW irradiation in a series of sono- and chemi-luminescence experiments. They observed surprising additional and multiplicative effects when an aqueous alkaline solution of luminol and pure water was treated with simultaneous MW (2.45 GHz) and US (20 kHz) irradiation.

Although the combination of US and MW sources involves technical and safety considerations, its outstanding potential in chemico-physical modifications has led to the publication of a number of papers and the technique's efficiency has now been well documented [2, 3].

4.1
Introduction

The physical activation capacities of acoustic cavitation [4] and dielectric heating [5] have been widely exploited in the synthesis of nanomaterials. High-quality nanostructures, nanocomposites containing carbon nanoforms, elemental metals, metal oxides, salts and complexes, as well as polymers, have all been generated with predictable and well defined size, shape, composition and crystallinity [6].

MW irradiation in chemical synthesis offers several advantages over conductive heating, namely improved reaction rates, higher yields, selective heating in the function of absorbing properties while avoiding direct contact with the heating source, a reduction in side reactions and improved reproducibility, a lack of a temperature gradient (volumetric heating), easier automatization, and higher throughput.

Microwaves in Nanoparticle Synthesis, First Edition. Edited by Satoshi Horikoshi and Nick Serpone.
© 2013 Wiley-VCH Verlag GmbH & Co. KGaA. Published 2013 by Wiley-VCH Verlag GmbH & Co. KGaA.

The only critical aspects of MW chemistry are the high costs of professional reactors, short penetration depth which limits reactor size and, therefore, process scale-up, and difficulty in monitoring nanoparticle formation *in situ*.

Sonochemistry is a well-established technique based on intense, nonhazardous ultrasonic irradiation (from 20 kHz to 1 MHz), commonly used in laboratories to promote chemical reactions, which causes both homogeneous and heterogeneous mixtures to be efficiently mixed and generates and disperses nanoparticles and colloids.

US affects reaction media via the acoustic cavitation phenomenon: the formation, growth and collapse of microscopic bubbles that generate elevated localized pressures and temperatures called "hot spots". These harsh localized conditions promote the formation of reactive radicals and nanomaterials. US waves convey energy into the reaction mixture without dramatically altering its overall physical properties, thus leaving the solution bulk relatively unperturbed. Nonhazardous acoustic waves dramatically enhance mass transfer processes, creating higher selectivity and yields and often reducing the number of synthetic steps. Water and aqueous mixtures are the typical media in sonochemistry. The most common US reactors are depicted in Figure 4.1.

Combined US/MW irradiation can be performed simultaneously with a non-metallic US-horn directly inserted into the MW cavity (Figure 4.2) or otherwise sequentially by circulating the reacting mixture in a loop reactor (Figure 4.4).

Quartz, pyrex®, ceramic and a specific type of PEEK® are the most common MW-transparent materials used by the authors, in collaboration with Danacamerini (Torino, Italy), to build suitable horns (Figure 4.2 on the left). Of these pyrex® is the cheapest material, it is highly transparent to MW and shows satisfactory resistance to US vibrations.

External cooling is often required to keep the temperature constant during US/MW irradiation and to avoid a drop in cavitation.

Efficient cooling can be achieved by circulating a refrigerating fluid that is transparent to MW, for example, silicone oil or Galden® (Solvay-Solexis), a perfluoropolyether with a high boiling point and low viscosity (Figure 4.3).

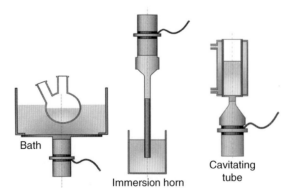

Figure 4.1 Example of sonochemical apparatus (Author's laboratory).

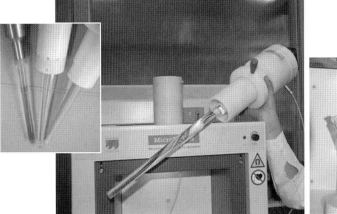

Figure 4.2 On the upper left the picture shows quartz, pyrex® and PEEK® US horns. Center, pyrex® horn insertion into a professional MW oven (MicroSYNTH, Milestone) for simultaneous US/MW irradiation and, on the right, a pear-shaped vessel set with a fiber optic temperature probe and US horn (Author's laboratory).

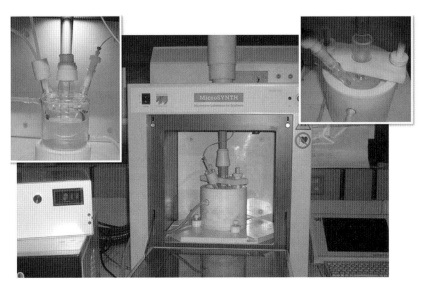

Figure 4.3 Simultaneous US/MW irradiation with cooling systems. On the right, a Teflon vessel with refrigerating fluid into which the reaction flask can be immersed; on the left, a glass vessel with an integrated external cooling chamber (Author's laboratory).

In a simpler method, US and MW can be applied in two separate steps as sequential treatments.

Sequential US/MW irradiation can be achieved via the use of flow loop-reactors. In these reactors, a pump circulates the reacting mixture through two separate

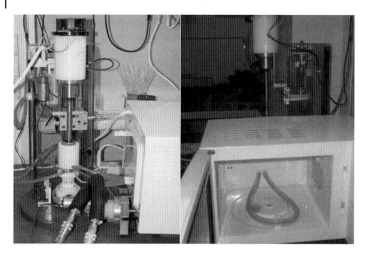

Figure 4.4 US/MW loop reactor with a peristaltic pump and cooling by coaxial tubing, a titanium US horn and a domestic MW oven (Author's laboratory).

reaction cells, one placed inside the MW oven and the other (fitted with a US probe) outside it (Figure 4.4) [7].

4.2
Simultaneous Ultrasound/Microwave Treatments

In the last ten years several papers have reported the advantageous use of combined US/MW irradiation for the preparation of nanomaterials.

In 2011, Li et al. [8] described the synthesis of well-defined flower-like zinc oxide (ZnO) nanostructures with a central petal and six symmetrical petals growing radially from the center. A solution of zinc acetate in aqueous sodium hydroxide was sonicated for 5 min at a power rating of 1000 W, and then treated with combined US/MW (pulsed US, 1 s irradiation and 2 s pause) at the power rating of 500 W for 30 min. This special flower-like feature of ZnO (Figure 4.5) showed stronger photocatalytic performance in Methylene Blue (MB) degradation than ZnO microrods.

Platinum nanoparticles were prepared by Ishikawa et al. [9] in a heterogeneous solid–liquid system and an excellent fine and homogeneous size distribution was obtained.

Tai and Guo's group developed an US/MW-assisted protocol for the synthesis of cadmium sulfide nanoparticles [10]. In 2008, they carried out a fast, high-yield, seedless and template-free preparation of flower-like hexagonal nanopyramids and nanoplates.

A powder mixture with $CdCl_2$, thiourea (Tu) or thioacetamide (TAA) and sulfur was placed in a combined US/MW reactor (Figure 4.6) and irradiated with MW

Figure 4.5 SEM images of the flower-like ZnO nanostructures. (Reprinted from Ref. [8], Copyright (2011), with permission from Elsevier).

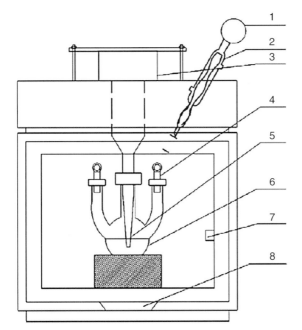

Figure 4.6 US/MW reactor. (1) Balloon, (2) condenser, (3) US transducer, (4) glass pipe, (5) US horn (from Xinzhi Co., China, JY92-2D, 20 kHz, pulsed mode with a duty cycle of 0.5 s), (6) reaction vessel, (7) IR thermometer and (8) MW generator. (Reprinted from Ref. [10], Copyright (2008), with permission from Elsevier).

and US power ratings of 400 W and 60 W cm^{-2}, respectively. Different irradiation times and temperatures were tested since they were critical points for the production of well-defined CdS nanoflowers; the optimal irradiation parameters were 45 min at 140 °C (Figure 4.7).

The authors also tested the synthesis of these cadmium nanostructures under simple MW irradiation and observed a large amount of residual sulfur powder in

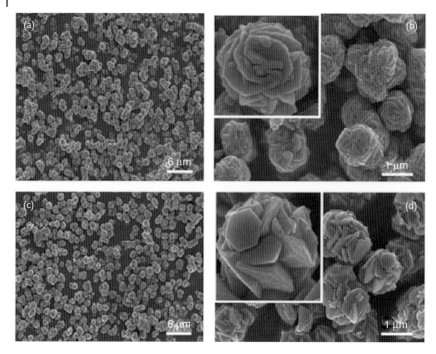

Figure 4.7 FESEM images of CdS nanoflowers synthesized with simultaneous US/MW irradiation at 140 °C for 45 min, mixing CdCl$_2$, sulfur powder and other different sulfur sources (Tu (a,b) and TAA (c,d)). (Reprinted from Ref. [10], Copyright (2008), with permission from Elsevier).

the resulting triangular and hexagonal pyramid nanostructures (Figure 4.8a). The combined US/MW protocol was able to fully transform sulfur powder into CdS, as shown by the XRD patterns (Figure 4.8b).

In this procedure, US irradiation reduced sulfur powder aggregation, thus promoting the effective reaction between the sulfur powders and cadmium ions. Moreover, US avoided or reduced the localized overheating that may have been caused by the MW irradiation and thus more uniform nanostructures were produced. Such synergistic effects provided the optimal experimental conditions needed to completely convert the sulfur sources into CdS hexagonal nanopyramids and nanoplates and to permit their regular assembly into 3D nanoflowers.

The same combined US/MW protocol was used by Guo's group [11] for the synthesis of Ag-doped CdS nanoparticles, which were made from cadmium acetate, thiourea and sulfur powders in the presence of AgNO$_3$ (15 min at 140 °C).

In an investigation into the individual contribution of US and MW, it was established that MW irradiation alone was only able to afford low quantities of products that were highly contaminated with unreacted sulfur powder. Simple US irradiation or mechanical stirring gave quite inhomogeneous products in much longer reaction times.

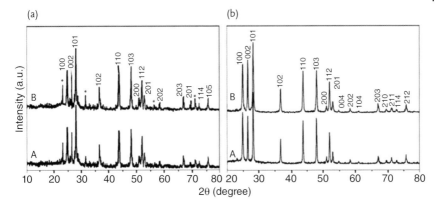

Figure 4.8 XRD patterns of CdS nanostructures prepared by MW alone (a) or by US/MW irradiation (b) at 140 °C for 45 min with reacting $CdCl_2$ and various sulfur sources. (A) TAA and sulfur powder, (B) Tu and sulfur powder. The peaks marked * are from sulfur powders. (Reprinted from Ref. [10], Copyright (2008), with permission from Elsevier).

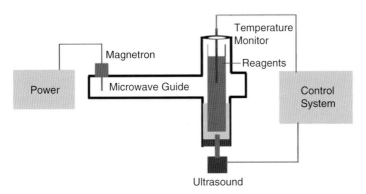

Figure 4.9 Schematic illustration of the laboratory-made US/MW system (MW power from 0–300 W, US from 0–50 W with a frequency of 38.4 kHz). (Reprinted from Ref. [13], Copyright (2009), with permission from Wiley).

Combined US/MW irradiation was, therefore, considered to be the method of choice for the production of noble metal-doped semiconductor nanostructures. Thanks to fast and volumetric heating and simultaneous pulsed sonication, the sulfur powder was finely dispersed and Ag easily penetrated the CdS matrix.

A further application of this combined US/MW procedure was the preparation of CdS nanoparticles from a zinc-blende to wurtzite structure [12].

In 2009, Shen [13] applied US/MW irradiation to the rapid synthesis of Pb(OH)Br nanowires, starting from $Pb(OAc)_2$ and 1-butyl-3-methylimidazolium bromide ([BMIM]Br) as reactant and structure-directing agent (see schematic diagram of the US/MW apparatus in Figure 4.9).

The synthesis was carried out under conventional heating in a water bath at 70 °C for 24 h for comparison and yielded multi-angular prisms (23.0%) of 20–30 μm in diameter and 2–3 mm in length (Figure 4.10a and b, Table 4.1).

MW irradiation, at 50 W for 10 min, gave uniform wires (35.1%) with a diameter of about 1800–2000 nm and a length of 60–80 μm (Figure 4.10c, Table 4.1), while US alone at 50 W for 10 min afforded finer but less uniform products (32.9%) (Figure 4.10d, Table 4.1). The product yield was higher (45.0%) under combined US/MW irradiation (both 50 W, 10 min) than those obtained under MW or US alone and the nanowires created (Figure 4.10e) were straight and smooth with diameters ranging from 80–800 nm and lengths of 50–100 μm.

A further increase in MW irradiation power to 250 W (50 W for US) led to a dramatic reduction in reaction time (80 s) and the formation of nanowires (48.2%) of much shorter diameter (100–500 nm) and length (10–30 μm) (Figure 4.10f, Table 4.1).

Figure 4.10 SEM images of Pb(OH)Br synthesized by (a) and (b) conventional heating at 70 °C for 24 h, (c) MW at 50 W for 10 min, (d) US at 50 W for 10 min, (e) simultaneous US/MW irradiation at 50 W/50 W for 10 min, and (f) at 250 W/50 W for 80 s. The scale bar in (a–f) is 20 μm and in (b) it is 10 μm. (Reprinted from Ref. [13], Copyright (2009), with permission from Wiley).

Table 4.1 Synthesis of Pb(OH)Br nanowires: comparison of conventional heating, MW, US and simultaneous US/MW irradiation.

	Reaction time	Temperature (°C)	Yield (%)	Diameter (nm)	Length (μm)
Conventional heating	24 h	70	23.0	20 000–30 000	2000–3000
MW (50 W)	10 min	89	35.1	1 800–2 000	60–80
US (50 W)	10 min	34	32.9	700–1 500	20–40
US/MW (50 W/50 W)	10 min	66	45.0	80–800	50–100
US/MW (250 W/50 W)	80 s	95	48.2	100–500	10–30

Reprinted from Ref. [13], Copyright (2009), with permission from Wiley.

Simultaneous US/MW irradiation demonstrated favorable effects on the synthesis of 1D nanostructures with a marked improvement in product yield and reaction rate.

4.3
Sequential Ultrasound and Microwaves

Sequential US/MW irradiation, used in several synthetic applications, has appeared on a great number of occasions in the literature. US and MW can be applied separately in different steps of the same reaction or, otherwise, in sequential reactions.

4.3.1
Sequential Steps of the Same Reaction

A fast and efficient MW-assisted solvothermal strategy [14] was used for the synthesis of metal nanoparticles (Pd, Ni, Sn) supported on sulfonated multi-walled carbon nanotubes (SF-MWCNTs). SF-MWCNTs obtained through the oxidation/sulfonation of pristine MWCNT were added to a $PdCl_2$ solution at pH 7.4 and sonicated for 1 h. The solution was transferred into a specific vessel (liner-rotor 16 F100 TFM) and then placed in a Multiwave 3000 MW oven (1400 W max, Anton Paar) and heated at a 1000 W power rating at 170 °C for 60 s (ramp of 13 min).

SF-MWCNT-Pd and its mixed bimetallic electrocatalysts were tested for ethanol oxidation in an alkaline medium. The observed behavior showed that the mixed Pd-based catalysts (obtained by simple US-mixing of the individual MWCNT-metal nanocomposites) gave better electrocatalytic activity than their alloy nanoparticles (obtained by the co-reduction of metal salts) or Pd alone.

Zinc phosphate ($Zn_3(PO_4)_2$) nanocrystals were prepared via a simple and efficient US/MW assisted route [15]; dodecahydrate $Na_3PO_4 \cdot$ and hexahydrate $Zn(NO_3)_2 \cdot$ aqueous solutions were added dropwise into deionized water in the presence of Triton X-100 and irradiated with US until precipitation was complete. The white solid was washed with deionized water and dried completely for 2 h in an MW oven at 400 W. US promoted separation between particles in the depositing process in this work, while MW avoided agglomeration in the drying step.

Hydroxyapatite (HAP) is a widely used biocompatible ceramic in many biomedical applications and devices. Recently, Poinern et al. [16] prepared ultrafine nano-HAP via a wet precipitation protocol using $Ca(NO_3)_2$ and KH_2PO_4 in the presence of NH_4OH, used as precipitation agent, under US irradiation. During wet milling, US caused the sample particles to be efficiently dispersed and de-agglomerated. Furthermore, the influence of two different thermal treatments was evaluated, these were conventional radiant tube furnace and MW heating.

A solution of tetrahydrate $Ca(NO_3)_2$ at pH 9 was exposed to US irradiation at 50 W for 1 h (UP50H, 30 kHz, MS7 Sonotrode, Hielscher, Germany). At the end of this first hour a KH_2PO_4 solution was slowly added to the first and sonicated for 1 h further. Portions of the white precipitate underwent various different thermal treatment methods, that is, MW at powers ranging from 20 to 100 W (1100 W at 2450 MHz; LG®, Australia) and conventional heating at temperatures ranging from 100 to 400 °C. Both thermal techniques effectively afforded ultrafine HAP nanoparticles with similar crystalline structure, morphology and particle sizes, 30 nm in diameter for conventional furnace and 36 nm for MW oven, respectively.

Ultrafine and highly crystalline hydroxyapatite particles prepared in this manner were examined as potential agents for the removal of fluoride from contaminated water and were proved to be very effective [17]. The solution pH after de-fluoridation was within the normal pH range for potable water and so there was no need for it to be adjusted for human consumption.

Hao et al. prepared magnetic composites of nickel-coated carbon nanofibers (CNFs) by employing a simple MW-assisted procedure [18].

In a typical experiment, CNFs and hexahydrate $NiCl_2$ were dispersed in ethylene glycol using US vibration for 0.5 h. The slow addition of monohydrate hydrazine (80 wt% solution) to the solution was followed by the addition of aqueous NaOH. The solution was then irradiated with MW for 5 min at a 425 W power rating. Upon such functionalization, the CNFs can be aligned in a relatively small external magnetic field and thus incorporated into metal matrixes to produce new composites with novel properties.

In 2011, Hu et al. [19] prepared Au/CNT hybrid nanostructures via a MW-assisted procedure. Pristine CNTs were sonicated in the mixed oleylamine and oleic acid and then irradiated with MW (WF-4000C system). The temperature was accurately controlled via the automatic adjusting of MW power. $HAuCl_4$ in ethylene glycol was added to the reaction system when the temperature reached 180 °C and the reaction was kept at the same temperature for a further 15 min.

Au NPs were well assembled on the surface of the CNTs without aggregation in solution suggesting the existence of highly selective Au NP nucleation on the functional CNT surface (Figure 4.11). Since Au/CNTs nanocomposites are

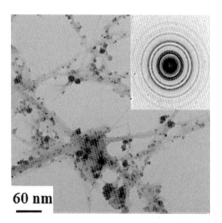

Figure 4.11 TEM image of Au/CNT composites (inset, corresponding SAED pattern) (Reprinted from Ref. [19], Copyright (2011), with permission of Elsevier).

conductive nanomaterials, they could be incorporated into the electrochemical systems to promote the electron exchange between electrode and electrochemical probe.

Yttria-stabilized zirconia (YSZ) nanocrystals were synthesized by Wang et al. using a US/MW assisted method [20]. Aqueous solutions of yttrium oxide and octahydrate zirconium oxychloride were irradiated with US in the presence of Triton X-100. The precipitate was washed with water reacted with hydrogen peroxide once the pH was adjusted to 9. The final precursor was dried under MW heating (600 W). YSZ nanoparticles mainly showed diameters of about 25 nm. Both US vibrations and MW heating promoted the collision probability of ions, enhancing the reaction rate.

Graphene nanosheets were synthesized from expandable graphite by MW heating and US assisted dispersion [21]. In a typical synthetic protocol, expandable graphite flake was put in an alumina melting pot and placed in a domestic MW oven (LG-MP700, power of 700 W, 90 s) to give a product that had expanded 150-fold. The graphite obtained was suspended in ethanol overnight and then sonicated at room temperature for 11 h, affording graphene nanosheets. The experimental results indicate that graphene nanosheets have great potential as a field emitter in various electron sources.

In 2008, Nyutu et al. [22] developed the INM process to obtain spinel metal oxides (nickel ferrites and zinc aluminates) using combined *in situ* mixing (I), US nozzle treatment (N), and MW heating (M) (Figure 4.12).

The authors investigated the effects of MW heating and US nozzle mixing on the interfacial properties, composition, particle size, purity and surface area of the resulting nanoparticles.

When the starting materials were simply flowed into the test tube without the US nozzle (sample 1–3, 9; Table 4.2) or when the US nozzle was used without

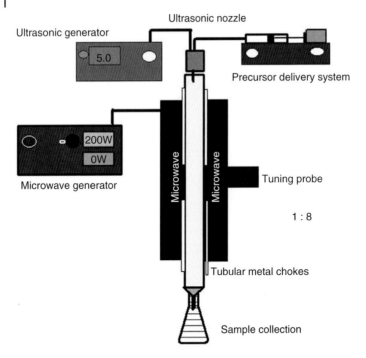

Figure 4.12 Scheme of continuous flow US-nozzle MW (INM) reactor, made up of an US nozzle (Sono-Tek Corp., 48 and 120 kHz) and a tunable, cylindrical single-mode cavity (Wavemat MW Processing System model CMPR 250, max power 1250 W) (Reprinted from Ref. [22], Copyright (2008), with permission from ACS).

Table 4.2 Experimental conditions of nickel ferrite preparation and results from XRD.

Sample	Nozzle (kHz)	MW (W)/ Temp. (°C)	Calcination temp. (°C)	XRD crystallite size (nm)	Present phases
1/9	–	–	800/600	39[a], 22[b]	$NiFe_2O_3$ + impurities
2	–	100 W	800	46[b]	$NiFe_2O_3$ + impurities
3	–	300 W	800	13[a]	$NiFe_2O_3$ + impurities
4	120	100 W	800	29[a], 18[b]	$NiFe_2O_3$ + impurities
5	120	–	800	7[a], 6[b]	$NiFe_2O_3$ + impurities
6	120	300 W	600	6[b]	$NiFe_2O_3$
7	120	300 W	800	9[a], 13[b]	$NiFe_2O_3$
8	120	400 W	800	19[b]	$NiFe_2O_3$
10	120	100 °C	600	11	$NiFe_2O_3$
11	120	150 °C	600	10	$NiFe_2O_3$
12	48	300 W	600	8	$NiFe_2O_3$
13	120	300 W	600	12	$NiFe_2O_3$

a) Refers to CO_3^{2-} as coprecipitating agent.
b) Refers to OH^- as coprecipitating agent (Reprinted from Ref. [22], Copyright (2008), with permission from ACS).

MW irradiation (sample 5, Table 4.2), the product obtained after calcination was impure (multiple phase) regardless of MW irradiation or MW power usage. No single-phase products could be found (sample 4, Table 4.2) when the nozzle was combined with MW at a power of <300 W. Pure products were obtained with a MW power setting of 300–400 W combined with the 120 kHz nozzle (sample 6–8, 12–13, Table 4.2).

The field emission scanning electron microscope (FESEM) images proved the existence of a lower degree of particle agglomeration when a low-frequency nozzle (4 kHz; Figure 4.13b, sample 12) was used. This is probably because high-frequencies (120 kHz; Figure 4.13c, sample 13) lead to the formation of smaller particles which tend to cluster. The control experiment (no MW, no US nozzle) (sample 9, Figure 4.13a) showed a huge amount of agglomerated particles.

Nyutu's group also tested also a process called INC (*in situ* nozzle, conventional heating) with the aim of comparing MW and conventional heating. In this case, pure $NiFe_2O_3$ was obtained (sample 10–11, Table 4.2), but particles were highly aggregated and less uniform, with lower surface areas than those obtained from the INM process (Figure 4.13d). Since the use of the US nozzle or MW separately did not result in a pure material, the sequential combination of the techniques was sufficient to provide single-phase products.

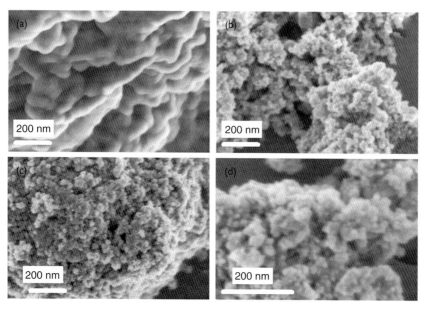

Figure 4.13 FESEM micrographs of (a) conventional mixing of nickel ferrite precursors (no nozzle and MW, sample 9 in Table 4.2). (b) INM 300 W MW, 48 kHz nozzle (sample 12), (c) INM 300 W MW, 120 kHz nozzle (sample 13), (d) INC 100 °C, 120 kHz nozzle (sample 10) (Reprinted from Ref. [22], Copyright (2008), with permission from ACS).

Kuznetsova et al. [23] described the formation of $CoFe_2O_4$ nanocrystals from $CoCl_2$ and $FeCl_3$ in the presence of NaOH under hydrothermal conditions (MW or conventional heating) and in doing so investigated the effects of preliminary US treatment.

$CoFe_2O_4$ nanocrystalline ferrite was synthesized at 130 °C for 24 h in a steel autoclave, and MW heating gave complete conversion in 2 h (MARS-5 MW oven). In the case of conventional heating, pre-sonication (3 min, in the presence of NaOH) did not affect particle formation and growth, while in the case of MW heating US pretreatment strongly reduced the reaction time and gave a much higher conversion percentage. US alone did not lead to $CoFe_2O_4$ ferrite formation.

Tapala et al. [24] reported a sonocatalyzed MW-heated hydrothermal preparation of tetragonal phase-pure lead titanate nanopowders. A mixture of $Pb(NO_3)_2$, TiO_2 and KOH was sonicated at 70 (±5) °C for 3 h in a US bath (Bandelin Electronic RK255H, 160/320 W, 35 kHz) then placed in a 95 (±5) °C water bath in a domestic MW oven (Whirlpool XT-25ES/S, 900 W) for 1–3.5 h at continuous MW power (720 W). MW heating reduced the reaction temperature from 180 °C for conventional heating to 95 (±5) °C. MW irradiation did not affect particle size and aggregation, which may be mainly governed by the US irradiation step, but reduced the tetragonal distortion and improved the regularity of the crystal habit.

A CuO nanofluid was prepared by Zhu et al. [25] by transforming an unstable copper(II) hydroxide precursor to copper(II) oxide in water under sonication and then heating using MW. Nanofluids were generated from a suspension of solid nanoparticles in traditional fluids, conferring upon them excellent heat transfer properties.

Copper(II) hydroxide was sonicated in the presence of ammonium citrate, used as a dispersant, for 30 min in a US disrupter (Autoscience, AS3120A, 120 W) and then irradiated with MW (3 min) (Galanz, WP750) to afford a black CuO nanofluid. The samples synthesized without US or MW irradiation gave unsatisfactory results. In particular, US alone did not provide complete precursor transformation, while MW irradiation led to the formation of irregular CuO flakes in large agglomerates when used in isolation.

US irradiation was employed as an efficient alternative mixing mechanism in the synthesis of Pt/C catalysts (platinum salt and carbon in acetone) which were then dried in a MW oven in a repeated 5 s on/off cycle [26]. The platinum oxide formed on carbon was then reduced with formic acid and dried again in a MW oven under Ar (20 s on/15 s off cycle) affording a uniformly dispersed Pt/C catalyst.

MW drying heats the sample bulk evenly, unlike conventional drying methods, making every particle reach the same temperature instantaneously and, thanks to the internal pressure formed by the expansion of moisture during evaporation, the lumped sample is broken into powder, resulting in smaller particles and subsequently a higher surface area.

4.3.2
Sequential Reactions

Recently, Hu et al. [27] synthesized graphene nanosheet–gold (GNS-Au) nanocomposites under MW irradiation in aqueous media. Graphene oxide (GO) was dispersed in deionized water and then sonicated for 1 h to generate exfoliated graphene oxide nanosheets (GONS). After the addition of aqueous $HAuCl_4$ to the GONS dispersion and further sonication for 15 min, the mixture was placed in a MW oven (MAS-I, Sineo MW Chemistry Technology, China) and then irradiated at 80 °C for 5 min (800 W). Meanwhile, hydrate hydrazine was dropped into the suspension. The final product was then dried to obtain the GNS–Au nanocomposites with a mean diameter of 9.3 nm. In this way, the GNS surfaces were strictly coated with Au nanoparticles in order to prevent GNS from agglomerating. GNS-Au nanocomposite-modified electrodes showed greatly enhanced electrochemical response values when compared with both simple and GNS-modified glassy carbon electrodes.

In 2011 Si et al. [28] synthesized nano-amorphous indium trioxide, a good photocatalyst in the degradation of acidic black dye. An indium hydroxide precursor was prepared by ultrasonication in the presence of carbamide and an indium salt; indium trioxide was obtained under MW irradiation, with an average grain size of 12 nm.

Tiwari's group described the synthesis of diamond films by MW plasma chemical vapor deposition (MPCVD) on a $Pt/SiO_2/Si$ substrate [29]. The platinum particles were sprayed onto the SiO_2/Si surface at room temperature, whereas adamantane was deposited onto the SiO_2/Si surface by ultrasonic treatment (10 min) (Figure 4.14).

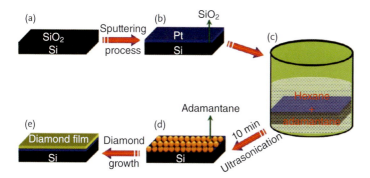

Figure 4.14 Schematic diagram showing diamond film synthesis in five steps: (a) silicon substrate with native oxide layer, (b) Pt coated on the SiO_2/Si substrate by a sputtering process, (c) $Pt/SiO_2/Si$ immersed into the hexane + adamantane solution, (d) adamantane deposited on the $Pt/SiO_2/Si$ surface by ultrasonication, and (e) diamond growth by MPCVD (Reprinted from Ref. [29], Copyright (2011), with permission of ACS).

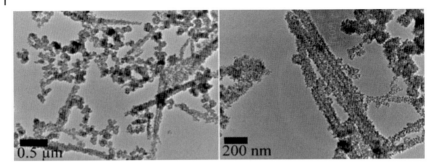

Figure 4.15 TEM micrographs of CuS nanotubes, prepared using Cu(OH)$_2$ nanowires by MW heating at 80 °C for 120 min (Reprinted from Ref. [30], Copyright (2011), with permission of Elsevier).

Finally, the adamantane-seeded Pt/SiO$_2$/Si substrates were placed in a MPCVD system (1.5 AsTeX-type) for diamond growth.

In this process, adamantane converts not only into the nanodiamond phase but also into other carbon phases which can then act as nuclei for diamond growth.

Copper sulfide (CuS) nanotubes were successfully synthesized via a MW-assisted solvothermal procedure using Cu(OH)$_2$ nanowires as the precursor and template (assembled with nanoparticles) [30]. Since CuS is an important transition metal semiconductor, it has a wide variety of applications in solar cells, photocatalysis, lithium-ion batteries and sensors. Cu(OH)$_2$ nanowires were prepared using US irradiation for 30 min from aqueous CuCl$_2$ and NaOH at room temperature. Successively, thiourea and Cu(OH)$_2$ nanowires were stirred in ethylene glycol and then irradiated with MW at 80 °C for 60 min in a 60 ml Teflon-lined sealed autoclave (MDS-6, Sineo, China) (Figure 4.15).

The MWCNTs surface was purified by MW-assisted acid purification then modified with HNO$_3$ under US irradiation and finally functionalized with platinum nanoparticles in methanol expanded with supercritical CO$_2$ [31]. Platinum nanoparticles on the functionalized MWCNTs shown in TEM images (Figure 4.16) were approximately 4 nm on the outer tube walls. The purification of MWCNTs was carried out under MW irradiation in 4.8 M H$_2$SO$_4$ at either 180 or 200 °C for 30 min (ETHOS SEL, Milestone). Purified MWCNTs were then sonicated in 1 HNO$_3$ at 25 °C (Branson 450) at 20 kHz for 10 min.

These sequential treatments attained both good MWCNT dispersion in solution and gave outer layer MWCNT surface modification with functional groups (i.e., –COOH and –OH) that was efficient enough to provide nucleation sites for the deposition of platinum nanoparticles. The surface modification of MWCNTs is necessary for metal deposition onto carbon. The functionalization of MWCNTs with PtCl$_2$ was obtained in supercritical CO$_2$ after sonication in methanol for 10 min.

Rangari et al. [32] studied the effect of MW on the curing behavior of an epoxy resin system in the presence and absence of MWCNTs. US irradiation was used

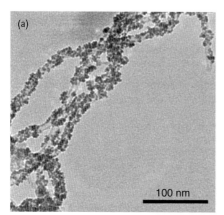 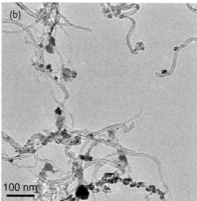

Figure 4.16 TEM images of CNTs. (a) purified at 200 °C in MW, ultrasonicated at 25 °C for 15 min and subsequently metalized with platinum at 80 °C and 120 bar in MeOH expanded with supercritical CO_2 for 30 min, (b) purified at 200 °C in MW for 30 min, refluxed for 24 h and subsequently metalized with platinum at 80 °C and 120 bar in MeOH expanded with supercritical CO_2 for 30 min (Reprinted from Ref. [31], Copyright (2010), with permission of Elsevier).

for the reinforcement of the EPON 862 Part A resin with 0.1–0.3 wt% MWCNTs (Ti-horn, 20 kHz, 100 W cm^{-2}) at an amplitude of 50% for 15 min. The mixture obtained was cured using a MW oven (Sharp BP210) for only 10 min instead of the 8 h needed with conventional heating (curing process is 4 h at 120 °C and post-curing at 170 °C for 4 h). Since the graphite carbon absorbs the maximum amount of MW radiation (1100 °C in 2 min), this process led to the rapid curing of EPON 862 resin in the presence of CNTs.

In 2011 the same authors evaluated the mechanical properties of neat EPON 862 and CNT infused EPON 862 nanocomposites [33]; the fast MW curing method was better than the prolonged conventional heating curing method and showed increases in compression modulus and strength.

Mediavilla et al. [34] described the synthesis of Pt/H-ZSM5 catalysts through a MW-assisted protocol followed by US irradiation. Platinum salts dissolved in ethylene glycol were irradiated at different power ratings (184 and 461 W) for time periods of 30 and 60 s (Samsung MW1050 oven) in order to form platinum nanoparticle suspensions, which were then mixed with the selected supports, sonicated for 30 min and finally reduced under H_2 at 500 °C for 6 h to afford the desired platinum catalysts.

Studies into MW irradiation time and power demonstrated the existence of some favorable effects on the reducibility of the catalysts, these included a strong (13-fold) increase in activity as irradiation time was raised from 30 to 60 s. However, a 90 s MW treatment resulted in the total loss of catalytic activity. The catalyst prepared at 461 W for 60 s showed the highest activity because of the smaller nanoparticle size.

Sequential US/MW irradiation has been applied, in several cases, to the preparation of nanofluids. For example, in 2010 Chandrasekar et al. [35] synthesized nanocrystalline alumina (Al_2O_3) via a MW-assisted chemical precipitation method and then the nanofluid by dispersing Al_2O_3 nanoparticles in water under US irradiation for 6 h (Toshiba, India, US pulses of 100 W, at 36 ± 3 kHz) in order to obtain a homogeneous dispersion and a stable suspension.

4.4 Conclusions

The impressive additive and even synergic effects of the combination of US and MW irradiation on the preparation of nanomaterials have been described. The combined irradiation is often the method of choice for the synthesis of noble metal-doped semiconductor nanostructures, and metal nanoparticles and nanowires in general.

US irradiation strongly enhances the mixing of reacting mixtures, a fundamental condition in nanoparticles synthesis. Even though cavitation bubbles cannot break up nanoparticles individually, they can break the nanoparticle cluster into smaller clusters, resulting in higher stability and generating nucleating new phase centers.

As has been widely demonstrated, MW provides fast, efficient and homogeneous heating which leads to significant improvements in product yields and reaction rates. The application of US irradiation during MW heating can reduce and avoid overheating in local areas and produce more uniform nanostructures. In several cases, the sequential application of US and MW has been chosen in order to benefit from distinct US and MW irradiation effects for the preparation of multi-component metal oxides, highly activated catalysts, functionalized MWCNTs and nanofluids.

In particular, it should be mentioned that US cannot be applied throughout MW heating when the reaction temperature is near to the boiling point of the solvent, since the boiling bubbles extinguish the cavitation bubbles, countering US effects. This method can also be useful when different steps of a reaction or different reactions in sequence require different treatments (e.g., heating or just mixing). These hybrid techniques will surely find further practical applications in nanoscience and nanotechnology.

References

1 Maeda, M., and Amemiya, H. (1995) Chemical effects under simultaneous irradiation by microwaves and ultrasound. *New J. Chem.*, **19**, 1023–1028.

2 Cravotto, G., and Cintas, P. (2007) The combined use of microwaves and ultrasound: new tools in process chemistry and organic synthesis. *Chem. Eur. J.*, **13**, 1902–1909.

3 Cintas, P., Cravotto, G., and Canals, A. (2012) Combined ultrasound–microwave technologies, in *Handbook on Applications*

of Ultrasound (eds D. Chen, S.K. Sharma, and A. Mudhoo), CRC Press, Taylor and Francis Group, New York, pp. 659–673.

4 Kharissova, O.V., Kharisov, B.I., Valdés, J.J.R., and Méndez, U.O. (2011) Ultrasound in nanochemistry: recent advances. *Synth. React. Inorg. Met.-Org. Nano-Met. Chem.*, **41** (5), 429–448.

5 Bilecka, I., and Niederberger, M. (2010) Microwave chemistry for inorganic nanomaterials synthesis. *Nanoscale*, **2** (8), 1358–1374.

6 Patete, J.M., Peng, X.H., Koenigsmann, C., Xu, Y., Karn, B., and Wong, S.S. (2011) Viable methodologies for the synthesis of high-quality nanostructures. *Green Chem.*, **13** (3), 482–519.

7 Cravotto, G., Beggiato, M., Penoni, A., Palmisano, G., Levêque, J.M., and Bonrath, W. (2005) High-intensity ultrasound and microwave, alone or combined, promote Pd/C-catalyzed aryl–aryl couplings. *Tetrahedron Lett.*, **46**, 2267–2271.

8 Li, H., Liu, E., Chan, F.Y.F., Lu, Z., and Chen, R. (2011) Fabrication of ordered flower-like ZnO nanostructures by a microwave and ultrasonic combined technique and their enhanced photocatalytic activity. *Mater. Lett.*, **65** (23–24), 3440–3443.

9 Ishikawa, D., Hayashi, Y., and Takizawa, H.J. (2008) Preparation of platinum nanoparticles in heterogeneous solid-liquid system by ultrasound and microwave irradiation. *Nanosci. Nanotechnol.*, **8**, 4482–4487.

10 Tai, G., and Guo, W. (2008) Sonochemistry-assisted microwave synthesis and optical study of single-crystalline CdS nanoflowers. *Ultrason. Sonochem.*, **15**, 350–356.

11 Ma, J., Tai, G., and Guo, W. (2010) Ultrasound-assisted microwave preparation of Ag-doped CdS nanoparticles. *Ultrason. Sonochem.*, **17**, 534–540.

12 Tai, G., Zhou, J., and Guo, W. (2010) Inorganic salt-induced phase control and optical characterization of cadmium sulfide nanoparticles. *Nanotechnology*, **21** (17), 175601–175607.

13 Shen, X.-F. (2009) Combining microwave and ultrasound irradiation for rapid synthesis of nanowires: a case study on Pb(OH). *Br. J. Chem. Technol. Biotechnol.*, **84**, 1811–1817.

14 Ramulifho, T., Ozoemena, K.I., Modibedi, R.M., Jafta, C.J., and Mathe, M.K. (2012) Fast microwave-assisted solvothermal synthesis of metal nanoparticles (Pd, Ni, Sn) supported on sulfonated MWCNTs: Pd-based bimetallic catalysts for ethanol oxidation in alkaline medium. *Electrochim. Acta*, **59**, 310–320.

15 Wang, J., Li, D., Liu, J., Yang, X., He, J., and Lu, Y. (2011) One-step preparation and characterization of zinc phosphate nanocrystals with modified surface. *Soft Nanosci. Lett.*, **1** (3), 81–85.

16 Poinern, G.J.E., Brundavanam, R., Le, X.T., Djordjevic, S., Prokic, M., and Fawcett, D. (2011) Thermal and ultrasonic influence in the formation of nanometer scale hydroxyapatite bio-ceramic. *Int. J. Nanomedicine*, **6**, 2083–2095.

17 Poinern, G.E.J., Ghosh, M.K., Ng, Y.-J., Issa, T.B., Anand, S., and Singh, P. (2011) Defluoridation behavior of nanostructured hydroxyapatite synthesized through an ultrasonic and microwave combined technique. *J. Hazard. Mater.*, **185** (1), 29–37.

18 Hao, C., Li, X., and Wang, G. (2011) Magnetic alignment of nickel-coated carbon fibers. *Mater. Res. Bull.*, **46** (11), 2090–2093.

19 Hu, Q., Gan, Z., Zheng, X., Lin, Q., Xu, B., Zhao, A., and Zhang, X. (2011) Rapid microwave-assisted synthesis and electrochemical characterization of gold/carbon nanotube composites. *Superlattices Microstruct.*, **49** (5), 537–542.

20 Wang, J.-D., Luo, C.-X., Liu, J.-K., Lu, Y., and Li, G.-M. (2010) Synthesis of yttria-stabilized cubic zirconia nanocrystals by ultrasonic-microwave route. *Nano*, **5** (5), 271–277.

21 Dong, J., Zeng, B., Lan, Y., Tian, S., Shan, Y., Liu, X., Yang, Z., Wang, H., and Ren, Z.F. (2010) Field emission from few-layer graphene nanosheets produced by liquid phase exfoliation of graphite. *J. Nanosci. Nanotechnol.*, **10** (8), 5051–5055.

22 Nyutu, E.K., Conner, W.C., Auerbach, S.M., Chen, C.-H., and Suib, S.L. (2008) Ultrasonic nozzle spray *in situ* mixing

and microwave-assisted preparation of nanocrystalline spinel metal oxides: nickel ferrite and zinc aluminate. *J. Phys. Chem. C*, **112**, 1407–1414.

23 Kuznetsova, V.A., Almjasheva, O.V., and Gusarov, V.V. (2009) Influence of microwave and ultrasonic treatment on the formation of $CoFe_2O_4$ under hydrothermal conditions. *Glass Phys. Chem.*, **35**, 205–209.

24 Tapala, S., Thammajak, N., Laorattanakul, P., and Rujiwatra, A. (2008) Effects of microwave heating on sonocatalyzed hydrothermal preparation of lead titanate nanopowders. *Mater. Lett.*, **62**, 3685–3687.

25 Zhu, H.T., Zhang, C.Y., Tang, Y.M., and Wang, J.X. (2007) Novel synthesis and thermal conductivity of CuO nanofluid. *J. Phys. Chem. C*, **111**, 1646–1650.

26 Tian, Z.Q., Xie, F.Y., and Shen, P.K. (2004) Preparation of high loading Pt supported on carbon by on-site reduction. *J. Mater. Sci.*, **39**, 1507–1509.

27 Hu, H., Wang, X., Xu, C., Wang, J., Wan, L., Zhang, M., and Shang, X. (2012) Microwave-assisted synthesis of graphene nanosheets-gold nanocomposites with enhancing electrochemical response. *Fuller. Nanotub. Car. N.*, **20** (1), 31–40.

28 Si, W., Yu, J.-Y., and Li, Q. (2011) Synthesis of nano amorphous In_2O_3 by microwave method and its photocatalysis. *Adv. Mater. Res.*, **299–300** (Pt. 1, Materials and Manufacturing), 542–545.

29 Tiwari, R.N., Tiwari, J.N., Chang, L., and Yoshimura, M. (2011) Enhanced nucleation and growth of diamond film on Si by CVD using a chemical precursor. *J. Phys. Chem. C*, **115** (32), 16063–16073.

30 Liu, X.-L., and Zhu, Y.-J. (2011) CuS nanotubes prepared using $Cu(OH)_2$ nanowires as self-sacrificial template. *Mater. Lett.*, **65** (7), 1089–1091.

31 Liu, J., Ebert, A., Variava, M.F., Dehghani, F., and Harris, A.T. (2010) Surface modification and Pt functionalization of multi-walled carbon nanotubes in methanol expanded with supercritical CO_2. *Chem. Eng. J.*, **165** (3), 974–979.

32 Rangari, V.K., Bhuyan, M.S., and Jeelani, S. (2010) Microwave processing and characterization of EPON 862/CNT nanocomposites. *Mater. Sci. Eng. B*, **168**, 117–121.

33 Rangari, V.K., Bhuyan, M.S., and Jeelani, S. (2011) Comparative study of microwave and thermal curing of high temperature epoxy/carbon nanotubes polymer nanocomposites and their properties. *Int. J. Nanosci.*, **10**, 1225–1230.

34 Mediavilla, M., Morales, H., Melo, L., Sifontes, A.B., Albornoz, A., Llanos, A., Moronta, D., Solano, R., and Brito, J.L. (2010) Microwave-assisted polyol synthesis of Pt/H-ZSM5 catalysts. *Micropor. Mesopor. Mater.*, **131**, 342–349.

35 Chandrasekar, M., Suresh, S., and Chandra Bose, A. (2010) Experimental investigations and theoretical determination of thermal conductivity and viscosity of Al_2O_3/water nanofluid. *Exp. Therm. Fluid Sci.*, **34**, 210–216.

5
Nanoparticle Synthesis through Microwave Heating
Satoshi Horikoshi and Nick Serpone

5.1
Introduction

Recent years have witnessed an exponential growth in the application of microwaves in organic syntheses. By contrast, syntheses of inorganic nanoparticles such as metallic, metal oxides and metal chalcogenides in general, among others, and in general microwave chemistry have yet to reach their full potential, despite a growing number of publications in nanoscience and nanotechnology that have made effective use of the dielectric heating provided by microwave radiation [1]. Microwave heating has been shown not only to reduce reaction times significantly, but also to minimize, if not to suppress side reactions, causing chemical yields to be enhanced and processes to be reproducible. The major role of microwaves in nanoparticle synthesis is rapid and uniform heating, a point that will be emphasized throughout this chapter.

A simple and straightforward method for bulk and shape-controlled synthesis of prisms, cubes, hexagonal and spherical nanostructures of such noble metals as Au, Ag, Pt, and Pd by microwave-assisted reduction of their corresponding salts in aqueous glucose, sucrose, and maltose was recently reported by Mallikarjuna and Varma [2], to fill the growing need for the development of ecofriendly processes that avoid the use of toxic chemicals in preparative protocols. Uniform nano and bulk particle sizes are typically obtained by uniform nuclear growth through microwave specific heating. In this regard, synthetic methods that utilize conventional convective heating, because of the need for high-temperature initiated nucleation followed by controlled precursor addition to the reaction media, rely on a conduction pathway to drive the synthetic process in which the reactor acts as the intermediary to transfer the thermal energy from the external heat source to the solvent, and ultimately to the reactants [3]. Such a pathway typically leads to thermal gradients throughout the bulk media, and to inefficient and non-uniform reactions, which can cause serious issues when attempting process scale-up and, more importantly, in nanomaterial syntheses where uniform nucleation and growth rates are essential in maintaining product quality, not to mention the formation of a metallic coating on the inner walls of the reactor.

Microwaves in Nanoparticle Synthesis, First Edition. Edited by Satoshi Horikoshi and Nick Serpone.
© 2013 Wiley-VCH Verlag GmbH & Co. KGaA. Published 2013 by Wiley-VCH Verlag GmbH & Co. KGaA.

Unlike convective heating, microwave internal heating can prevent formation of a metallic coating on the synthesis reactor walls in nanoparticle synthesis, and plays an important role in the wash-free, continuous flow production of metallic nanoparticles [4]. There is a clear need to clarify some of the features of the microwaves in nanoparticle syntheses so as to develop novel microwave methodologies for possible process scale-ups. In this chapter, nanoparticle synthesis employing microwave heating efficiently is introduced, focusing on the authors' own research efforts.

5.2
Microwave Frequency Effects

In a previous chapter, we noted that microwave radiation has become one more laboratory tool in the available arsenal of the chemist to carry out syntheses. In this regard, recent years have witnessed an increasing number of microwave-assisted organic syntheses in batch reactors and in continuous flow reactors to prepare organic substances for electronic applications, solid catalysts and nanoparticles [5]. However, studies on the significance of the microwave frequency factor in such syntheses have been rather scant for possible scale-up, except in the field of food heating, whereby heating efficiencies relative to the microwave frequency have been reported [6]. The optimal frequency was typically chosen from the viewpoint of the absorption efficiency of food, the penetration depth of the microwaves, and the size and composition (solid or liquid) of the foodstuffs. Chemists have come to expect features from the microwaves in a manner otherwise similar to the wavelength effects in photochemical processes, and, therefore, have begun to examine the frequency effect of the microwave radiation.

For food defrosting and wood drying, the microwave frequency used most in Japan is 915 MHz, even though it is not permitted as an ISM band in Japan. Earlier, we reported some preliminary work that showed that both the chemical structure of the substrates and the frequency effect of the microwave radiation may be important parameters in organic syntheses [7]. Despite the worldwide preponderance of microwave ovens (including commercial systems used by research chemists) that operate at 2.45 GHz, the use of other microwave frequencies by researchers should further the development of microwave chemistry, in particular, and the possible scale-up of microwave-assisted chemical syntheses, in general.

Our overall strategy in microwave-assisted chemistry studies has always been to clarify some of the characteristics of the microwave methodology in nanoparticle synthesis. In this section, two microwave applications have been improved taking advantage of the characteristics of the different frequencies of the microwave radiation: (i) nanoparticle synthesis with non-polar solvents [8], and (ii) nanoparticle synthesis in a reactor larger than the penetration depth of the microwaves.

5.2.1
Synthesis of Ag Nanoparticles through the Efficient Use of 5.8-GHz Microwaves

The advantages of the 5.8-GHz microwave frequency are the prompt heating and superheating resulting from the shallow penetration depth of these microwaves, and the greater heating efficiencies of nonpolar solvents at this frequency relative to the more commonly used 2.45-GHz frequency [9]. The synthesis of gold nanoparticles in the nonpolar oleylamine solvent ($C_{18}H_{35}NH_2$) was carried out using hydrogen tetrachloroaurate(III) tetrahydrate ($HAuCl_4 \cdot 4H_2O$) dissolved in oleylamine. The 5.8-GHz frequency microwave generator was a Panasonic M5801 with maximal power of 700 W. The sample solution was contained in a Pyrex glass cylindrical reactor (length 160 mm, 37 mm i.d., maximum volume, 150 mL) and was positioned in a metal pipe connected to the waveguide (see Figure 5.1). The reactor was sealed with two Byton O-rings and a stainless steel cap through which was inserted a thermocouple and a condenser. Prior to utilization of the apparatus, the position of the reactor relative to the waveguide was tested through a computer simulation using the Comsol Inc. Multiphysics software.

Initial rates of temperature increase for a 6-min irradiation period were 16.5 °C min^{-1} for the 5.8-GHz and 2.79 °C min^{-1} for the 2.45-GHz microwaves, respectively, under a continuous applied microwave power of 45 W. The rate of increase of temperature was nearly fivefold faster for the 5.8-GHz microwaves than for the 2.45-GHz microwaves, with the latter increasing gradually with irradiation time. The surface plasmon resonance (SPR) absorption spectra at irradiation times of 7 and 15 min for the 5.8-GHz microwaves and 30 min for the 2.45-GHz microwaves are displayed in Figure 5.2a. An SPR band occurred at 528 nm on irradiation with the 5.8-GHz microwaves, in accord with the absorption spectra of gold

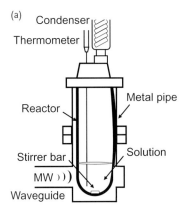
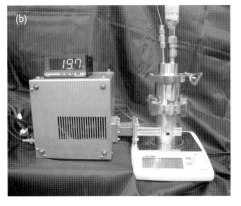

Figure 5.1 Experimental set-up of the 5.8 GHz microwave batch reactor apparatus with a single-mode applicator used in the synthesis of gold nanoparticles. (a) Cartoon and (b) photo of the equipment. Reproduced from [8] with permission.

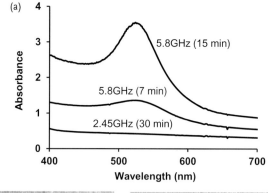
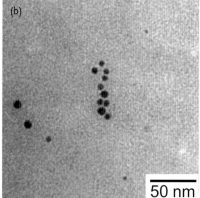
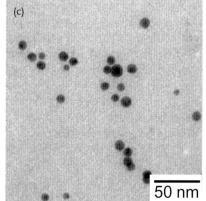

Figure 5.2 (a) UV–visible absorption spectra of the product from irradiating the oleylamine/aurochloric acid solution with 5.8-GHz and 2.45-GHz microwaves at the times indicated; microwave power levels, 45 W. Spectra were obtained after a twofold dilution of the respective colloidal sols. TEM images of the gold nanoparticles produced after (b) 7 min, and (c) 15 min of dielectric heating by the 5.8-GHz microwave radiation. Note the larger gold nanoparticles produced at the longer irradiation time. Reproduced from [8] with permission.

nanoparticles reported earlier by Mohamed and coworkers [10]. By contrast, no absorption bands were seen after 15-min (not shown) and 30-min heating times with the 2.45-GHz microwaves, indicating that no gold nanocolloids are formed under these conditions. Evidently, the Au^{3+} ions in $HAuCl_4$ are reduced by the pure oleylamine under the 5.8-GHz microwaves and the product displays only one SPR band at 528 nm, an indication of the formation of spherical gold nanoparticles. The oleylamine solvent acted as both the reducing agent and, possibly, as the capping material.

TEM images of the resulting Au nanoparticles produced subsequent to the 5.8-GHz microwave heating for irradiation times of 7 and 15 min are illustrated in Figure 5.2b and c. A fairly uniform particle size distribution of spherical nanocolloids was observed from the TEM images: a random sampling of 35 particles gave a size distribution of 6 ± 2 nm after 7 min and 9 ± 2 nm after the 15-min heating time. By contrast, the TEM images (not shown) from the 2.45-GHz microwave heating of the oleylamine/aurochloric acid mixture resulted in no formation of Au nanoparticles, even after an irradiation period of 3 min, which confirms the results observed from the SPR spectra. We inferred from these observations that the lack of formation of gold nanoparticles under the 2.45-GHz microwaves was the result of not reaching the appropriate reaction temperature.

5.2.2
Metal Nanoparticle Synthesis through the Use of 915-MHz Microwaves

The advantage of the 915-MHz microwave frequency is in the deeper penetration depth of the microwaves. Moreover, in using the conventional 2.45-GHz frequency, the heating of large-scale samples has to use equipment that includes a circulating reactor with either a single-mode applicator or a multi-mode applicator, but is unsuitable for reactor sizes of several centimeters in diameter.

An important factor in scale-up of microwave chemistry is the penetration depth of the microwaves into the samples (D_p; in cgs units), which is the depth that microwaves pervade into the material at which the power flux has fallen to $1/e$ (= 36.8%) of its surface value; D_p can be estimated from Eq. 5.1 [11]. Neglecting this factor makes it impossible to ascertain whether the center of the reactor is heated or not. That is, if the solution is heated unevenly, microwave irradiation of the sample cannot generate nanoparticles of uniform size. In order to acquire the optimal experimental conditions, the penetration depth of the microwaves into the sample, the reactor size and the stirring rate factors must all be examined. In this regard, the penetration depths of microwaves into samples of 23 common solvents were determined at the microwave frequencies of 5.8 and 2.45 GHz and 915 MHz at 25 ± 1 °C; the results are reported in the histograms of Figure 5.3.

$$D_p = \frac{\lambda}{4\pi}\left[\frac{2}{\varepsilon'\left(\sqrt{1+(\varepsilon''/\varepsilon')^2}-1\right)}\right]^{1/2} \tag{5.1}$$

The 915 MHz microwave frequency is well suited for heating a larger reactor since it displays the deeper penetration depth. The synthesis of Ag nanoparticles using the polyol method was performed in a 500-mL Pyrex glass reactor (i.d. 86 mm) on addition of polyvinylpyrrolidone, which acted as the dispersing and protective agent of the nanoparticles, to an ethylene glycol medium containing the diaminesilver(I) complex, $Ag(NH_3)_2^+$ [12].

A picture of the experimental set-up for the single-mode microwave apparatus used, operating at a microwave frequency of 915 MHz with the sample located in the waveguide is illustrated in Figure 5.4. The incident microwaves were generated using a 915-MHz microwave apparatus that consisted of a semiconductor generator (Fuji Electronic Ind. Co FSU-301VP-01). The temperature of the water was

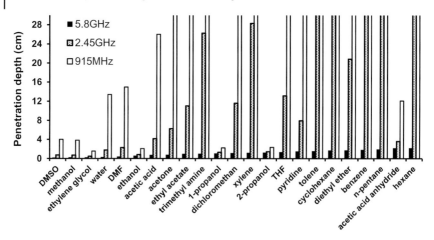

Figure 5.3 Penetration depth of the microwaves (cm) at 5.8 and 2.45 GHz, and 915 MHz for water and some 22 pure organic solvents at ambient temperature. Volume of the sample was 100 ml. Reproduced from [5] with permission.

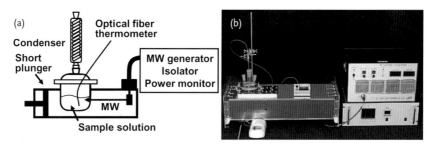

Figure 5.4 (a) Cartoon of the experimental set-up and (b) photo of the desktop single-mode semiconductor microwave generator apparatus operating at a frequency at 915 MHz [12].

measured with an optical fiber thermometer (Anritsu Meter Co., Ltd. FL-2000). The same reactor and thermometer were used in all the experiments carried out at both the 915-MHz and 2.45-GHz frequencies. Although the reaction temperature of each sample was 197 °C (reflux conditions), about 120 W of microwave power was needed for the 915-MHz microwaves, whereas 720 W was needed for the 2.45-GHz microwaves. The sample temperature conditions were adjusted for each frequency.

The TEM images of the silver nanoparticles produced at the 915 MHz and 2.45 GHz microwave frequencies are reported in Figure 5.5. Generation of the more uniform nanoparticle was checked by using 915 MHz (Figure 5.5a). On the other hand, although bigger particles were generated in 2.45 GHz, the particle size distribution was uneven compared with 915 MHz (Figure 5.5b). In oil bath heating, wide distribution of particle size was observed (Figure 5.5c) and as a consequence of the reaction a silver mirror resulted in the inner reactor wall.

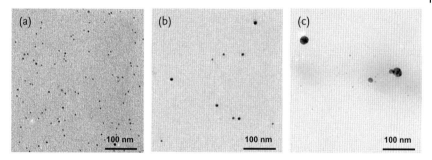

Figure 5.5 TEM images of the Ag nanoparticles produced by 7-min microwave irradiation at frequency of (a) 915 MHz and (b) 2.45 GHz; (c) TEM image of the Ag nanoparticles from oil bath heating [12].

The inner walls of the reactor were inspected after the Ag nanoparticle reaction synthesis was complete. At the 915 MHz frequency the synthesis of Ag nanoparticles left no signs of anything unusual on the reactor wall. By contrast, on carrying out the synthesis of Ag nanoparticles at the 2.45 GHz frequency the reactor wall showed signs of particles adhering to it. This is caused by the difference in the penetration depth of the 2.45-GHz microwaves into the sample solution which was three times less than for the 915-MHz microwaves (penetration depth = 0.5 cm for the 2.45-GHZ microwaves, 1.6 cm for the 915-MHz microwaves). Consequently, the 2.45-GHz microwaves cannot heat the solution uniformly to the center of the sample in a reactor that has a diameter of 8.6 cm. Microwaves at a higher power (720 W) were needed to heat the solution in the 500-mL reactor, and in order to achieve a suitable temperature distribution in solution.

5.3
Nanoparticle Synthesis under a Microwave Magnetic Field

Cheng and coworkers [13] reported significant differences in the heating behavior of solids by the microwaves' electric field (*E*-field) and magnetic field (*H*-field). In fact, in that article, and later in a patent [14], they showed that the *E*- and *H*- fields in a 2.45-GHz microwave reactor interact very differently with matter. They demonstrated for the first time that the microwave field itself, independent of temperature, profoundly affected the thermodynamics of any system in which electrons have unpaired spins. This important effect was realized using single-mode microwave radiation where the points of maximal *E*-field and *H*-field were spatially separated. Moreover, they showed [13] that using a 2.45 GHz waveguide cavity, operating in single-mode TE$_{103}$ excitation, they could physically position compacted 5 mm pellets of the samples separately at the *H*- node (where the *E*-field is nearly zero), or at the *E*-node (where the *H*-field is nearly zero). It must be emphasized that they [13] were able to physically separate the maximal density of the *E*-field and the *H*-field. They further noted that for the general theory of energy loss in various materials when these are placed in a microwave field, the effect of the magnetic field component could no longer be ignored, particularly for conductor and semiconductor materials. They also suggested that contributions to the magnetic loss mechanism could be hysteresis, eddy currents, magnetic resonance, and domain wall oscillations. Their empirical data re-opened the matter of microwave–material interaction to incorporate more detailed consideration of the effects of the microwaves' magnetic field. The *H*- field was effective in the heating of iron, but failed to heat a metal oxide such as ZnO.

In an earlier report, the heating characteristics of aqueous electrolyte solutions (NaCl, KCl, $CaCl_2$, $NaBF_4$, and NaBr) of varying concentrations in ultrapure water by 2.45-GHz microwave radiation from a single-mode resonance microwave device and a semiconductor microwave generator were examined under conditions where the *E*-field was dominant and where the *H*-field dominated [15]. Although magnetic field heating is not generally used in microwave chemistry, the electrolyte solutions were heated almost entirely by the microwaves' *H*-field. These results showed that a solid and a liquid can be heated by the *H*-field component of the microwave radiation. Next, we describe recent work in which we synthesized Ag and Ni nanoparticles by means of microwave magnetic field heating.

In using the polyol method, polyvinylpyrrolidone (PVP), the dispersing and protective agent of the nanoparticles, was added to an ethylene glycol medium containing the $Ag(NH_3)_2^+$ complex, after which the solution was introduced into the quartz reactor (i.d. 4 mm). When the sample was heated by the microwaves' magnetic field irradiation, the solution reached a temperature of only 60 °C after 110 s. This heating is considered to be Joule heating of ethylene glycol containing the silver(I) complex ions. However, the reaction temperature of the polyol method needs to be over 110 °C for silver nanoparticles to form. Accordingly, no silver nanoparticles were generated under magnetic field irradiation. Since nickel(II) is paramagnetic, some improvement in the heating efficiency of the Ag(I) solution was expected if a nickel(II) salt was added to the solution. Hence, nickel(II) nitrate hexahydrate was then added to the sample solution, and microwave magnetic field irradiation was re-tried. Contrary to anticipation, the temperature still did not reach above 60 °C. However, the color of the solution changed to a light yellow on heating for 160 s at a temperature below 60 °C. No generation of Ni nanoparticles particles was observed when an aqueous nickel(II) nitrate hexahydrate solution, that did not contain the diaminesilver(I) complex, was heated by the magnetic field component of the microwaves. In this case, no change in color due to plasmon absorption was observed at a temperature of 60 °C. Silver nanoparticles were generated only when a solution containing both the diaminesilver(I) complex and nickel(II) nitrate hexahydrate was heated by the electric field component of the microwave radiation. The temperature of the solution needed to be over 110 °C, as occurred upon being heated by the microwaves' electric field component, for the solution to change color from colorless to light yellow. The TEM images of Ag nanoparticles formed when the Ag(I) aqueous solution also contained nickel(II) nitrate are displayed in Figure 5.6. In the

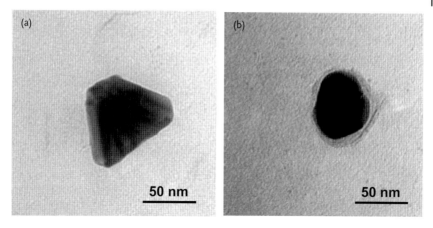

Figure 5.6 TEM images of the Ag nanoparticles produced under irradiation by the microwaves' (a) *E*-field and (b) *H*-field.

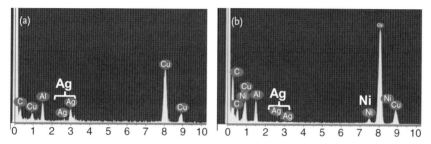

Figure 5.7 EDX observation of the silver nanoparticles produced under irradiation by (a) *E*-field and (b) *H*-field.

synthesis of silver nanoparticles generated under electric field irradiation (at 110 °C), nanoparticles of a trapezoidal nature were observed (Figure 5.6a). By contrast, under magnetic field heating (at ca. 60 °C), particles of a somewhat spherical nature were generated, as shown in Figure 5.6b; the particle surface displayed different contrast. Furthermore, a thin film could be observed on the surface of the nanoparticle generated by magnetic field heating.

Energy dispersive X-ray spectroscopy (EDX) measurements of the particles observed by TEM are shown in Figure 5.7. No nickel nanoparticles formed from the synthesis involving the microwaves' electric field component. However, under magnetic field heating of the mixed solution of $Ag(NH_3)_2^+$ and $Ni(NO_3)_2$, the EDX spectra displayed a peak attributed to nickel in addition to the Ag peaks, indicating the possible formation of Ni/Ag composite nanoparticles. We deduced that, under the experimental conditions used, the nickel may have acted as a reducing agent in generating the Ni/Ag nanoparticles under magnetic field irradiation.

5.4
Synthesis of Metal Nanoparticles by a Greener Microwave Hydrothermal Method

The reducing agent currently used conventionally was not needed in the synthesis of Ag nanoparticle by the microwave hydrothermal synthesis method. Water is a low-boiling-point solvent compared with ethylene glycol presently used in the typical polyol method; it is thus considered a green solvent. The role of water as a solvent in microwave-assisted organic syntheses has risen dramatically because of the interest in ecofriendly processes germane to Green Chemistry, as recently described by Polshettiwar and Varma [16]. In green chemistry, microwaves represent an ecofriendly source of thermal energy, and since no reducing agent was used in the microwave hydrothermal synthesis of Ag nanoparticles, no impurities were expected and none formed. The precursor salt used for the silver nanoparticles was silver(I) nitrate ($AgNO_3$) and PVP in an aqueous medium. The sample solution was heated using an Anton Paar Monowave 300 apparatus. The temperature distribution in the solution was measured using a ruby fiber optic thermometer and a radiation thermometer. In a closed system, the solution was added to an Anton Paar high-pressure Pyrex cylinder as the reactor, subsequent to which the sample was heated rapidly within 10 s by the microwaves to the reaction temperature, followed by rapid cooling immediately thereafter.

The SPR spectra at different temperatures are shown in Figure 5.8. Even without the use of a reducing agent, the growth of Ag nanoparticles was observed at a reaction temperature above 190 °C. In this regard, the reaction temperature reached 200 °C when the solution was exposed to microwave radiation for 100 s. The walls of the reactor were free of any particles. By contrast, when heating was done using the conventional oil bath method, a silver mirror formed on the inner walls of the reactor.

TEM images of the silver nanoparticles formed at 190 and 200 °C are shown in Figure 5.9. Close inspection of the TEM results indicated that the nanoparticles

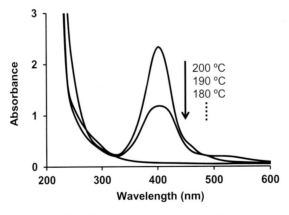

Figure 5.8 UV–visible absorption spectra at various temperatures in the microwave hydrothermal process for the generation of Ag nanoparticles.

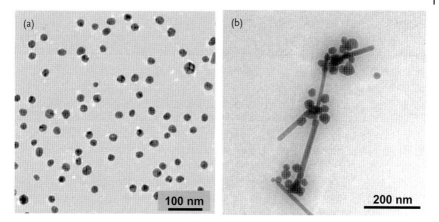

Figure 5.9 TEM images of the silver nanoparticles produced by the microwave hydrothermal method. (a) 190 °C, (b) 200 °C.

were of relatively uniform size (about 20–40 nm) when the synthesis was carried out at 190 °C, while at the higher temperature of 200 °C some of the nanoparticles took on the shape of nanorods, in addition to some spherical nanoparticles.

5.5
Nanoparticle Synthesis with Microwaves under Cooling Conditions

Recent studies have examined external cooling of reactions subjected to microwave dielectric heating as a special case of microwave-assisted organic syntheses [17, 18]. For instance, cooling microwave-assisted reactions of radical species caused removal of excess heat and maintained reaction temperature at near ambient. This led to relatively high product yields and significant minimization of generated side-products (impurities), two important and attractive outcomes of microwave heating when coupled to a cooling methodology. Cooling allows a higher level of microwave power to impact directly on the reaction mixture, thus potentially enhancing nonthermal microwave effects that rely on the electric field strength, while at the same time preventing overheating by the continuous removal of heat [19].

The characteristics of cooling in the microwave-assisted nanoparticle synthesis were examined using an integrated Pyrex flask reactor equipped with a cooling system consisting of a cooling coil and a silicone oil bath fabricated with microwave non-absorbing materials. Microwave irradiation (frequency, 2.45 GHz) used a multi-mode microwave applicator. The reaction temperature was controlled with a Tokyo Rikakikai Co., Ltd. MWO-1000 chemical reaction system for microwave heating and cooling using a circulation system [4]. Different protocols were thus examined in the synthesis of silver nanoparticles: (i) irradiation of the sample solution with 390 W microwaves under cooling conditions (MW-390W/Cool

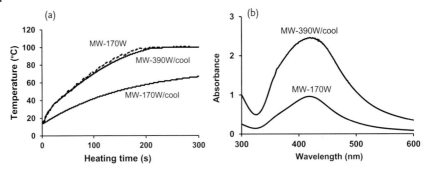

Figure 5.10 (a) Temperature–time profiles of the aqueous CMC/diaminesilver(I) solution using the MW heating method with 170 W microwaves and by the MW/Cooling hybrid methods with 390 and 170 W microwaves. (b) Plasmon resonance absorption of Ag nanoparticles using the MW heating method with 170 W microwaves and by the MW/cooling hybrid method with 390 W microwaves for 5 min. Note that the sample solution of the MW/cooling method was diluted 8-fold with water, whereas for the MW-170 method the solution was diluted 4-fold. Reproduced from [4] with permission.

protocol), (ii) microwave irradiation alone with 170 W microwaves (MW-170W protocol), and (iii) irradiation with 170 W microwaves under cooling conditions (MW-170W/Cool protocol). Rates of temperature rise with the MW-390W/Cool and MW-170W protocols were 0.50 and 0.51 °C min^{-1}, respectively, reaching a temperature of 100 °C in about 3.5 min. By contrast, the rate of temperature rise for the MW-170W/Cool protocol was 0.27 °C min^{-1} and the temperature reached after around 4 min was 67 °C (Figure 5.10a). Thus the coolant inhibited the temperature rise by 46% in the MW-170W/Cool protocol.

Maximum absorption of the SPR band of the silver colloids produced by the MW-390W/Cool and MW-170W protocols occurred at 420 nm, with the band intensity being 5.5-fold greater for the colloids formed with the former protocol (Figure 5.10b). Note that the samples were diluted 8-fold and 4-fold, respectively, prior to recording the spectra. The rate of formation of silver nanoparticles was enhanced with the 390 W microwaves under cooling conditions (MW-390W/Cool protocol; temperature, 100 °C) relative to the use of 170 W microwaves, even though the temperature was also 100 °C.

TEM images of Ag nanoparticles produced by the MW-390W/Cool and MW-170W protocols are displayed in Figure 5.11. In the former case, the size distribution of the nanoparticles spanned the 1–2 nm range, whereas for the latter protocol the sizes ranged from 3 to 5 nm. Clearly, smaller silver nanoparticles formed at a higher concentration under cooling conditions and at the higher microwave radiation power; the weaker 170 W microwaves (MW-170W protocol) led to formation of somewhat larger size nanoparticles (compare the TEMs of Figure 5.11a and b).

No silver nanoparticles were produced with the MW-170W/Cool protocol as evidenced by both UV–visible absorption spectroscopy and by TEM microscopy,

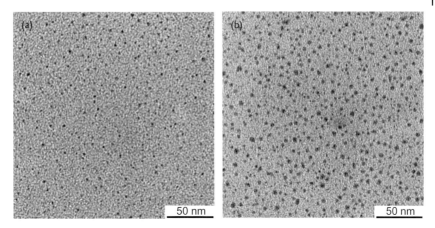

Figure 5.11 TEM images of Ag nanoparticles generated from (a) the 390 W MW/Cool hybrid protocol, and (b) the 170 W microwave heating protocol for a 5 min reaction period. Reproduced from [4] with permission.

as the maximum temperature reached after 5 min was only 67 °C (Figure 5.4a). Recall that a reaction temperature of about 100 °C is necessary to nucleate the growth of the silver nanoparticles. Accordingly, a certain threshold of the thermal energy was a necessary condition in the synthesis of silver nanoparticles.

5.6
Positive Aspects of Microwaves' Thermal Distribution in Nanoparticle Synthesis

This section discusses some of the thermal features of the 2.45-GHz microwave radiation (MW) used in the synthesis of silver nanoparticles in comparison with the conventional heating method that uses an oil bath (CH), with a special emphasis on the temperature effects and the characteristics (if any) of different microwave synthesizers [4].

The following materials were used in the synthesis of silver nanoparticles. An aqueous solution of the diaminesilver(I) complex and carboxymethylcellulose (CMC) was introduced into a 150-mL Pyrex glass batch-type cylindrical reactor. Continuous microwave irradiation (power, 64 W) was provided with a microwave single-mode resonance apparatus. The reactor was positioned such that irradiation was achieved by the microwaves' electric field at its maximum intensity by adjustment of the short plunger and the three-stub tuner. The conventional heating source was a silicone oil bath. After heating the sample solution, either with microwaves or conventionally with an oil bath, the reactor contents were rapidly cooled in an ice/water bath to arrest the synthesis of the Ag nanoparticles, which would have otherwise continued as a result of residual heat from the microwave and oil-bath heating.

The microwave-assisted synthesis of nanoparticles is characterized by rapid and homogeneous heating in contrast to a conventional heat-assisted synthesis, even though the thermal effects are similar to those of other heating methods [20].

The observed rapid heating by microwaves was controlled and closely monitored so as to extract the characteristic thermal features of the microwaves. The temperature conditions of both heating methods were closely matched in the nanoparticle synthesis. This was achieved by soaking the cylindrical reactor in the oil bath pre-heated to 190 °C, followed by determination of the temperature rise of the solution (Figure 5.12, dashed line). The rise in temperature of the solution exposed to microwave radiation was subsequently matched to the temperature rise in the oil-bath heating using microwave power levels in the range 60–70 W at 1 W increments controlled by the proportional-integral-derivative (PID) device available in the microwave apparatus. The solid line in Figure 5.12 represents the temperature rise in the solution exposed to the 64-W microwaves.

Following the temperature match between microwave heating and oil-bath heating, the formation of silver nanoparticles was monitored at heating times of 0.5, 1, 2, 3, 4 and 5 min by the SPR band around 420 nm in the UV–visible absorption spectra. Silver nanoparticles formed after a 3-min heating period by both MW and CH methods. In both cases, the synthesis of Ag nanoparticles necessitated a temperature of 100 °C reached only after this time period. Figure 5.13 shows the corresponding TEM images of the resulting silver nanoparticles after the 5-min heating time. A uniform particle size distribution was observed from the TEM image of the colloids obtained from microwave heating under otherwise identical temperature conditions as from conventional heating. The light scattering results indicate that Ag nanoparticles prepared by MW heating were in the range 1–2.3 nm (Figure 5.13a inset), whereas the CH heating method yielded polydipsersed nanoparticles mostly in the 3–5.7 nm range (Figure 5.13b inset), with some up to 30 nm. Thus, even though the reaction temperatures for the microwave and oil-bath

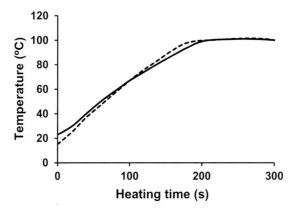

Figure 5.12 Temperature–time profiles of the aqueous CMC/diaminesilver(I) solution by microwave heating (applied power, 64 W; solid line) and oil bath heating (consumed power, 400 W; dashed line). Reproduced from [4] with permission.

methods were identical, there was an otherwise noticeable difference in the rate of formation of the silver nanoparticles and their size distributions.

Variations in the two heating methods are clearly evident in the results illustrated in Figure 5.14. After 5 min microwave irradiation of the aqueous CMC/diaminesilver(I) solution a slightly yellow-colored sol was formed in the reactor (Figure 5.14a), and its concentration did not change even after 60 min of irradiation. Silver nanoparticles that adsorbed on the reactor walls were easily removed by simple washing with water. By contrast, the silver film (mirror) coating the inner reactor walls formed after 4 min by CH heating (Figure 5.14b), under

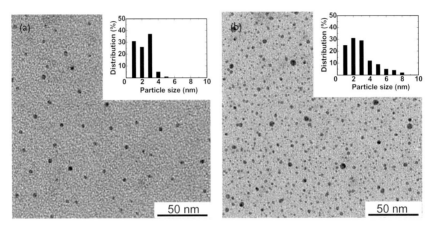

Figure 5.13 TEM images of the silver nanoparticles produced by (a) microwave and (b) oil bath heating methods (inset: particle size distribution from light scattering experiments). Reproduced from [4] with permission.

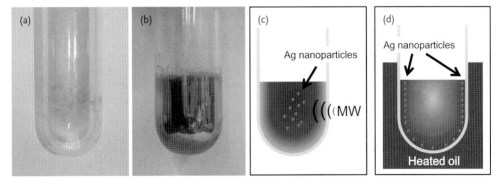

Figure 5.14 Photograph of the reactor after the sample discharge (a) after 4 min of microwave irradiation, and (b) after 5 min of oil bath heating. Cartoon representing the temperature distribution in the reactor (c) after 4 min of microwave heating, and (d) after 4 min of oil bath heating. Note that the photographs in (a) and (b) were taken immediately after the 4-min heating time. Experiments were carried out under non-stirring conditions. Reproduced from [4] with permission.

otherwise identical temperature conditions as the MW method, resisted such washings.

The initial temperature distributions and formation of silver nanoparticles produced under non-stirring conditions by both MW and CH heating methods are illustrated, respectively, in the cartoons of Figure 5.14c and d. In the former case, microwave radiation penetrated the CMC/diaminesilver(I) solution causing the temperature to rise by dielectric loss, and to some extent by Joule heating, followed by subsequent loss of heat to the surroundings through the reactor walls. As such, the temperature near the inner walls of the reactor tended to be lower than at the center of the reactor. That is, the synthesis of Ag nanoparticles by the microwave-assisted process progresses outward from the center of the reactor to the inner reactor wall owing to a temperature gradient that gave rise to a concentration gradient (thermophoretic migration, i.e., the Ludwig–Soret effect [21]). On the other hand, heat from the oil bath was most prominent at the reactor wall and was subsequently transmitted to the solution by thermal conduction and convection mechanisms. This ultimately led to formation of a silver film on the reactor's inner wall as the concentration of the nanoparticles was greatest at this position. Germane to the present discussion, Schanche [22] has described temperature distributions produced from MW and CH heating methods in various organic syntheses. In a later study, the Kappe group noted that the temperature distribution in the MW method may not always be uniform [19]. That is, the location where heat is generated may be different from the position where the endothermic process occurs. This calls attention to the fact that even completely homogeneous solutions must be stirred/agitated when using single-mode microwave reactors so as to avoid temperature gradients from developing, a consequence of the inherent field inhomogeneities that exist inside the single-mode microwave cavity.

5.7
Microwave-Assisted Nanoparticle Synthesis in Continuous Flow Apparatuses

Scaled-up production of monodisperse colloidal nanoparticles has become an important research subject in recent years. In this regard, continuous flow reactors are generally favored over batch reactors. Nonetheless, batch reactors still constitute the most frequently used system in the wet synthesis of nanoparticles. This is so in spite of several obvious drawbacks such as: (i) heterogeneous distribution of reactants and temperature in the reactor, (ii) insufficient mixing, (iii) variations in the physicochemical characteristics of the resulting products among different batches, (iv) inherent discontinuity, (v) difficulty of scaling-up, and (vi) frequent need for post-synthesis purification steps [23]. The advantage of the microwave method can best be summarized as rapid heating. Moreover, the rapid growth of nanoparticles can lead to particles with a narrow size distribution. A microwave continuous flow system is more suitable for scale-up of nanoparticle syntheses than the batch system. Uniform heating of a batch system consisting of a large

5.7 Microwave-Assisted Nanoparticle Synthesis in Continuous Flow Apparatuses

size reactor is difficult from the viewpoint of the penetration depth of the microwaves. Moreover, to the extent that microwaves cause rapid heating of the solutions, the continuous flow and/or circulation system is definitely to be preferred. Next, we describe some continuous flow apparatuses for microwave nanoparticle synthesis.

5.7.1 Microwave Desktop System of Nanoparticle Synthesis in a Continuous Flow Reactor

The proposed continuous-flow reactor system consisted of a Pyrex pipe (length 135 mm, i.d 8 mm) placed horizontally in the microwave waveguide through which the aqueous CMC/diaminesilver(I) solution was circulated by means of a peristaltic pump [4]. Irradiation of the reactor contents was performed with 1200 W microwaves obtained from a microwave generator (maximum power, 3000 W). In the photograph of Figure 5.15a, the microwaves emanate from the back toward the reactor with the maximum of the microwaves' electric field positioned at the center of the reactor. A metal mesh closed the waveguide to prevent microwave leakages and to observe events occurring in the reactor. For maximum heating

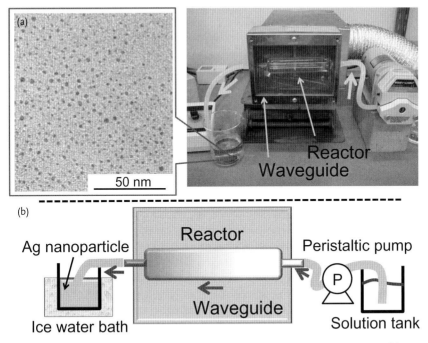

Figure 5.15 (a) Photograph of the continuous flow reactor system for the microwave-assisted synthesis of silver nanoparticles; the TEM image of the resulting silver colloids is also displayed. (b) Schematic image of the overall experimental set-up. Reproduced from [4] with permission.

efficiency the flow rate of the peristaltic pump was set at 600 ml min^{-1}. The solution temperature increased rapidly to 100 °C under microwave irradiation to yield a colloidal sol of Ag nanoparticles which were then collected in the receiver flask and rapidly cooled with an ice/water bath to stop any further reaction.

The TEM image of the nanoparticles displayed in Figure 5.15a confirmed the generation of a fairly narrow size distribution of silver nanoparticles (range, 1–4 nm) with this reactor set-up. The cartoon in Figure 5.15b summarizes the reactor set-up and our observations. Interestingly, the internal reactor walls showed no visible evidence of a yellow stain (seen above in the batch reactor) by this continuous-flow synthesis of silver nanoparticles even after 7 min of microwave irradiation. With microwave power levels greater than 1200 W, we had some difficulty in controlling the temperature, which could (in principle) be corrected by cooling the reactor externally with a cooling device and by a faster solution flow rate to enhance synthesis efficiency. However, addition of any cooling device (e.g., cooling jacket, cold wind, etc.) would unnecessarily complicate the reactor set-up. In the present instance, with the 1200-W microwaves there was no necessity for cooling and as such this microwave power level was maintained throughout the synthesis. Understandably, a similar continuous-flow rapid synthesis of silver nanoparticles of similar size distribution (i.e., 1–4 nm) would be somewhat difficult to achieve in a reactor system that relied on the conventional heating method.

5.7.2
Synthesis of Metal Nanoparticles with a Hybrid Microreactor/Microwave System

Microfluidic reactors have recently been used in the synthesis of various organic, inorganic, and biological materials [24]. Such microreactors are also successful in the synthesis of colloidal nanoparticles [25]. The microreactor method as well as the microwave method can lead to nanoparticles of uniform size. A uniform seed crystal is the principal feature of the microreactor. Combining the microwave method, in which the microreactor can make a uniform seed crystal such that uniform crystal growth is possible, results in nanoparticles with a narrower diameter distribution. A novel hybrid system that combines a microreactor and microwave radiation for the continuous nanoparticle synthesis has been proposed using the formation of Ag nanoparticles as an example [26]. For instance, addition of polyvinylpyrrolidone (PVP) as the dispersing agent, and glucose as the reducing agent of the nanoparticles to an aqueous medium will be referred to as "solution A" (Figure 5.16a); "solution B" consists of a diaminesilver(I) complex aqueous solution. These two solutions were then mixed in the microreactor (YMC. Co. Ltd Deneb mixer) using a syringe driver. The mixed solution was subsequently added to a quartz spiral reactor using an HPLC pump. The reactor was then heated rapidly using a microwave resonator (Saida FDS Inc). The sample solution was adjusted to a temperature of 162 °C and a pressure of 1 MPa using a back pressure regulator.

When the mixed solution in the microreactor was heated with the microwaves, uniform size Ag nanoparticles of 5 nm resulted (Figure 5.17a). The maximum

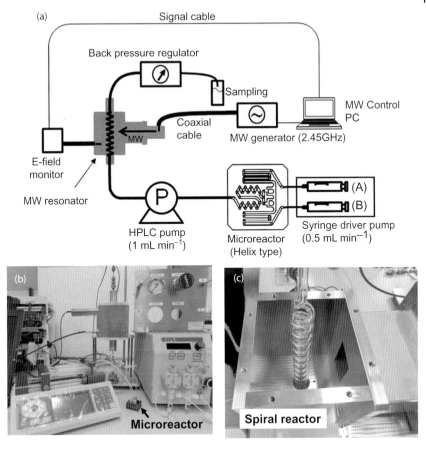

Figure 5.16 Experimental set-up of a hybrid microreactor/microwave system (a) detailed schematic, (b) photograph, (c) quartz spiral reactor in applicator [26].

absorption of the SPR band of the silver colloids produced by the microreactor/microwave combination occurred at around 420 nm (Figure 5.17b).

Control experiments were carried out in which both solutions A and B were mixed in the microreactor system, yet no silver nanoparticles formed under these conditions,and in the presence of glucose as the reducing agent; no doubt due to a very slow reducing rate in the absence of microwaves. On the other hand, although the mixed solution comprising solution A and B was heated by the microwave system but without the microreactor, no generation of silver nanoparticles formed, at least none were observed, in spite of having arranged the appropriate temperature conditions under microwave irradiation. Silver nanoparticles did form, however, with the hydrothermal synthesis method utilizing a continuous-flow system and a synthesis apparatus that combined both the microwaves and the microreactor.

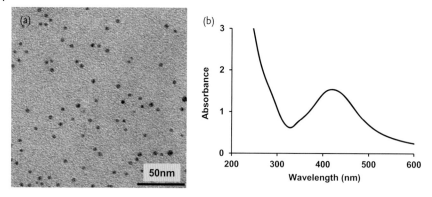

Figure 5.17 (a) TEM images and (b) plasmon resonance absorption of the silver nanoparticles produced by the hybrid microreactor/microwave system [26].

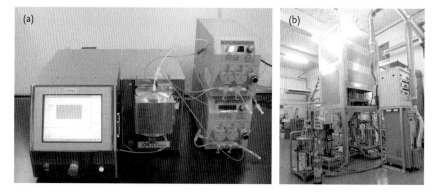

Figure 5.18 Continuous-flow microreactor nanoparticle synthesis systems. (a) Unit type microwave heating device; (b) Industrial scale device comprising an ultrasonic atomization type microwave nanoparticle continuous-flow synthesis equipment. Photograph (a) was obtained through the courtesy of Dr. Nishioka and photograph (b) was kindly provided by the Kojundo Chemical Lab. Co. Ltd.

5.7.3
Other Examples of Continuous Microwave Nanoparticle Synthesis Equipment

Nishioka and coworkers reported the synthesis of $Ag@SiO_2$ core–shell particles by a two-step reaction with microwave systems [27] in which a polytetrafluoroethylene (PTFE) tube (i.d. 1 mm) was mounted coaxially in the center of the TM_{010} single-mode cavity for use as the flow-type reactor (see Figure 5.18a). Since the microwave system is a unit type microwave heating device, the interesting feature of this equipment is that it can connect equipment perpendicularly. Homogeneous heating of the long reactor can be carried out with this microwave unit.

The large scale ultrasonic atomization type microwave nanoparticle continuous-flow synthesis equipment system (Kojundo Chemical Lab. Co. Ltd) is shown in

Figure 5.18b. A 3.5-m reactor is contained at the center of the equipment; the sample solution is sprayed by an ultrasonic wave into the apparatus, following which resistance heating and microwave heating lead to formation of the nanoparticles. This type of system has been used for the synthesis of ferrite, a material that absorbs electromagnetic waves.

5.7.4
Microwave Calcination Equipment for the Fabrication of Nanometallic Inks

Microwave calcination equipment for the fabrication of nanometallic inks is displayed in Figure 5.19. The lower melting point of the metal nanoparticles compared with bulk metal makes the former useful in new electronic wiring, among other applications. Since differential heating can be carried out in the open atmosphere under microwave irradiation, plastics such as PET can form the base of the electronic wiring. When the pattern on the plate is written with a paste of either silver or gold, sparks would result if microwave irradiation were to be carried out in a regular microwave oven. However, a significant advantage of the microwaves' magnetic field heating is that no sparks occur when heating the metal nanoparticles. Such applications as low-temperature calcination of an ITO film or a ZnO film, and recovery of a rare earth (among others) are possible with such a methodology.

An additional advantage of the microwaves' magnetic field heating method is in the coating of a PET plastic base to form Ag, Cu, and CuO nanoparticle ink (Figure 5.19). Generally, oxidation of the surface of copper occurs when heated by a conventional heating method in an open atmosphere. However, with microwave magnetic field heating, even if heating were carried out in the presence of an air atmosphere, the copper surface is not oxidized, as such oxidation would drastically change the electrical properties of the ink. Clearly the microwave method that avoids formation of an oxide film is an effective methodology in such material processing.

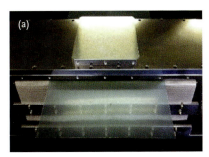
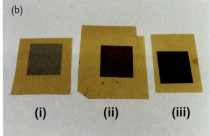

Figure 5.19 (a) Microwave nanometallic ink continuous processor system, and (b) microwave processing of (i) nano-Ag ink, (ii) nano-Cu ink, and nano-CuO ink on a PET plastic base. Photographs provided by courtesy of Fuji Electronic Industrial Co. Ltd.

5.7.5
Synthesis of Metal Nanoparticle Using Microwave Liquid Plasma

In physics and chemistry, plasma is a substance similar to a gas in which a certain portion of the particles is ionized. Plasma is typically defined as an electrically conducting medium, generally consisting of negatively charged electrons, positively charged ions, and neutral atoms or molecules or both. Plasma gas applications are readily found in various industrial fields, such as surface treatments and etching of microelectronics, among others [28]. In environmental remediation, plasma is widely used in certain air cleaners to sterilize and to decompose various volatile organic contaminants (VOCs) [29]. By contrast, reports of plasma generated in liquids have been rather scarce compared to gas-phase plasma. In addition, research in the use of plasma in liquids has been rather limited in environmental remediation [30].

Sato and coworkers reported that the synthesis of gold, silver and platinum nanoparticles could be mass produced by a liquid plasma system (see Figure 5.20) [31]. A feature of this technique is that the nanoparticle synthesis from a salt precursor is possible in the absence of a reducing agent. Metal nanoparticles were produced by vaporization in pure water. Because this process requires no reducing agent, contamination of the nanoparticles is minimized, if not eliminated altogether, as they are produced directly from the metal salt.

5.7.6
Compendium of Microwave-Assisted Nanoparticle Syntheses

Research into nanoparticles of various types and shapes through the use of microwave radiation had some remarkable but limited success. Table 5.1 lists several examples of metallic, metal oxide and metal chalcogenide nanoparticles from their corresponding precursor salts, together with the dispersing agents, reducing agents, solvent media and reaction temperature conditions [32–103].

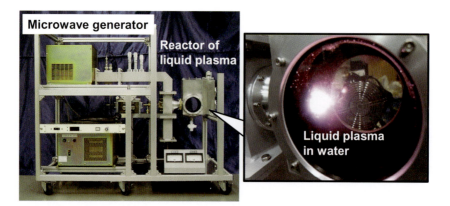

Figure 5.20 Liquid phase plasma synthesis of metal nanoparticles using microwave liquid plasma. Photograph provided by courtesy of ARIOS Ltd.

Table 5.1 Synthesis of nanoparticles under a microwave heating method.

Particles	Shape (size)	Precursor	Dispersing agent	Reducing agent/solvent	Reaction temp. and time	Ref.
Ag	Spheres (50 nm)	$AgNO_3$	PVP	EG	150 °C, 15 min	[32]
Ag	Rods (10–2 nm × 50–200 nm)	$AgNO_3$	Na_3citrate	H_2O	100 °C, 10 min	[33, 34]
Ag	Spheres and cubes (5–10 nm)	$AgNO_3$	Dodecylthiol	EG, toluene, thiourea	160–170 °C, 3 h	[35]
Ag	Spheres (26 ± 3 nm)	$AgNO_3$	starch	L-lysine, L-arginine, H_2O	150 °C, 10 s	[36]
Ag	Spheres, cubes, rods, wires (30–100 nm)	$AgNO_3$	PVP	Na_2S, EG	130–170 °C, 20 s	[37]
Ag	Rods (40–120 nm × 1–8 μm)	Ag_2O	1,2-ethanedithiol	1,2-ethanedithiol	80–140 °C, 10 min	[38]
Ag	Spheres (1–2.3 nm)	$Ag(NH_3)_2$	Carboxymethylcellulose	Carboxymethylcellulose, H_2O	100 °C, flow rates = 600 ml min^{-1}	[4]
Ag	Spheres (5 nm)	$Ag(NH_3)_2$	PVP	Glucose, H_2O	100 °C, flow rates = 1 ml min^{-1}, 1 MPa	[26]
Au	Wires (20 nm × 1–3 μm)	$HAuCl_4$	DNA	DNA, EDTA, H_2O	<100 °C, 3 min	[39]
Au	Spheres, triangular shapes, rods	$HAuCl_4$	CTAB	2,7-DHN, NaOH, H_2O	30–90 s	[40]

(Continued)

Table 5.1 (Continued)

Particles	Shape (size)	Precursor	Dispersing agent	Reducing agent/solvent	Reaction temp. and time	Ref.
Au	Plates (30 nm)	$HAuCl_4$	PVP	EG	196 °C, 1 min	[41]
Au	Spheres (5–25 nm)	$HAuCl_4 \cdot 3H_2O$	PVP	HCl, NaOH	1000 W, 5–60 s	[42]
Au	Spheres (9 ± 2 nm)	$HAuCl_4$	Oleylamine	Oleylamine	<1 min, 2.45 or 5.8 GHz	[8]
Ni	Spheres (6 ± 3 nm)	$Ni(OH)_2$	PVP	EG, H_2O	150 °C, 5 min	[43]
Ni	Polyhedrons (40–100 nm)	$Ni(CH_3CO_2)_2$	PVP, DDA	EG	195 °C, 45 min	[44]
Pt	Spheres (1.5–6.5 nm)	$H_2[PtCl_6]\cdot 6H_2O$	PVP	EG, glycerol, or 1,3-propanediol	>160 °C, flow rates = 3–200 ml h^{-1}	[45]
Pt	Spheres (nm)	H_2PtCl_6	PVP	EG	150 °C, 15 min	[46]
Pd	10–30 nm	$PdCl_2$	PEG	D-glucose	800 W, 20 s	[47]
Pd	Cubes (25 nm)	$PdCl_2$	PVP, CTAB, CTAC	TEG	900 W, 80 s	[48]
Te	Rods, wires (15–40 nm × ~700 nm)	TeO_2	PVP	(Bpy)[BF$_4$], NaBH$_4$	180 °C, 10 min	[49]
Cu	Spheres (10 nm)	$CuSO_4$	PVP	NaH_2PO_2, EG	5 min	[50]
Cu	Spheres (90–240 nm)	$Cu(ac)_2$	PVP	ascorbic acid, DEG	<170 °C	[51]
Cu	Spheres (2–6 nm)	$Cu(octa)_2$, $Cu(myri)_2$, $Cu(AOT)_2$, $Cu(EHP)_2$	—	1-propanol, 1-hexanol, 1-heptanol	<170 °C, 20–120 min	[52]

Product	Morphology (size)	Precursor	Additive	Solvent	Conditions	Ref.
Cr, Mo, W, Mn, Re, Fe, Ru, Os, Co, Rh, Ir	Spheres (<13 nm)	Mn(CO)m (n = 1–6, m = 6–16)	–	[Bmim][BF$_4$]	250 °C, 3 min	[53]
CuO, Cu$_2$O	Spheres (30 nm), cubes (300 nm)	Cu(OAc)$_2$·H$_2$O	–	EG, H$_2$O	15 min	[54]
CuO	Spheres (4 nm)	Cu(OAc)$_2$	PEG	NaOH, EtOH	10 min	[55]
Cu$_2$O	Ellipsoids (300 nm)	Cu(OAc)$_2$	–	glucose, H$_2$O	70 °C, 30 min	[56]
CuS	5–10 nm	Cu(OAc)$_2$, thioacetamide	–	Formaldehyde, H$_2$O	Reflux, 20 min	[57]
TiO$_2$	Cubes (5 nm)	TTIP	–	EtOH, [bmim][BF$_4$]	3–40 min	[58]
TiO$_2$	Cubes (25 nm), rods (4 × 17 nm), spheres (8 nm)	TiCl$_4$	–	NH$_3$ or NaCl or NH$_4$Cl, H$_2$O	20–60 min	[59]
TiO$_2$	Spheres (1 nm), rods 10 × 100 nm)	TiOCl$_2$	–	H$_2$O	195 °C, 5–60 min	[60]
Fe$_3$O$_4$	3–10 nm	FeCl$_3$	poly acrylic acid	NaOH, diethylene glycol	10 min	[61]
Fe$_2$O$_3$	Cubes (240 nm), ellipsoids (300 nm), spheres (300 nm)	FeCl$_3$	–	NH$_4$H$_2$PO$_4$, H$_2$O	220 °C	[62]
Fe$_2$O$_3$	<20 nm	FeCl$_3$	–	EtOH, AcOH	150 °C, 15 min	[63]
Fe$_3$O$_4$	Wires (30–50 nm × 1 μm)	FeSO$_4$	PEG	H$_2$O	180 °C, 8 min	[64, 65]

(Continued)

Table 5.1 (Continued)

Particles	Shape (size)	Precursor	Dispersing agent	Reducing agent/solvent	Reaction temp. and time	Ref.
Fe_3O_4	Spheres (6 nm)	$Fe(acac)_3$	Oleylamine, 1,2-hexadecandiol	[bmin][BF_4], Phenyl ether	250°C, 5–10 min	[66]
Tl_2O_3	Spheres (20 nm)	$TlCl_3$	—	NH_3/H_2O	60 min	[67]
SnO_2	Spheres (5 nm)	$SnCl_4$	—	HCl, H_2O	180°C, 45–120 min	[68]
CeO_2	Cubes (50 nm)	$Ce(NO_3)_3$	Oleic acid	EG, t-butylamine, H_2O	60 min	[69]
CeO_2	Spheres (3 nm)	$Ce(NO_3)_3$	Hexamine	Diethylene glycol	60 min, 10 bar	[70]
$PbTiO_3$	50 nm	$Pb(Ac)_2$, $Ti(OPr^i)_4$	—	EG	1 min	[71]
BiO_2	Rods (2 × 30 µm)	$Bi(NO_3)_3$	PVP	NaOH, H_2O	6 min reflux	[72]
$Cd(OH)_2$	Wires (5–30 nm × 0.3 µm)	$Cd(NO_3)_2$	—	NH_3, H_2O	5 min	[73]
$BaTiO_3$	50–100 nm	$BaCl_2$, $Ti(OPr^i)_4$	—	EG	5 min	[74]
$BaTiO_3$	Cube particles	$BaCl_2 \cdot 2H_2O$, titanium (IV) isopropoxide	—	NaOH, HCl, H_2O	2–40 h, 2.45–5 GHz	[75]
CdSe	Spheres (200 nm)	Cd(Ac), Se(powder)	—	EG	60 min	[76]
CdS	Spheres (<5 nm)	$Cd(OAc)_2$	—	Thiourea, DMF	18 s	[77]
CdS	Spheres (2.8 nm)	$Cd(CH_3COO)_2$	1-thioglycerol	DMF, thiourea	900 W, 25–50 s	[78]
CdTe	Spheres (4.5 nm)	CdO, TeTBP	—	ODE, TDPA	180–280°C, 1–90 s	[79]
CdTe	Spheres (5 nm)	$CdCl_2$, NaHTe	MPA	H_2O	300 W, 10–150 min	[80]

Material	Morphology	Precursors	Additives	Solvent	Conditions	Ref.
PbS	Spheres (10 nm)	$Pb(OAc)_2$, S	–	NaOH, EtOH	Reflux, 20 min	[81]
PbSe	Cubes (5–15 nm)	$Pb(OAc)_2$, selenourea	Olaylamine	Phenyl ether, DMF	160 °C, 1 min, 3 bar	[82]
ZnS, CdS	6 nm	$ZnSO_4$, $CdSO_4$	Sodium citrate	Thioacetamide, KOH	905 W, 1 min	[83]
ZnO	Rods, wires, spindles	$Zn(NO_3)_2 \cdot 6H_2O$	–	NaOH, H_2O	100–200 °C, 10–20 min, 0.5–4.0 MPa	[84]
ZnO	Spheres (57–210 nm)	$Zn(OAc)_2 \cdot 2H_2O$	–	DEG	180 °C, 10 min	[85]
ZnS	Spheres (7 nm)	$Zn(OAc)_2$, thiourea	PVP	DMF	1–12 min	[86]
ZrO_2	Spheres (2 nm)	$Zr(NO_3)_4 \cdot 5H_2O$	–	PVA, NaOH	6 min	[87]
ZnSe	Spheres (2.5–4 nm)	$ZnCl_2$, Se(powder)	MPA	$NaBH_4$, H_2O	140 °C, 5 min	[88]
WO_3	Wires (20–30 nm × >1 μm)	$Na_2WO_4 \cdot 2H_2O$	$(NH_4)_2SO_4$	HCl, H_2O	150 °C, 20–180 min	[89]
WO_3	Aggregates	WCl_6	–	$C_6H_5CH_2OH$	210 °C, 5–15 min	[90]
$ZnWO_4$	Aggregates	H_2WO_4, $Zn(OAc)_2$	–	Citric acid, H_2O	30 min	[91]
Bi_2Se_3	50 × 30 nm	$BiONO_3$, Se	–	NaOH, EG	Reflux, 30 min	[92]
$MoSe_2$	Aggregates	$Mo(CO)_6$, Se	–	EG	Reflux, 60 min	[93]
$NiFe_2O_4$ $CoFe_2O_4$ $MnFe_2O_4$	Spheres (5 ± 2 nm)	$Fe(NO_3)_3 \cdot 9H_2O$ $Ni(NO_3)_2 \cdot 6H_2O$ $Co(NO_3)_2 \cdot 6H_2O$ $Mn(NO_3)_2 \cdot 4H_2O$	–	Oleic acid Toluene	500 W, 160 °C, 1 h	[94]

(Continued)

Table 5.1 (Continued)

Particles	Shape (size)	Precursor	Dispersing agent	Reducing agent/solvent	Reaction temp. and time	Ref.
$M_xZn_{1-x}O$ (M=V, Co, Fe, Ni, Mn)	7–19 nm	$Zn(CH_3COO)_3$ $VO[OCH(CH_3)_2]_3$ $Co(CH_3COO)_2$ $Fe(CH_3COO)_2$ $Ni(CH_3COO)_2 \cdot 4H_2O$ $Mn(CH_3COO)_2$	—	Benzyl alcohol	160 °C, 3 min	[95]
$MnZnFe_2O_4$	Aggregates	$Fe(NO_3)_3$, $Mn(NO_3)_2$, $Zn(NO_3)_2$	—	NaOH, H_2O	100 °C, 5–30 min	[96]
CdZnSe	Aggregates	$Cd(OAc)_2$, $Zn(OAc)_2$, Se	—	NaOH, EG	Reflux, 60 min	[97]
GdF_3:Eu^{3+}	Aggregates	$Gd(TFA)_3$, $Eu(TFA)_3$	—	[bmin][BF_4], EG	120 °C, 15 min	[98]
Cu—Ag	Spheres (20 ± 5 nm)	$Cu(NO_3^-)_2 \cdot 3H_2O$, $AgNO_3$	—	Ascorbic acid, H_2O	90 s	[99]
Au@Pd	12 nm	$HAuCl_4 \cdot 6H_2O$, $PdCl_2$	—	EG	10–60 s	[100]
Pd@Pt	Cubes (10–30 nm)	$K_2(PtCl_4)$, $PdCl_2$	CTAB	Ascorbic acid, H_2O	100 °C, 3 min	[101]
CdTe@ CdS@ZnS	Spheres (3–5 nm)	$CdCl_2$, NaHTe, $ZnCl_2$	MPA	Na_2S, H_2O	100 °C, 1–5 min	[102]
CdSe@ZnS	Aggregates	$CdCl_2$, Se, $ZnCl_2$	MPA	NaOH, NH_4OH, $NaBH_4$, H_2O	140–170 °C, 45–120 min	[103]

EG: ethylene glycol; FA: formaldehyde; DHN: dihydroxynaphthalene; DDA: dodecylamine; MPA: 3-mercaptopropionic acid; ODE: 1-octadecane; PVP: polyvinylpyrrolidone; TDPA: tetradecylphosphonic acid; TTIP: titanium tetraisopropoxide; PVA: polyvinyl alcohol.

References

1. Bilecka, I., and Niederberger, M. (2010) *Nanoscale*, **2**, 1358–1374.
2. Mallikarjuna, N.N., and Varma, R.S. (2007) *Cryst. Growth Des.*, **7**, 686–690.
3. Gerbec, J.A., Magana, D., Washington, A., and Strouse, G.F. (2005) *J. Am. Chem. Soc.*, **127**, 15791–15800.
4. Horikoshi, S., Abe, H., Torigoe, K., Abe, M., and Serpone, N. (2010) *Nanoscale*, **2**, 1441–1447.
5. Horikoshi, S., and Serpone, N. (2012) Microwave frequency effects in organic synthesis, in *Microwaves in Organic Synthesis*, 3rd edn (eds A. de la Hoz and A. Loupy), Wiley-VCH Verlag GmbH, Weinheim, Germany, pp. 377–423.
6. Meredith, R. (ed.) (1998) *Engineers' Handbook of Industrial Microwave Heating*, IEE power series, vol. 25, Institute of Electrical Engineers, London, UK.
7. Horikoshi, S., Matsuzaki, S., Mitani, T., and Serpone, N. (2012) *Radiat. Phys. Chem.*, **81**, 1885–1895.
8. Horikoshi, S., Abe, H., Sumi, T., Torigoe, K., Sakai, H., Serpone, N., and Abe, M. (2011) *Nanoscale*, **3**, 1697–1702.
9. Horikoshi, S., Iida, S., Kajitani, M., Sato, S., and Serpone, N. (2008) *Org. Proc. Res. Dev.*, **12**, 257–263.
10. Mohamed, M.B., Abouzeid, K.M., Abdelsayed, V., Aljarash, A.A., and El-Shall, M.S. (2010) *ACS Nano*, **4**, 2766–2772.
11. Metaxas, A.C., and Meredith, R.J. (1983) *Industrial Microwave Heating*, Peter Peregrinus, London, UK.
12. Horikoshi, S., et al. (2012) *J. Jpn. Soc. Color Mater.*, to be submitted for publication.
13. Cheng, J., Roy, R., and Agrawal, D. (2002) *Mater. Res. Innovat.*, **5**, 170–177.
14. Cheng, J., Roy, R., and Agrawal, D. (2002) Microwave processing in pure H fields and pure E fields, U.S. Patent No. 6,365,885, Issued April 2.
15. Horikoshi, S., Sumi, T., Abe, M., and Serpone, N. (2012) *J. Microwave Power Electromagnetic Energy*, **46**, 215–228.
16. Polshettiwar, V., and Varma, R.S. (2010) *Aqueous Microwave Assisted Chemistry: Synthesis and Catalysis*, RSC publishing, London, UK.
17. Horikoshi, S., Ohmori, N., Kajitani, M., and Serpone, N. (2007) *J. Photochem. Photobiol. A: Chem.*, **189**, 374–379.
18. Horikoshi, S., Tsuzuki, J., Kajitani, M., Abe, M., and Serpone, N. (2008) *New J. Chem.*, **32**, 2257–2262.
19. Herrero, M.A., Kremsner, J.M., and Kappe, C.O. (2008) *J. Org. Chem.*, **73**, 36–47.
20. Tu, W., and Liu, H. (2000) *J. Mater. Chem.*, **10**, 2207–2211.
21. Wiegand, S. (2004) *J. Phys. Condens. Matter.*, **16**, R357–R379.
22. Schanche, J.-S. (2003) *Mol. Divers.*, **7**, 293–300.
23. Chang, C.-H., Paul, B.K., Remcho, V.T., Atre, S., and Hutchison, J.E. (2008) *J. Nanopart. Res.*, **10**, 965–980.
24. Ehrfeld, W., Hessel, V., and Löwe, H. (2001) *Microreactors: New Technology for Modern Chemistry*, Wiley-VCH Verlag GmbH, Weinheim, Germany.
25. Zhao, C.-X., He, L., Qiao, S.Z., and Middelberg, A.P.J. (2011) *Chem. Eng. Sci.*, **66**, 1463–1479.
26. Horikoshi, S., Sumi, T., and Serpone, N. (2012) *Chem. Eng. J.*, submitted.
27. Nishioka, M., Miyakawa, M., Kataoka, H., Koda, H., Sato, K., and Suzuki, T.M. (2011) *Chem. Lett.*, **40**, 1204–1206.
28. National Research Council (1991) *Plasma Processing of Materials: Scientific Opportunities and Technological Challenges*, National Academies Press, Washington, DC.
29. Urashima, K., and Chang, J.-S. (2000) *IEEE Trans. Dielectr. Electr. Insul.*, **7**, 602–614.
30. Horikoshi, S., Sato, S., Abe, M., and Serpone, N. (2011) *Ultrason. Sonochem.*, **18**, 938–942.
31. Sato, S., Mori, K., Ariyada, O., Hyono, A., and Yonezawa, T. (2011) *Surf. Coat. Technol.*, **206**, 955–958.
32. Tsuji, M., Hashimoto, M., and Tsuji, T. (2002) *Chem. Lett.*, **31**, 1232–1233.
33. Liu, F.-K., Huang, P.-W., Chang, Y.-C., Ko, C.-J., and Chu, T.-C. (2005) *J. Cryst. Growth*, **273**, 439–445.

34 Liu, F.-K., Huang, P.-W., Chu, T.-C., and Ko, F.-H. (2005) *Mater. Lett.*, **59**, 940–944.

35 Gao, F., Lu, Q., and Komarneni, S. (2005) *Chem. Mater.*, **17**, 856–860.

36 Hu, B., Wang, S.-B., Wang, K., Zhang, M., and Yu, S.-H. (2008) *J. Phys. Chem. C*, **112**, 11169–11174.

37 Zhao, T., Fan, J.-B., Cui, J., Liu, J.-H., Xu, X.-B., and Zhu, M.-Q. (2011) *Chem. Phys. Lett.*, **501**, 414–418.

38 Zhu, Y.-J., and Hu, X.-L. (2004) *Mater. Lett.*, **58**, 1517–1519.

39 Kundu, S., and Liang, H. (2008) *Langmuir*, **24**, 9668–9674.

40 Kundu, S., Peng, L., and Liang, H. (2008) *Inorg. Chem.*, **47**, 6344–6352.

41 Liu, F.K., Ker, C.J., Chang, Y.C., Ko, F.H., Chu, T.C., and Dai, B.T. (2003) *J. Appl. Phys. Part 1*, **42**, 4152–4158.

42 Kundu, S., Wang, K., and Liang, H. (2009) *J. Phys. Chem. C*, **113**, 5157–5163.

43 Zhu, Y.-J., and Hu, X.-L. (2003) *Chem. Lett.*, **32**, 1140–1141.

44 Fernández-García, M., Martínez-Arias, A., Hanson, J.C., and Rodriguez, J.A. (2004) *Chem. Rev.*, **104**, 4063–4104.

45 Nishioka, M., Miyakawa, M., Daino, Y., Kataoka, H., Koda, H., Sato, K., and Suzuki, T.M. (2011) *Chem. Lett.*, **40**, 1327–1329.

46 Komarneni, S., Li, D., Newalkar, B., Katsuki, H., and Bhalla, A.S. (2002) *Langmuir*, **18**, 5959–5962.

47 Mehta, S.K., and Gupta, S. (2011) *J. Appl. Electrochem.*, **41**, 1407–1417.

48 Yu, Y., Zhao, Y., Huang, T., and Liu, H. (2010) *Mater. Res. Bull.*, **45**, 159–164.

49 Zhu, Y.-J., Wang, W.-W., Qi, R.-J., and Hu, X.-L. (2004) *Angew. Chem. Int. Ed.*, **43**, 1410–1414.

50 Zhu, H.-T., Zhang, C.-Y., and Yin, Y.-H. (2004) *J. Cryst. Growth*, **270**, 722–728.

51 Blosi, M., Albonetti, S., Dondi, M., Martelli, C., and Baldi, G. (2011) *J. Nanopart. Res.*, **13**, 127–138.

52 Nakamura, T., Tsukahara, Y., Sakata, T., Mori, H., Kanbe, Y., Bessho, H., and Wada, Y. (2007) *Bull. Chem. Soc. Jpn.*, **80**, 224–232.

53 Vollimer, C., Redel, E., Abu-Shandi, K., Thomann, R., Manyar, H., Hardacre, C., and Janiak, C. (2010) *Chem. Eur. J.*, **16**, 3849–3858.

54 Zhao, Y., Zhu, J.-J., Hong, J.-M., Bian, N., and Chen, H.-Y. (2004) *Eur. J. Inorg. Chem.*, 4072–4080.

55 Wang, H., Xu, J.-Z., Zhu, J.-J., and Chen, H.-Y. (2002) *J. Cryst. Growth*, **244**, 88–94.

56 Volanti, D.P., Orlandi, M.O., Andrés, J., and Longo, E. (2010) *Cryst. Eng. Commun.*, **12**, 1696–1699.

57 Liao, X.-H., Zhu, J.-J., and Chen, H.-Y. (2001) *Mater. Sci. Eng. B*, **85**, 85–89.

58 Ding, K., Miao, Z., Liu, Z., Zhang, Z., Han, B., An, G., Miao, S., and Xie, Y. (2007) *J. Am. Chem. Soc.*, **129**, 6362–6363.

59 Suprabha, T., Roy, H.G., Thomas, J., Kumar, K.P., and Mathew, S. (2009) *Nanoscale Res. Lett.*, **4**, 144–152.

60 Corradi, A.B., Bondioli, F., Focher, B., Ferrari, A.M., Grippo, C., Mariani, E., and Villa, C. (2005) *J. Am. Ceram. Soc.*, **88**, 2639–2641.

61 Xiao, W., Gu, H., Li, D., Chen, D., Deng, X., Jiao, Z., and Lin, J. (2012) *J. Magn. Magn. Mater.*, **324**, 488–494.

62 Hu, X., and Yu, J.C. (2008) *Adv. Funct. Mater.*, **18**, 880–887.

63 Li, Y., Li, H., and Cao, R. (2009) *J. Am. Ceram. Soc.*, **92**, 2188–2191.

64 Muraliganth, T., Murugan, A.V., and Manthiram, A. (2009) *Chem. Commun.*, 7360–7362.

65 Hu, X., and Yu, J.C. (2006) *Chem. Asian J.*, **1**, 605–610.

66 Hu, H., Yang, H., Huang, P., Cui, D., Peng, Y., Zhang, J., Lu, F., Lian, J., and Shi, D. (2010) *Chem. Commun.*, **46**, 3866–3868.

67 Patra, C.R., and Gedanken, A. (2004) *New J. Chem.*, **28**, 1060–1065.

68 Jouhannaud, J., Rossignol, J., and Stuerga, D. (2008) *J. Solid State Chem.*, **181**, 1439–1444.

69 Tao, Y., Gong, F.-H., Wu, H.-P., and Tao, G.-L. (2008) *Mater. Chem. Phys.*, **112**, 973–976.

70 Zawadzki, M. (2008) *J. Alloys Compd.*, **454**, 347–351.

71 Palchik, O., Zhu, J.-J., and Gedanken, A. (2000) *J. Mater. Chem.*, **10**, 1251–1254.

72 Anandan, S., and Wu, J.-J. (2009) *Mater. Lett.*, **63**, 2387–2389.

73 Raj, D.S., Krishnakumar, T., Jayaprakash, R., Donato, N., Latino, M., and Neri, G. (2010) *Sci. Adv. Mater.*, **2**, 432–437.

74 Wang, W.-W. (2008) *Mater. Chem. Phys.*, **108**, 227–231.

75 Nyutu, E.K., Chen, C.-H., Dutta, P.K., and Suib, S.L. (2008) *J. Phys. Chem. C*, **112**, 9659–9667.

76 Palchik, O., Kerner, R., Gedanken, A., Weiss, A.M., Slifkin, M.A., and Palchik, V. (2001) *J. Mater. Chem.*, **11**, 874–878.

77 Wada, Y., Kuramoto, H., Anand, J., Kitamura, T., Sakata, T., Mori, H., and Yanagida, S. (2001) *J. Mater. Chem.*, **11**, 1936–1940.

78 Karan, S., and Mallik, B. (2007) *J. Phys. Chem. C*, **111**, 16734–16741.

79 Song, Q., Ai, X., Topuria, T., Rice, P.M., Alharbi, F.H., Bagabas, A., Bahattab, M., Bass, J.D., Kim, H.-C., Scott, J.C., and Miller, R.D. (2010) *Chem. Commun.*, **46**, 4971–4973.

80 Li, L., Qian, H., and Ren, J. (2005) *Chem. Commun.*, 528–530.

81 Ding, T., and Zhu, J.-J. (2003) *Mater. Sci. Eng. B*, **100**, 307–313.

82 Sliem, M.A., Chemseddine, A., Bloeck, U., and Fischer, R.A. (2011) *Cryst. Eng. Comm*, **13**, 483–488.

83 Ortíz, S., Gómez, I., Elizondo, P., and Cavazos, J. (2010) *Phys. Status Solidi*, **C 7**, 2683–2687.

84 Huang, J., Xia, C., Cao, L., and Zeng, X. (2008) *Mater. Sci. Eng. B*, **150**, 187–193.

85 Hu, X., Gong, J., Zhang, L., and Yu, J.C. (2008) *Adv. Mater.*, **20**, 4845–4850.

86 He, R., Qian, X.-F., Yin, J., Xi, H.-A., Bian, L.-J., and Zhu, Z.-K. (2003) *J. Colloids Surf. A*, **220**, 151–157.

87 Liang, J., Deng, Z., Jiang, X., Li, F., and Li, Y. (2002) *Inorg. Chem.*, **41**, 3602–3604.

88 Qian, H., Qiu, X., Li, L., and Ren, J.J. (2006) *J. Phys. Chem. B*, **110**, 9034–9040.

89 Phuruangrat, A., Ham, D.J., Hong, S.J., Thongtem, S., and Lee, J.S. (2010) *J. Mater. Chem.*, **20**, 1683–1690.

90 Houx, N.L., Pourroy, G., Camerel, F., Comet, M., and Spitzer, D. (2010) *J. Phys. Chem. C*, **114**, 155–161.

91 Ryu, J.H., Lima, C.S., Oh, W.C., and Shim, K.B. (2004) *J. Ceram. Process. Res.*, **5**, 316–320.

92 Harpeness, R., and Gedanken, A. (2003) *New J. Chem.*, **27**, 1191–1193.

93 Harpeness, R., Gedanken, A., Weiss, A.M., and Slifkin, M.A. (2003) *J. Mater. Chem.*, **13**, 2603–2606.

94 Baruwati, B., Nadagouda, M.N., and Varma, R.S. (2008) *J. Phys. Chem.*, **C 112**, 18399–18404.

95 Bilecka, I., Luo, L., Djerdj, I., Rossell, M.D., Jagodič, M., Jagličić, Z., Masubuchi, Y., Kikkawa, S., and Niederberger, M. (2011) *J. Phys. Chem. C*, **115**, 1484–1495.

96 Zhenyu, L., Guangliang, X., and Yalin, Z. (2006) *Nanoscale Res. Lett.*, **2**, 40–43.

97 Grisaru, H., Palchik, O., Gedanken, A., Palchik, V., Slifkin, M.A., Weiss, A.M., and Hacohen, Y.R. (2001) *Inorg. Chem.*, **40**, 4814–4815.

98 Lorbeer, C., Cybinska, J., and Mudring, A.-V. (2010) *Chem. Comm.*, **46**, 571–573.

99 Valodkar, M., Modi, S., Pal, A., and Thakore, S. (2011) *Mater. Res. Bull.*, **46**, 384–389.

100 Harpeness, R., and Gedanken, A. (2004) *Langmuir*, **20**, 3431–3434.

101 Zhang, H., Yin, Y., Hu, Y., Li, C., Wu, P., Wei, S., and Cai, C. (2010) *J. Phys. Chem. C*, **114**, 11861–11867.

102 He, Y., Lu, H.-T., Sai, L.-M., Su, Y.-Y., Hu, M., Fan, C.-H., Huang, W., and Wang, L.-H. (2008) *Adv. Mater.*, **20**, 3416–3421.

103 Schumacher, W., Nagy, A., Waldman, W.J., and Dutta, P.K. (2009) *J. Phys. Chem. C*, **113**, 12132–12139.

6
Microwave-Assisted Solution Synthesis of Nanomaterials
Xianluo Hu and Jimmy C. Yu

6.1
Introduction

The development of new synthetic strategies is a highly active field in nanotechnology research [1–5]. The ability to fabricate and process nanostructured materials is the first cornerstone in nanotechnology, paving the way for understanding novel properties and realizing potential applications of nanostructured materials [6–10]. To date, many methods have been explored to synthesize nanostructured materials. These technical approaches can be essentially grouped in two paradigms: top-down and bottom-up [11, 12]. The top-down approach could not achieve the manufacture of structures with infinitesimal dimensions due to technological and economic reasons. This approach often needs to be carried out in a clean room or a vacuum environment, and large machinery is required in all cases. Another disadvantage is that the conventional top-down approaches, such as lithography, may cause significant crystallographic damage to the processed patterns, and thus additional defects may be introduced. In contrast, versatile bottom-up methods based on chemistry produce a diverse range of nanostructures at relatively low cost, with high throughput and the potential for large-scale industrial production [13–17]. Since bottom-up processes are driven mainly by the reduction of Gibbs free energy, the as-formed nanostructures are in a state closer to a thermodynamic equilibrium state. In comparison to top-down methods, therefore, bottom-up approaches based on chemistry promise a better chance to design, fabricate, and manipulate nanostructures with less defects, more homogenous chemical composition, and better short- and long-range ordering [18–20].

It is reasonable that we can take systematic, logical procedures to fabricate nanostructures with desired dimensions and morphologies by carefully controlling experimental conditions (e.g., temperature, concentration, and chemical interactions) or by introducing additives (e.g., heterogeneous nuclei to lower nucleation barriers). In the past decades, a wide variety of chemical techniques, such as co-precipitation, sol–gel processes, microemulsions, freeze drying, hydrothermal processes, laser pyrolysis, ultrasound and microwave irradiation, templates, and

Microwaves in Nanoparticle Synthesis, First Edition. Edited by Satoshi Horikoshi and Nick Serpone.
© 2013 Wiley-VCH Verlag GmbH & Co. KGaA. Published 2013 by Wiley-VCH Verlag GmbH & Co. KGaA.

chemical vapor deposition, have been developed to simultaneously control size, morphology, and uniformity of nanostructures [21–27]. The successful implementation of the "bottom-up" strategy requires, in the end, the controlled growth of nanostructures. The development of new strategies for the controlled synthesis of nanostructured materials with specified architectures/properties, particularly soft wet chemical routes, is therefore critical in the field of nanotechnology. Among various techniques, solution-based methods have several significant advantages for the synthesis of nanostructured materials including: (i) low reaction temperatures, (ii) size-selective growth, (iii) morphological control, and (iv) large-scale production [28–31]. Liquid-phase techniques, including coprecipitation, sol–gel processing, microemulsions, hydrothermal/solvothermal processing, and template-engaged and biomimetic methods for the synthesis of inorganic nanostructures have been recently reviewed [32–34].

Conventional heating in solution-based chemical processes relies on thermal conduction of black-body radiation to drive chemical reactions, whereby the reaction vessel serves as an intermediary for energy transfer from the heating source to the solvent, and finally to reactant molecules [35]. It inevitably suffers from several disadvantages, especially at relatively low temperatures, including sharp thermal gradients throughout the bulk solution, slow reaction kinetics, and nonuniform reaction conditions [35]. In particular, as the area of nanotechnology becomes more of a commercial reality, the industrialization of nanomaterials is limited by the need for new material compositions and the development of high throughput automation for materials preparation. For large-scale reactions, inhomogeneous effects and thermal gradients may be magnified severely during the processes of nanomaterials preparation, resulting in poor nucleation and broadened size distributions [36, 37]. Therefore, it is highly desirable to develop new synthetic strategies to avoid the existing specific problems encountered in the preparation of nanomaterials by conventional heating.

Microwave heating methods can address the problem of heating inhomogeneity. In fact, microwave irradiation has been becoming an increasingly popular method of heating samples for nanomaterials synthesis, since the first reports of microwave-assisted synthesis in 1986 [38]. The technique has found a valuable place in synthetic chemists' and materials scientists' toolboxes. This is evidenced by the large number of papers and reviews appearing in the literature in the last decades [25, 39–43]. Because of its high efficiency in energy utilization, microwave irradiation offers a clean, cheap, and convenient method of heating which often results in a higher yield and shorter reaction time. Fundamentally, microwaves heat reacting species differently from the conventional means. Some effects generated under microwave irradiation cannot be duplicated by conventional heating. Microwave dielectric heating is also unique in providing scaled-up processes with a uniform nucleation environment, thus leading to a potential advancement in large-scale industrial production of high-quality nanocrystals [44–49].

Microwave-promoted chemistry is founded on the fact that materials, be they solvents or reagents, can absorb microwave energy and convert it to heat [50]. Microwaves are a form of electromagnetic energy with frequencies in the range of 300 MHz to 300 GHz. The commonly used frequency is 2.45 GHz. Interactions between materials and microwaves are based on two specific mechanisms: dipole interactions and ionic conduction. Both mechanisms require effective coupling between components of the target material and the rapidly oscillating electrical field of the microwaves. Dipole interactions occur with polar molecules. The polar ends of a molecule tend to re-orient themselves and oscillate in step with the oscillating electrical field of the microwaves. Heat is generated by molecular collision and friction. Generally, the more polar a molecule, the more effectively it will couple with the microwave field. Figure 6.1 shows a schematic illustration for the microwave interaction with polar H_2O molecules. Ionic conduction is only minimally different from dipole interactions. Ions in solution do not have a dipole moment. They are charged species distributed in solution and can couple with the oscillating electrical field of microwaves. The concentration of ions in solution often significantly affects the efficiency of microwave heating of an ionic solution [51].

Much work in the field of microwave-assisted synthesis has been conducted using domestic microwave ovens. However, some problems exist in a domestic microwave system. For example, reactions are hard to control. It is particularly

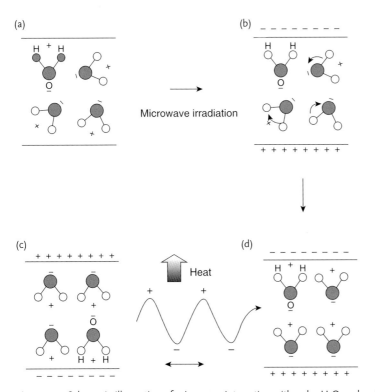

Figure 6.1 Schematic illustration of microwave interaction with polar H_2O molecules. Reproduced from reference [25].

hazardous if flammable solvents and reagents are used. Also, it is very difficult to accurately measure the temperature and microwave power in a domestic microwave system. This is the reason why the reaction conditions published using domestic microwave apparatus are poorly reproduced. With the advent of scientific microwave systems, these problems can be overcome [52–57]. In such a system, temperature can be measured either using an IR probe located outside the reaction vessel or a fiber-optic probe inserted into the reaction vessel. Pressure is measured using a load cell. The system software and electronics make it possible to heat a reaction mixture to a desired bulk temperature and hold it there by a predetermined program. The automatic temperature-control system allows continuous monitoring and control of the internal temperature of the reaction systems. The preset profile (desired time, temperature, pressure) could be followed automatically by continuously adjusting the applied microwave power. The temperature, pressure, and power profiles of each reaction can also be recorded easily. Since modern microwave systems possess the capabilities to program time and temperature of consecutive steps, several parallel reactions can be conducted in one run. This allows us not only to significantly reduce the preparation time from days to minutes, but also to screen a wide range of experimental conditions in order to optimize and scale up materials preparation. Depending on the apparatus used, reactions can be carried out either in sealed vessels of varying volumes or in open vessels under atmospheric pressure. This chapter reviews some recent progress in the design, fabrication, and modification of nanomaterials by microwave-assisted solution synthesis.

6.2
Synthesis of ZnO Nanocrystals

Zinc oxide is a wide-bandgap (3.37 eV) semiconductor with large exciton binding energy (60 meV). To date, many approaches have been reported for fabricating well-defined ZnO nanostructures with not only diverse morphologies but also novel properties [36, 58–61]. Owing to the intrinsic nature of polar hexagonal-phase ZnO with an $a:c$ axial ratio of 1:1.6, most of the reported ZnO nanoarchitectures show the one-dimensional (1D) or branched morphology. It should be noted that the fabrication of complex ZnO colloids with secondary microstructures is often difficult because of the lack of appropriate and generalized synthetic methodologies [62]. A high-temperature nonhydrolytic sol–gel route based on the ester-elimination reaction between zinc acetate and alcohol has been proposed for growing uniform hierarchically self-assembled ZnO spheres composed of anisotropic cone-like nanocrystals, as well as soluble ZnO nanocrystals with tetrapod and spiked-cluster morphologies [63, 64]. Zeng and Liu fabricated three-tiered organization of ZnO nanobuilding blocks into hollow spherical conformations by a template-assisted method [65]. Interestingly, aggregation of ZnO nanocrystallites involving secondary structures has been proven to be an effective way to generate light scattering within the photoelectrode film of dye-sensitized solar cells (DSSCs) while retaining the desired specific surface area for dye-molecule adsorption [66].

6.2.1
Synthesis of Colloidal ZnO Nanocrystals Clusters

Recently, we have developed an efficient microwave-polyol process to synthesize uniform-sized ZnO colloidal nanocrystal clusters (CNCs) in an open polyol system [36]. Fine tuning of ZnO CNCs with precise size control ranging from about 57 to 274 nm can be achieved. It is anticipated that the as-formed ZnO CNCs may provide flexible ZnO building blocks for potential three-dimensional (3D) photonic crystals, but are ideal candidates for systematically studying their nanoarchitecture-dependent performance in optical, catalytic, and sensing applications. Microwave heating at 180 °C promotes the hydrolysis of $Zn(OAc)_2$ in DEG to form zinc glycolates or alkoxide derivatives. Then, the Zn-complexes transform into ZnO nanocrystals through dehydration under microwave irradiation. By optimizing the experimental conditions, these fresh-formed ZnO nanocrystals spontaneously aggregate to form raspberry-like 3D clusters with tunable sizes.

Figure 6.2 shows the scanning electron microscopy (SEM) and transmission electron microscopy (TEM) images of the products. The monodispersed aggregates are spherical in shape and have rough surfaces. Close observation confirms that these monodisperse colloidal particles consist of small primary particles. The size of the ZnO CNCs could be tuned from ~57 to 274 nm by simply reducing the amount of the Zn-complex precursor while keeping all other parameters constant. The realized size-tuning capability should originate from the slight difference in the amount of H_2O and crystal nuclei induced by varying additions of Zn-complex solution. Both higher H_2O content and higher Zn-complex concentration could accelerate the hydrolysis of $Zn(OAc)_2$ and dehydration of the newly formed Zn-complexes, thus leading to more nuclei in the bulk solution and finally smaller assembled clusters. We propose that the growth of ZnO CNCs follows the well-documented two-stage growth model [67]. The primary ZnO nanocrystals (~8 nm) first nucleate in a supersaturated solution and then aggregate into larger raspberry-like assemblies (~57–274 nm in diameter). The X-ray diffraction (XRD) pattern show a pure phase of wurtzite-type hexagonal ZnO. High-magnification and high-resolution TEM images suggest the single-crystal-like nature of ZnO CNCs. All of the primary nanocrystals align crystallographically with adjacent ones in the same crystal orientation.

As an n-type semiconductor, a variety of ZnO nanostructures, such as nanowires and nanotetrapods, have been applied in humidity sensors [68]. Indeed, these materials show a considerable increase in sensitivity compared to their bulk counterparts. Prompted by the unique secondary and complex nanoarchitectures, we expect that the as-formed ZnO CNCs from our microwave-polyol synthesis would be advantageous for fabricating humidity sensors. Our preliminary results demonstrate that the sensors made of ZnO CNCs exhibit high sensitivity for humidity measurement at room temperature. The thin-film sensor made of monodispersed

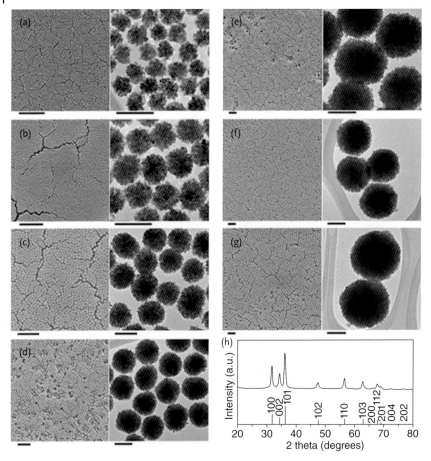

Figure 6.2 (a) High-magnification TEM image of a single ZnO cluster of 86 nm. (b) ED pattern indicating the single-crystal-like nature of the CNCs. (c) Typical HRTEM image taken from a 57-nm cluster and the corresponding fast Fourier-transformation (FFT) pattern (inset). (d) Corresponding intensity profile for the line scan across the lattice fringes. (e) EDX spectrum of a single ZnO CNC, where the signal of Cu is generated from the Cu grids. Reproduced from reference [36].

ZnO nanocrystal clusters, possesses a loose-film feature analogous to a highly porous architecture and a network of interconnected hierarchical pores. The network of hierarchical inter-nanocrystallite and inter-cluster pores should contribute to the high sensitivity, since it allows the target molecules more access to all the surfaces of ZnO CNCs included in the sensing unit. Furthermore, we found that thin-film sensors made of ZnO CNCs with different diameters exhibited the size-dependent sensitivity. When the diameter of the clusters is decreased from 274, 210, 183, 162, to 125 nm, humidity sensitivity increases from 2.6, 6.3, 114, 180, to 2476. This is easily understood as higher surface areas of smaller clusters contribute to the distinct increase in sensitivity. When further decreasing the diameter to 86 and 57 nm, however, the sensitivity does not improve but actually drops slightly. This suggests that the clusters of an appropriate size between 87 and 125 nm possess the optimal inter-cluster porosity and internal surface area in the sensing unit, in favor of achieving high performance of ZnO CNCs for humidity sensing.

6.2.2
Controlled Growth of Basic and Complex ZnO Nanostructures

Lee *et al.* reported a chemical method to synthesize ZnO crystals from basic to complex structures by a microwave irradiation method [69]. The experimental conditions, such as the precursor chemicals, the capping agents, and the aging times. were controlled. At a low temperature (90 °C) with low power microwave-assisted heating (about 50 W) and a subsequent aging process, a variety of basic ZnO structures, including nanorods, nanocandles, nanoneedles, nanodisks, nanonuts, microstars, microUFOs, and microballs were simply synthesized (Figure 6.3). Moreover, more complex ZnO structures, including ZnO bulky stars, cakes, and jellyfishes, were achieved by microwave treating a mixture of the as-prepared basic ZnO structures and the different mixture solution containing the Zn source and the surfactants. The growth mechanisms for those ZnO structures were discussed. It was found that the key factor of the morphology-controlled growth of ZnO is controlling the crystal growth and dissolution rates in specific directions and the concentration of growth units.

6.2.3
Synthesis of ZnO Nanoparticles in Benzyl Alcohol

Niederberger's group recently carried out a detailed study of kinetic and thermodynamic aspects in the microwave-assisted synthesis of ZnO nanoparticles from zinc acetate and benzyl alcohol [70]. It is demonstrated that the formation of ZnO nanoparticles from the reaction of zinc acetate with benzyl alcohol proceeds along several consecutive steps that involve organic as well as inorganic reactions (Figure 6.4). Inseparable and fundamental processes occurring during the nanoparticle formation involve (i) the esterification reaction producing benzyl acetate and the monomeric species on the organic side, and (ii) nucleation and growth of ZnO on the inorganic side. Also, the dissolution rate and coarsening effects contribute greatly to the kinetics of the ZnO formation. However, the organic esterification reaction represents the key step as well as the bottleneck of the nucleation process. The monomers, presumably zinc hydroxo species, were generated from the esterification reaction. If the monomer concentration reaches supersaturation then ZnO clusters start to nucleate. Microwave irradiation prompts the esterification reaction. It is also observed that the monomer formation (as a result of the esterification reaction) does not influence the rate of crystal growth, but the coarsening is greatly affected by microwave irradiation. In addition to the esterification reaction

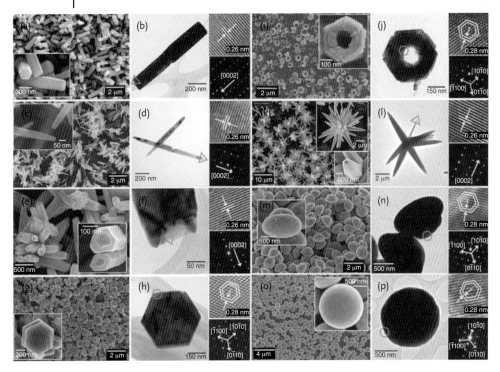

Figure 6.3 SEM (a,c,e,g,i,k,m,o) and TEM (b,d,f,h,j,l,n,p) images of the basic ZnO structures synthesized by the microwave irradiation: (a) and (b) nanorods, (c) and (d) nanoneedles, (e) and (f) nanocandles, (g) and (h) nanodisks, (i) and (j) nanonuts, (k) and (l) microstars, (m) and (n) micro-UFOs, (o) and (p) microballs. HRTEM images and SAED patterns were inserted as upper and lower insets in TEM images, respectively. Reproduced from reference [69].

and the crystal growth, microwave irradiation could accelerate the ZnO formation by improving the thermal ramping, resulting in a fast heating of the reaction mixture, and the dissolution of the precursor in benzyl alcohol. This work may provide a possible way to directly correlate the kinetics of the organic side reaction with the growth kinetics of the ZnO nanoparticles, by monitoring both the formation of the organic species as well as ZnO crystal size.

6.3
Synthesis of α-Fe$_2$O$_3$ Nanostructures

α-Fe$_2$O$_3$ (hematite), an n-type semiconductor ($E_g = 2.1$ eV), is the most stable iron oxide under ambient conditions. It has been extensively investigated because it has novel electrical and catalytic properties. Up to now, many α-Fe$_2$O$_3$ structures, such as 0D (particles), 1D (rods, wires, tubes, and belts), 2D/3D (disks, dendrites,

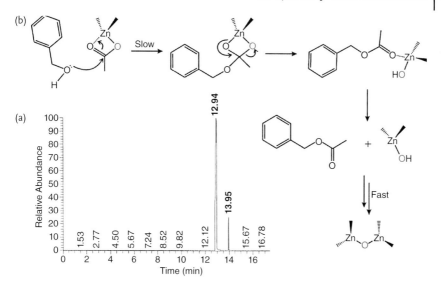

Figure 6.4 (a) Gas chromatogram of the final organic reaction solution with two elution peaks at 12.94 and 13.95 min corresponding to benzyl alcohol and benzyl acetate, respectively. (b) Ester elimination reaction as a result of a nucleophilic attack of benzyl alcohol on zinc acetate, finally resulting in the formation of benzyl acetate and zinc oxide. Reproduced from reference [70].

flowers, and mesopores), and hybrids, have been synthesized by a variety of methods [71–78]. As expected, the various α-Fe$_2$O$_3$ structures lead to interesting shape-dependent properties and a wide variety of potential applications, including sensors [54, 72, 74, 77], catalysts [78, 79], electrode materials in lithium secondary batteries [72, 80, 81], and water splitting [82], and water treatment [83].

6.3.1
α-Fe$_2$O$_3$ Hollow Spheres

Zhu et al. synthesized α-Fe$_2$O$_3$ hierarchically nanostructured hollow spheres through a microwave-solvothermal route combined with subsequent thermal decomposition [83]. The Fe-based precursor was formed by microwave-heating a mixture of FeCl$_3 \cdot$6H$_2$O, NaOH, and sodium dodecylbenzenesulfonate (SDBS) in ethylene glycol in a closed Teflon autoclave at 200 °C for 30 min. The Fe-based precursor exhibits hierarchically nanostructured hollow spheres that are assembled with nanosheets. After a further treatment at 500 °C in air, the Fe-based precursor transforms topotactically into α-Fe$_2$O$_3$, and the morphology is well maintained. The α-Fe$_2$O$_3$ product with hierarchical hollow nanostructures is demonstrated to have a good ability for photocatalytic degradation of salicylic acid and removal of Methylene Orange in water treatment.

6.3.2
Monodisperse α-Fe$_2$O$_3$ Nanocrystals with Continuous Aspect-Ratio Tuning and Precise Shape Control

Recently, designed synthesis of uniform α-Fe$_2$O$_3$ nanocrystals in aqueous solution has been successfully developed through the microwave-hydrothermal method, taking advantage of microwave irradiation and hydrothermal effects [37, 54]. Continuous aspect-ratio tuning and precise shape control over the resulting α-Fe$_2$O$_3$ nanocrystals could be easily realized. The capability of continuous aspect-ratio tuning and precise shape control of Fe$_2$O$_3$ nanocrystals should provide an ideal platform for systematically exploring the size/shape-related optical, catalytic, magnetic, and lithium-storage properties. Microwave-induced hydrolysis of the iron precursor (FeCl$_3$) in a closed aqueous NH$_4$H$_2$PO$_4$ system leads to monodisperse α-Fe$_2$O$_3$ nanocrystals with controlled size and shape. The experimental conditions for typical samples and their morphologies are listed in Table 6.1.

It is found that the shape of the α-Fe$_2$O$_3$ nanocrystals depends on the amount of NH$_4$H$_2$PO$_4$ used, and evolves from pseudocubes (S1, see Figure 6.5a), ellipsoids (S2 and 3, see Figure 6.5b and c), to gradually elongated spindles (S4-6, see Figure 6.5d–f). When the NH$_4$H$_2$PO$_4$ content was increased, the aspect ratios (c/a) of the as-formed Fe$_2$O$_3$ particles were enlarged from 1.1, 1.5, 2.0, 2.6, 3.8, to 6.3. The average aspect ratio (c/a) and length (c) of the α-Fe$_2$O$_3$ products as a function of phosphate content is shown in Figure 6.6, revealing a clear trend of morphology evolution.

In addition to NH$_4$H$_2$PO$_4$, the amount of the iron source (Fe^{3+}) plays an important role in the morphology of α-Fe$_2$O$_3$ nanocrystals. When 0.02 mmol of phosphate and a reaction temperature of 220 °C were used, the shape of α-Fe$_2$O$_3$ nanocrystals evolved from nanosheets (S7) to nanorings (S8) and then to ellipsoids

Table 6.1 Experimental conditions for the preparation of samples.

Sample No.	FeCl$_3$ (mmol)	NH$_4$H$_2$PO$_4$ (mmol)	Molar ratio	T (°C)	Size (c/a) (nm)	Aspect ratio	Morphology
S1	1.6	0	–	220	240/220	1.10	Pseudocubes
S2	1.6	0.008	200	220	280/190	1.5	Ellipsoids
S3	1.6	0.016	100	220	400/200	2.0	Ellipsoids
S4	1.6	0.02	80	220	500/195	2.6	Spindles
S5	1.6	0.03	53	220	710/185	3.8	Spindles
S6	1.6	0.04	40	220	885/140	6.3	Spindles
S7	0.25	0.02	12.5	220	90	–	Nanosheets
S8	0.5	0.02	25	220	60	–	Nanorings
S9	2.5	0.02	125	220	770/407	1.9	Ellipsoids
S10	0.5	0.01	50	220	225/50	4.5	Nanospindles
S11	1.6	0.05	32	220	300	–	Spheres

Reproduced from reference [37].

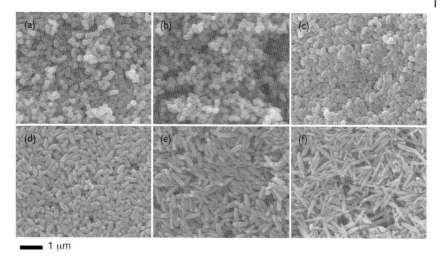

Figure 6.5 SEM images of sample (a) 1, (b) 2, (c) 3, (d) 4, (e) 5, and (f) 6. Reproduced from reference [37].

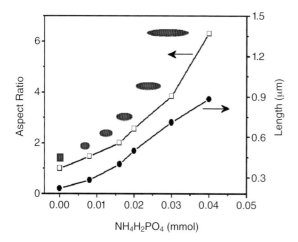

Figure 6.6 Aspect ratio (c/a) and length (c) of the α-Fe$_2$O$_3$ crystals as a function of phosphate content Reproduced from reference [37].

(S9) upon gradually increasing the initial Fe^{3+} amount from 0.25 to 2.50 mmol. If 0.01 mmol PO$_4^{3-}$ and 0.5 mmol Fe^{3+} were used, monodisperse nanospindles with an aspect ratio of about 4.5 (S10) were obtained. Spherical particles of ~300 nm were formed when both the initial amounts of PO$_4^{3-}$ and Fe^{3+} were further increased to 0.05 and 1.6, respectively (S11).

Based on a systematic analysis, it is concluded that the initial molar ratio of FeCl$_3$ to NH$_4$H$_2$PO$_4$ had a significantly direct relationship with the aspect ratio of the final products. There are two types of growth mechanism proposed for the controllable formation of $\alpha\text{-}Fe_2O_3$ nanocrystals. One is "nucleation–aggregation–recrystallization" for a molar Fe^{3+} to PO_4^{3-} ratio higher than 40, and the other is "nucleation–aggregation–dissolution–recrystallization" over a range of Fe^{3+} to PO_4^{3-} ratio of 10 to 35.

The infrared optical properties of a powdered ionic solid depend critically on the particle size and shape [84]. The size- and shape-dependent properties result from the polarization change induced at the particle surfaces by an external electromagnetic field [85]. Interestingly, the resulting $\alpha\text{-}Fe_2O_3$ nanocrystals that have good uniformity, continuous aspect-ratio tuning, and fine shape control may serve as the ideal model crystals, and they offer important and high-value opportunities for directly evidencing or systematically exploring the shape-dependent infrared optical properties.

6.3.3
Self-Assembled Hierarchical $\alpha\text{-}Fe_2O_3$ Nanoarchitectures

Single-crystal self-assembled hierarchical $\alpha\text{-}Fe_2O_3$ nanoarchitectures were synthesized by a microwave-enhanced hydrothermal process [86]. As shown in Figures 6.7 and 6.8, well-defined hierarchical $\alpha\text{-}Fe_2O_3$ assemblies were achieved by microwave-treating an aqueous $K_3[Fe(CN)_6]$ solution for less than 30 min. The $\alpha\text{-}Fe_2O_3$ formation proceeds via three steps [71]:

$$[Fe(CN)_6]^{3-} \leftrightarrow Fe^{3+} \rightarrow FeOOH/Fe(OH)_3 \rightarrow \alpha\text{-}Fe_2O_3 \qquad (6.1)$$

Under microwave irradiation, the $[Fe(CN)_6]^{3-}$ ions dissociate into Fe^{3+} ions and the available Fe^{3+} source becomes quickly depleted once the $\alpha\text{-}Fe_2O_3$ seeds start to grow. Because of the excellent microwave-absorbing characteristics of $[Fe(CN)_6]^{3-}$ ions and iron oxide, "hot spots" in solution and "hot surfaces" on newly-formed hematite crystals may be created under microwave irradiation. Thus, the crystal growth and hierarchical self-assembly would be speeded up. Under nonequilibrium conditions, $\alpha\text{-}Fe_2O_3$ dendrites are rapidly formed by hierarchical self-assembly along the six crystallographically equivalent $\langle 10\bar{1}0 \rangle$ directions, which may also be related to the intrinsic crystal nature of rhombohedral $\alpha\text{-}Fe_2O_3$ [71]. The heating time for fabricating self-assembled hierarchical nanoarchitectures can be shortened by more than two orders of magnitude by using microwave heating instead of conventional heating. The shorter microwave-heating time leads to enhanced efficiency and significant energy savings.

6.4
Element-Based Nanostructures and Nanocomposite

6.4.1
Silver Nanostructures

Nanophase metal powders which have numerous technological applications have been prepared by various techniques, including chemical precipitation from

6.4 Element-Based Nanostructures and Nanocomposite | 119

Figure 6.7 (a) Low-magnification and (b, c) high-magnification SEM images of the hierarchical α-Fe$_2$O$_3$ nanoarchitectures. Reproduced from reference [86].

aqueous or organic solutions [30, 87–91]. There has been a recent innovation to prepare metal powders using liquid polyols such as ethylene glycol [2, 92, 93], which was named the polyol process. The polyol method is a low-temperature process and is environmentally benign because the reactions are carried out under closed system conditions.

Bimetallic nanoparticles have received much attention in recent years because of their novel optical, electronic, magnetic, and catalytic properties, superior to the individual metals [89, 90, 94, 95]. Tsuji et al. reported a microwave-polyol method to prepare Au–Ag core–shell nanocrystals [96]. The formation process of Au–Ag core–shell nanocrystals involves two steps. First, Au nanocrystals are synthesized by reduction of HAuCl$_4$·4H$_2$O in the presence of poly(vinylpyrrolidone) (PVP) in ethylene glycol (EG) under microwave irradiation for 2 min. Second, a solution of AgNO$_3$ in EG is added to the as-formed Au suspension with a AgNO$_3$/HAuCl$_4$ molar ratio ranging from 3 to 30, and treated for another 2 min under microwave irritation. Au–Ag core–shell nanocrystals with different shapes were easily obtained (Figure 6.9). Inverted triangular Ag shells were prepared via a triangular Au core,

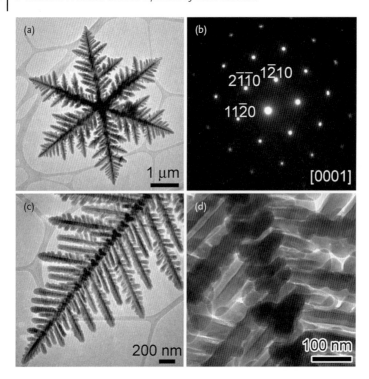

Figure 6.8 (a) TEM image of a sixfold-symmetric dendritic structure. (b) Electron diffraction pattern taken from a single dendritic structure, which shows the single-crystal nature of the entire dendritic structure. (c,d) High-magnification TEM images of the dendritic structure. Reproduced from reference [86].

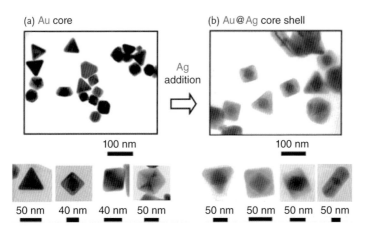

Figure 6.9 (a) TEM photographs of Au core nanocrystals prepared by MW heating of $HAuCl_4 \cdot 4H_2O$ (2.4 mM)/PVP(1 M)/EG for 2 min. (b) TEM photographs of Au@Ag nanocrystals prepared by addition of $AgNO_3$ (23 mM) to the solution obtained in (a) and MW heating for 2 min. Reproduced from reference [96].

square or rectangular Ag shells were prepared via square or rhombic Au cores, respectively, and Ag rods are formed via rhombic Au cores.

Recently, Zhu's group reported the synthesis of highly fluorescent Ag nanoclusters in aqueous solution with the help of a polyelectrolyte, polymethacrylic acid sodium salt (PMAA-Na) by microwave irradiation [97]. The Ag nanoclusters are monodispersed and uniform, and their average size is 2.0 nm (Figure 6.10). Importantly, the Ag nanoclusters can be used as a novel fluorescence probe for the determination of Cr^{3+} ions with high sensitivity and excellent selectivity.

Yu and coworkers demonstrated a one-pot solution-phase route, namely microwave-assisted hydrothermal reduction/carbonization (MAHRC), for the rapid synthesis of coaxial Ag/amorphous carbon (a-C) nanocables [98]. The as-grown Ag/C nanocables could self-assemble in an end-to-end fashion into a novel interlinked chain, even in the absence of any molecular connectors. More

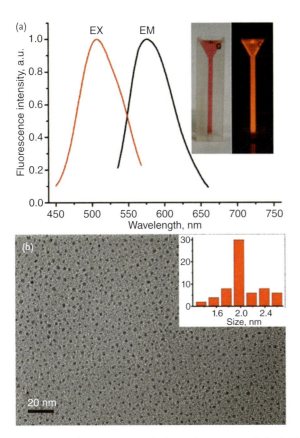

Figure 6.10 The maximum excitation and emission spectra (a) and a typical TEM image (b) of the highly fluorescent Ag nanoclusters. Inset of (a) shows photographs of the solution of Ag nanoclusters under room light (left) and illuminated by a UV lamp with excitation at 365 nm (right); inset of (b) shows the diameter histogram of Ag nanoclusters. Reproduced from reference [97].

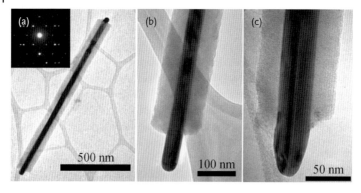

Figure 6.11 (a) TEM image of an individual coaxial Ag/C nanocable. The inset gives an electron microdiffraction pattern recorded by focusing the beam on the nanocable. The diffraction pattern indicates a bicrystalline structure for the Ag core; (b) and (c) Higher magnification TEM images taken from each end of the nanocable shown in (a). Reproduced from reference [98].

interestingly, this route does not need any seeds, surfactants or templates. As shown in Figure 6.11, dark/light contrast is clearly observed along the radial direction. The different contrast suggests a different phase composition, indicating the core–sheath cable structure. The dark contrast suggests Ag of larger mass thickness in the core region. Outside the core region, the light contrast suggests that C is in the sheath layer. More TEM observations reveal Ag/C nanocables with an average diameter of about 100 nm and lengths ranging from 50 nm to 2 μm. Each nanocable is straight and has a uniform diameter along its entire length. The thickness ratios of core to sheath are in the range 1–5. Interestingly, a novel 1D chain topology of interconnected Ag/C nanocables can be obtained (Figure 6.12). The Ag/C nanocables can self-assemble into chains in an end-to-end fashion. Most of the intersections between two straight segments form obtuse angles.

6.4.2
Te Nanostructures

Elemental Te is a p-type semiconductor exhibiting many intriguing and useful properties, such as narrow band gap (~0.35 eV), ultrafast electronic excitation on the A_1 phonon frequency, and high piezoelectric, thermoelectric, or nonlinear optical responses [99–104]. It offers great potential in fabricating electric and photoelectric devices [105, 106]. Moreover, the availability of 1D Te nanomaterials may bring about new applications, or greatly enhance the performance of the currently existing devices based on Te as a result of quantum confinement [107]. So far, many 1D t-Te nanostructures (wires, rods, tubes, and belts) have been synthesized [43, 106, 108–111].

Room temperature ionic liquids (RTILs) for chemical synthesis have attracted increasing interest worldwide because of their high fluidity, low melting

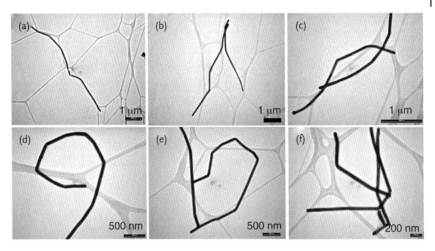

Figure 6.12 Representative TEM images of interconnected chains of nanocables. Such coaxial Ag/C nanocables are self-assembled in an end-to-end fashion without using any molecular connectors. Reproduced from reference [98].

temperature, and extended temperature range in the liquid state, air and water stability, low toxicity, nonflammability, high ionic conductivity, ability to dissolve a variety of materials, and, importantly, no measurable vapor pressure [112]. By combining the advantages of both RTILs and microwave heating, Zhu et al. developed a microwave-assisted ionic liquid (MAIL) method for the fast controlled synthesis of tellurium (Te) nanorods and nanowires [43]. By controlling experimental parameters, exclusively Te nanorods or nanowires are synthesized (Figure 6.13). It is demonstrated that both the ionic liquid and the microwave heating play an important role in the formation of Te nanorods and nanowires.

Yu et al. reported the rapid synthesis of uniform ultralong Te nanowires by microwave-assisted reaction in the presence of PVP using hydrazine hydrate as reducing agent and Na_2TeO_3 as tellurium source [113]. It is found that the formation process of high-quality Te nanowires is strongly dependent on the reaction conditions, such as the amount of PVP, pH of the initial solution, reaction time, and the choice of surfactant. Ultralong Te nanowires with a diameter of 2 nm and length of tens of micrometers can be rapidly synthesized (Figure 6.14). The hydrophilic Te nanowires display a broadened luminescent emission from the shortwave ultraviolet to the visible region excited by vacuum-ultraviolet (VUV) under synchrotron radiation at room temperature.

6.4.3
Selenium/Carbon Colloids

A rapid microwave-assisted route was developed for the synthesis of selenium/carbon colloids using starch and H_2SeO_3 as the reactants [114]. Dandelion-like Se

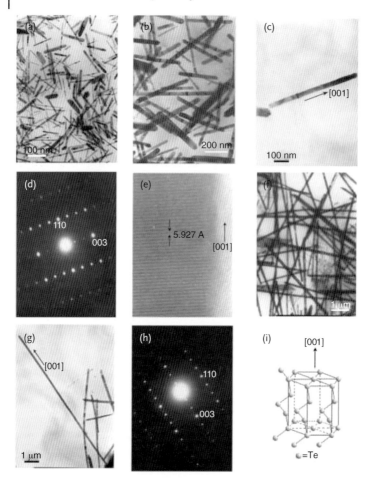

Figure 6.13 TEM micrographs of two typical samples prepared by the MAIL method. (a) and (b) sample 1; (c) a single Te nanorod from sample 1; (d) electron diffraction pattern of the same Te nanorod as shown in (c); (e) high-resolution transmission electron microscopy (HRTEM) image of the same Te nanorod as shown in (c); (f) and (g) sample 2; (h) electron diffraction pattern of a single Te nanowire (the longest one in (g)); (i) a schematic illustration showing the hexagonal crystal structure of Te and the [001] zone axis (c axis of the lattice). Reproduced from reference [43].

architectures are completely encapsulated in the C intestines (Figure 6.15). The entire structure of Se "dandelions" is built from a spherical assembly of centrally oriented Se nanorods with diameters mostly ranging from 5 to 15 nm and with lengths up to 100 nm. The intestine-like or spherical shells of C have an average wall thickness of ~40 nm. The average outer diameter of the C intestines along the radial direction is about 200 nm. Two main reactions occur simultaneously during the one-pot microwave-induced hydrothermal process. One is the reduction of selenious acid by starch and the other is the carbonization of starch. The 1D Se

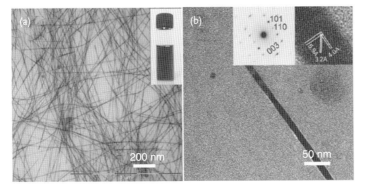

Figure 6.14 TEM images of Te nanowires. (a) TEM image of the Te nanowires. (b) TEM image of a single nanowire. The inset of (a) displays a digital photograph of Te nanowires and inset of (b) shows the corresponding HRTEM and the corresponding SAED pattern obtained from the single nanowire. The electron beam was focused along the [010] axis. Reproduced from reference [113].

nanorods probably grow from a precursor particle, as in the case of the sonochemical formation of Se nanowires [115]. The growth of the core–shell Se–C composites includes two stages. In the first stage, starch-encapsulated spherical amorphous Se (or t-Se or both) particles are produced by reducing selenious acid under microwave irradiation. Meanwhile, the shell layer of starch is carbonized, due to localized "hot surfaces" and "hot spots" created by the microwave energy. In the second stage, 1D Se nanostructures grow from the surface of the Se cores, where the newly-formed t-Se crystallites may serve as seeds. The growth process continues until all the spherical Se particles in the core region are depleted, leaving behind a pure assembly of t-Se nanorods encapsulated by an external layer of amorphous carbon. Furthermore, the resultant Se–C composite can be easily transformed into hollow carbon capsules by a simple thermal treatment. Zhu *et al.* reported a microwave-polyol method to prepare Se nanorods and nanowires [116]. SeO_2 and EG were used as the starting reagents. After microwave-heating the mixture at 195 °C for 30 min, Se nanorods with diameters ranging from 40 to 120 nm and lengths of 1–2.5 μm were obtained. Also, Se nanowires with higher aspect ratios (length to diameter ratio > 100) were yielded. This method is simple, fast and low-cost for large-scale production of 1D Se nanostructures.

6.5 Chalcogenide Nanostructures

6.5.1 Cadmium Chalcogenides

Colloidal semiconductor nanocrystals (NCs), due to their unique optical properties, are currently attracting extensive attention for potential applications in

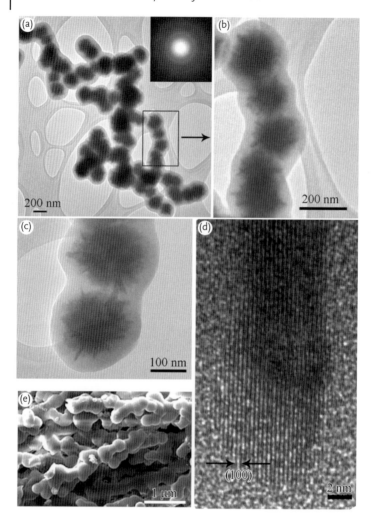

Figure 6.15 (a) Low-magnification bright-field TEM image of Se nanorod assemblies encapsulated in C; (b) and (c) high-magnification TEM images of the core-shell structured Se/C composite; (d) HRTEM image of an individual Se nanorod embedded in the carbon matrix; (e) SEM image of the as-prepared composite. Reproduced from reference [114].

optoelectronics and biolabeling [44, 117, 118]. In recent years, many high quality semiconductor nanocrystals (such as CdS, CdSe, and CdTe) with high quantum yield (QY~50–80%) and sharp emission spectra have been successfully prepared in organic or water phases [119–122].

The selective nature of dielectric heating for semiconductor nanocrystal growth is fascinating. The selectivity offered by microwave heating demonstrates the "specific microwave effect." This depends on the ability to selectively heat molecular precursors that are highly polarizable in the presence of molecules that are less polarizable. Such a specific microwave effect is beneficial in controlling the batch-to-batch variation in nanocrystal production, particularly if selective heating allows the nucleation event to be specifically triggered by the microwave and thereafter allows growth to proceed. Strouse and coworker demonstrate that controlling nanomaterial growth via the "specific microwave effect" can be achieved by selective heating of the chalcogenide precursor [123]. The high polarizability of the precursor allows instantaneous activation and subsequent nucleation, leading to the synthesis of CdSe and CdTe in non-microwave-absorbing alkane solvents. Selective absorption by the chalcogenide results in the isolation of elliptical (aspect ratio 1.2) CdSe and elliptical (aspect ratio 1.7) CdTe in the size range 2.5–8 nm. The materials are prepared in less than 3 min with a typical out-of-reactor dispersity of 6% for CdSe (12% for CdTe). More importantly, the product has a standard deviation in size for the CdSe (CdTe) reaction of 4.2 ± 0.14 nm (4.25 ± 0.3 nm) from batch-to-batch (averaged over 10 individual runs). The strategy of a stopped-flow synthesis is very useful for large-scale industrial production, allowing the production of high-quality materials within minutes instead of hours, or continuously at a specific size and composition.

He *et al.* reported a microwave synthesis of water-dispersed and highly luminescent CdTe–CdS core–shell nanocrystals [124]. The as-prepared CdTe–CdS core–shell nanocrystals without any post-preparative treatment possessed a high photoluminescence quantum yield (up to 75%) and a narrow size distribution (fwhm 35 nm) (Figure 6.16 and 6.17). Microwave irradiation is extremely suitable for accelerating epitaxial growth of the CdS shell. It is demonstrated that merely 5 min irradiation was required to form an optimum thickness of the CdS shell in the microwave synthesis, in comparison to several days expended in the illumination method. Moreover, compared with thioglycolic acid exclusively being available for synthesizing CdTe–CdS core–shell nanocrystals in the illumination method, the author showed that the microwave synthesis can be applied for synthesizing core–shell nanocrystals capped with other thiols, such as 3-mercaptopropionic acid.

Yanagida and coworkers prepared CdS nanocrystallites with controlled size and size distribution by the microwave-assisted reaction of cadmium acetate with thiourea in N,N-dimethylformamide (DMF) [125]. Photoluminescence of CdS nanocrystallites can give us much information on their size and surface structures. The absorption onset of the CdS nanocrystallites shifts to longer wavelength with increasing irradiation time, indicating particle size growth under prolonged irradiation. When the microwave irradiation of the solution is periodically interrupted and then repeated, keeping the solution at ambient temperature before each irradiation, the absorption onset remains at the same wavelength and only the optical density of the absorption band is increased. This suggests that the particle growth occurs only during the continuous irradiation and stops once the system is cooled. Photoluminescence for the resulting CdS nanocrystallites changes with the irradiation time and the repetition of the irradiation, suggesting a structural change of the nanocrystallite surface due to the consumption of excess Cd^{2+}. Mallik and

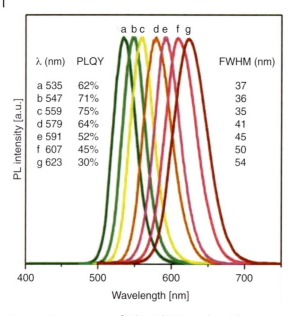

Figure 6.16 PL spectra of CdTe/CdS NCs with serial maximum emission wavelength, and related photoluminescence quantum yield and fwhm values are presented. Reproduced from reference [124].

Figure 6.17 Photograph of the wide spectral range of bright luminescence from a sample of CdTe/CdS NCs aqueous solution without any post- preparative treatment under irradiation with 365-nm ultraviolet light from a UV lamp. Reproduced from reference [124].

Karan prepared CdS nanocrystallites by the microwave-assisted reaction of cadmium acetate with thiourea in DMF in the presence of 1-thioglycerol as a capping agent [126]. The particle size can be tuned by changing the irradiation time. The peak position of the absorption band of the CdS nanocrystallites that are dispersed in chloroform shifts toward longer wavelength with increasing irradiation time, indicating the growth of particle size under prolonged irradiation.

Zedan et al. developed a facile, fast, and scalable microwave irradiation method for the synthesis of graphene and CdSe nanocrystals of controlled size, shape, and crystalline structure dispersed on graphene sheets [127]. Graphite oxide was reduced to graphene in DMSO within 2 min under microwave irradiation, in contrast to 12 h by conventional thermal heating at 180 °C. By using a variety of capping agents, the reduction of graphite oxide and the nucleation and growth of CdSe nanocrystals occured simultaneously. Cubic and hexagonal CdSe nanocrystals with average sizes of 2–4 and 5–7 nm, respectively, can be obtained by using a suitable capping agent within a few minutes under microwave irradiation. High-quality nearly monodisperse CdSe nanocrystals are distributed on the surface of graphene (Figure 6.18). This work should provides a new approach for exploring the size-tunable optical properties of CdSe nanocrystals supported on graphene. They may have important implications for energy conversion applications, such as photovoltaic cells where CdSe quantum dots, the light-harvesting material, are supported on the highly conducting flexible graphene electrodes.

Ren and coworkers prepared high-quality CdTe nanocrystals in the aqueous phase by microwave irradiation using the reaction between Cd^{2+} and NaHTe solution [128]. At a relatively lower temperature (below or equal to 160 °C vs. 300 °C used in organic phase synthesis), various sizes of CdTe NCs with high quantum yield were controllably synthesized by simply varying the temperature and reaction time. The as-prepared CdTe NCs possessed strong luminescence (QY 40%–60%) and pH-independent luminescence. Huang et al. reported the program process of microwave irradiation to prepare high-quality CdTe nanocrystals in aqueous solution [129]. A series (diameters 2–4 nm) of highly luminescent (PLQY 30–68%) CdTe nanocrystals were rapidly prepared (reaction time 1–30 min).

6.5.2
Lead Chalcogenides

Lead chalcogenides (PbS, PbSe, and PbTe) are very promising materials for thermoelectric (TE) applications [130]. In particular, PbS, as an important binary π–π semiconductor with small band gap (0.41 eV) and large exciton Bohr radius (18 nm), has attracted increasing attention because of its wide potential applications in optical devices such as optical switches, near-IR (NIR) communication, thermal and biologic images, and photovoltaic solar cells [131–134]. Qian et al. synthesized nanostructured lead chalcogenides (PbS, PbSe, and PbTe) by microwave irradiation using EG as the solution. PbS nanoarchitectures with a variety of morphologies including cubic particles, six arms, and bugle-like dendrites could be simply fabricated by adjusting the [Pb^{2+}] and sulfur sources (Figure 6.19) [131]. In addition, the dendrites of PbSe and PbTe have also been fabricated by the assistance of appropriate reductive reagent in a similar method.

Gedanken et al. prepared nanoparticles of various dimensions of II–VI binary chalcogenides CdSe, PbSe, and $Cu_{2-x}Se$ by a very simple fast reaction between acetates or sulfates of Cd, Pb, and Cu and Na_2SeSO_3 in the presence of a complexing agent, potassium nitrilotriacetate ($N(CH_2COOK)_3$-NTA) [135]. Chemseddine

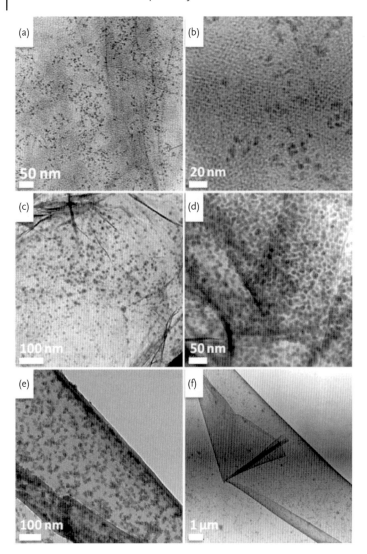

Figure 6.18 TEM images of (a,b) cubic and (c–f) hexagonal CdSe nanocrystals supported on graphene. Reproduced from reference [127].

and coworkers developed a microwave-assisted route for the preparation of highly crystalline PbSe nanocubes [136]. The shape of PbSe NCs evolves from truncated to spherical and finally to cubic. As shown in Figure 6.20a, truncated-shaped PbSe NCs of average particle size ~5.8 ± 2.2 nm are formed at the first stage of growth as a result of heating to 100 °C under microwave irradiation for 1 min with stirring. When the reaction mixture was heated to 160 °C for 1 min, spherical PbSe nanocrystals were obtained with an average diameter of ~7.3 ± 1.2 nm. These spherical particles form monolayer assemblies with hexagonal ordering (shown in Figure

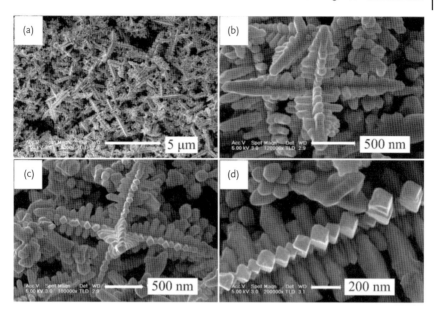

Figure 6.19 FESEM images of PbS at $[Pb^{2+}] = 0.25$ M for 10 min. (a) Low-magnification SEM image; (b) high-magnification SEM image of 3D dendrites from side view; (c) high-magnification SEM image of 3D dendrites from top view; (d) high-magnification SEM image of branches. Reproduced from reference [131].

6.20b). Their corresponding HRTEM image verifies the high crystallinity of PbSe. When the growth time was prolonged to 5 min, locally ordered and monodispersed PbSe NCs with an average size of ~14.4 ± 1.6 nm were obtained (Figure 6.20e) at 160 °C. The HRTEM image indicates a cubic PbSe NC. The distance of 3.06 Å between the adjacent lattice planes corresponds to the surface of the rock-salt structure. This work demonstrates that microwave irradiation can affect the selectivity of the nucleation and the growth rates of different compounds.

6.5.3
Zinc Chalcogenides

Kulkarni reported template-free, microwave-irradiation-assisted growth of ZnS nanorods, approximately 50–100 nm in diameter and more than 1 μm in length, by using zinc chloride and thiourea as the starting reagents dissolved in ethylenediamine [137]. Ren's group prepared highly luminescent ZnSe(S) alloyed quantum dots (QDs) in the aqueous phase [138]. A series of nanocrystals with different sizes and optical properties was produced within 1 h, which was several times faster than conventional aqueous synthesis. The as-formed ZnSe QDs were water-soluble and had high crystallinity, and their photoluminescence (PL) quantum yield was up to 17%. The results show that these properties were remarkably enhanced as compared with ZnSe QDs prepared by conventional aqueous

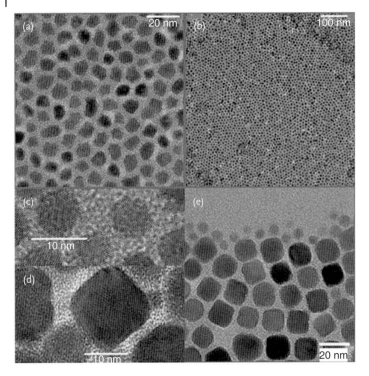

Figure 6.20 (a) TEM image of truncated PbSe NCs. (b) TEM image of hexagonally ordered, monolayer assembled PbSe NCs. (c) HRTEM of a spherical PbSe NC. (d) HRTEM of a cubic PbSe NC. (e) TEM image of cubic PbSe NCs. Reproduced from reference [136].

synthesis. It is worth noting that the reaction did not need rigorous reaction conditions, and some hazardous, expensive, environmentally unfriendly solvents such as trioctylphosphine (TOP), tributylphosphine (TBP), and trioctylphosphine oxide (TOPO). Further characterizations confirm that the as-prepared nanocrystal was an alloy ZnSe(S) shell formed on the surface of the ZnSe core, whereby the sulfur ions came from the decomposition of stabilizer by microwave irradiation.

6.6
Graphene

Graphene has stimulated wide interest because of the properties associated with the 2D crystal structure formed by sp^2 hybridized carbon [139, 140]. Considerable studies have shown great promise for a wide variety of applications, such as electronics, sensors, batteries, supercapacitors, and catalysts [141–144]. However, realization of several of these applications is still not feasible because large-scale

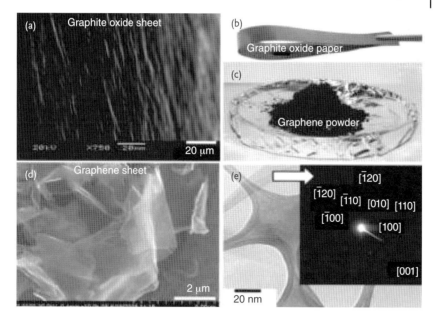

Figure 6.21 (a) SEM image of the cross-section of the GO paper, (b) photographic image of GO paper, (c) bulk quantity of GNS powder produced by the MW-ST process, (d) FE-SEM image of large micrometer-size single paper-like GNS, and (e) bright field TEM image with the inset showing the electron diffraction pattern of optically transparent GNS. Reproduced from reference [145].

production of graphene nanosheets remains a huge challenge. The production of large quantities of high-purity graphene sheets will allow the development of large-scale applications of this unique material. In recent years, various synthetic approaches have been pursued to prepare graphene nanosheets, including top-down and bottom-up approaches. Among the various strategies, the reduction of graphite oxide (GO) is one of the most promising methods. Manthiram et al. have recently developed a microwave-assisted solvothermal route based on reduction of exfoliated GO [145]. GO that was prepared from graphite by the Hummer's method was dispersed in high boiling tetraethylene glycol, and a yellow-brown colloidal suspension was formed. After the heating treatment by microwave irradiation within a short reaction time of 5–15 min at relatively low temperatures (180–300 °C), the colloidal solution turned black, indicating the reduction of GO to graphene. As shown in Figure 6.21, the resulting graphene exhibits the large (~15–5 μm) paper-like nanosheet morphology with slightly folded edges. The rapid absorption of microwave radiation by GO in polar solvents with a subsequent increase in temperature and pressure should greatly facilitate the reduction of GO to graphene. In addition to polyol glycol, graphene nanosheets can be obtained in other polar solvents such as DMF, ethanol, 1-butanol, and water. Recently, Yan and coworkers prepared micrometer-sized graphene nanosheets in a domestic

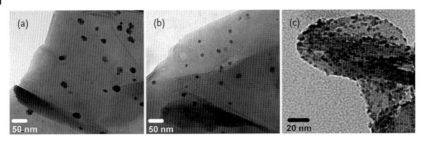

Figure 6.22 TEM images of the chemically converted graphene sheets containing (a) Pd, (b) Cu and (c) CuPd nanoparticles prepared by the simultaneous reduction of GO and the appropriate metal salt in water using hydrazine hydrate under MW irradiation. Reproduced from reference [147].

microwave oven, whereby GO was reduced in a mixed solution of N, N-dimethylacetamide (DMAc) and water [146]. It is expected that the rapid microwave-assisted synthesis of high-quality graphene nanosheets would provide the high-value opportunities for various successful applications, including microelectronics, photovoltaic, and energy storage and conversion.

Owing to the small thickness of a single carbon atom, graphene sheets possess unique physicochemical properties, including large surface area, good flexibility, high electrical conductivity, high chemical stability, and tunable surface functional groups. In particular, their large surface area and tunable surface properties allow them to be a competitive host substrate for the heterogeneous growth of desired active guest materials. To date, the designed synthesis of graphene-based nanocomposites has received much attention worldwhile. El-Shall's group developed a microwave-assisted route to prepare chemically converted graphene sheets and metal nanoparticles dispersed on the graphene sheets [147]. The method allows rapid chemical reduction of exfoliated GO using a variety of reducing agents in either aqueous or organic media. Also, the simultaneous reduction of GO and a variety of metal salts can be achieved, leading to the dispersion of metallic (Pd, Cu) and bimetallic (PdCu) nanoparticles supported on graphene sheets (Figure 6.22). Without using any surfactants, ultrafine Pt nanoparticles (2–3 nm) on graphene were prepared by the co-reduction of graphene oxide and Pt salt using ethylene glycol under microwave irradiation [148]. Figure 6.23 shows the Pt/graphene hybrid. Importantly, the Pt/graphene hybrids exhibit excellent activity for methanol oxidation coupled with good long-term stability and also show enhanced catalytic activity for the hydrogen combustion reaction. Besides the above-mentioned hybrids, many other graphene- based nanocomposites (e.g., Fe_3O_4/graphene, Co_3O_4/graphene, CdSe/graphene, ZnO/graphene, CdS/graphene. TiO_2/graphene, SnO_2/graphene, CuO/graphene, and Mn_3O_4/graphene) have been synthesized by microwave irradiation and show enhanced catalytic or lithium-storage properties [128, 149–157].

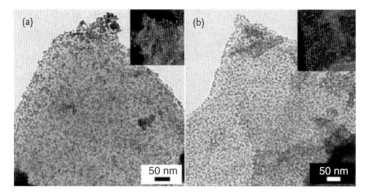

Figure 6.23 TEM images of the Pt/graphene nanohybrids. Reproduced from reference [148].

6.7
Summary

Microwave-assisted synthesis has become a very appealing tool in the preparation of nanostructured materials. This chapter documents some representative attempts in this direction. Microwave chemistry research is creating numerous opportunities for devising new strategies to fabricate nanostructured materials. It may be expected that continuing breakthroughs in microwave chemistry will bring novel nanomaterials with fascinating properties and exciting applications. In particular, modern scientific microwave systems possess the capabilities to program the duration and temperature of consecutive steps, conduct several parallel reactions at once, and enable chemical reactions to occur in a closed high-temperature system, similar to a hydrothermal condition. This allows one not only to reduce the preparation time significantly from days to minutes, but also to easily screen a wide range of experimental conditions in order to optimize and scale-up material preparation. However, fundamental principles in microwave chemistry have not yet been well established. Many microwave-assisted syntheses are mainly based on trial-and-error experiments. So far, there are still many controversial and challenging topics in the field of microwave-induced synthesis. For instance, *in situ* exploration of the microwave-induced reaction processes may be a good choice to elucidate the nucleation and growth of nanocrystals in solutions. We anticipate significant developments in this research area to set the basis for creating new routes for nanomaterials synthesis.

References

1 Park, J., An, K.J., Hwang, Y.S., Park, J.G., Noh, H.J., Kim, J.Y., Park, J.H., Hwang, N.M., and Hyeon, T. (2004) Ultra-large-scale syntheses of monodisperse nanocrystals. *Nat. Mater.*, **3**, 891–895.

2 Sun, Y.G., and Xia, Y.N. (2002) Shape-controlled synthesis of gold and silver nanoparticles. *Science*, **298**, 2176–2179.

3 Tian, N., Zhou, Z.Y., Sun, S.G., Ding, Y., and Wang, Z.L. (2007) Synthesis of tetrahexahedral platinum nanocrystals with high-index facets and high electro-oxidation activity. *Science*, **316**, 732–735.

4 Wang, X., Zhuang, J., Peng, Q., and Li, Y.D. (2005) A general strategy for nanocrystal synthesis. *Nature*, **437**, 121–124.

5 Yin, Y., and Alivisatos, A.P. (2005) Colloidal nanocrystal synthesis and the organic-inorganic interface. *Nature*, **437**, 664–670.

6 Mai, H.X., Zhang, Y.W., Si, R., Yan, Z.G., Sun, L.D., You, L.P., and Yan, C.H. (2006) High-quality sodium rare-earth fluoride nanocrystals: controlled synthesis and optical properties. *J. Am. Chem. Soc.*, **128**, 6426–6436.

7 Sanchez-Iglesias, A., Pastoriza-Santos, I., Perez-Juste, J., Rodriguez-Gonzalez, B., de Abajo, F.J.G., and Liz-Marzan, L.M. (2006) Synthesis and optical properties of gold nanodecahedra with size control. *Adv. Mater.*, **18**, 2529–2534.

8 Xia, Y.N., and Halas, N.J. (2005) Shape-controlled synthesis and surface plasmonic properties of metallic nanostructures. *MRS Bull.*, **30**, 338–344.

9 Xiong, Y.J., McLellan, J.M., Chen, J.Y., Yin, Y.D., Li, Z.Y., and Xia, Y.N. (2005) Kinetically controlled synthesis of triangular and hexagonal nanoplates of palladium and their SPR/SERS properties. *J. Am. Chem. Soc.*, **127**, 17118–17127.

10 Yan, R.X., Sun, X.M., Wang, X., Peng, Q., and Li, Y.D. (2005) Crystal structures, anisotropic growth, and optical properties: controlled synthesis of lanthanide orthophosphate one-dimensional nanomaterials. *Chem.-Eur. J.*, **11**, 2183–2195.

11 Alivisatos, A.P. (1996) Semiconductor clusters, nanocrystals, and quantum dots. *Science*, **271**, 933–937.

12 Lu, W., and Lieber, C.M. (2007) Nanoelectronics from the bottom up. *Nat. Mater.*, **6**, 841–850.

13 Ariga, K., Hill, J.P., and Ji, Q. (2007) Layer-by-layer assembly as a versatile bottom-up nanofabrication technique for exploratory research and realistic application. *Phys. Chem. Chem. Phys.*, **9**, 2319–2340.

14 Li, M., Bhiladvala, R.B., Morrow, T.J., Sioss, J.A., Lew, K.K., Redwing, J.M., Keating, C.D., and Mayer, T.S. (2008) Bottom-up assembly of large-area nanowire resonator arrays. *Nat. Nanotechnol.*, **3**, 88–92.

15 Lieber, C.M. (2003) Nanoscale science and technology: building a big future from small things. *MRS Bull.*, **28**, 486–491.

16 Magasinski, A., Dixon, P., Hertzberg, B., Kvit, A., Ayala, J., and Yushin, G. (2010) High-performance lithium-ion anodes using a hierarchical bottom-up approach. *Nat. Mater.*, **9**, 353–358.

17 Srinivasan, S., Praveen, V.K., Philip, R., and Ajayaghosh, A. (2008) Bioinspired superhydrophobic coatings of carbon nanotubes and linear π systems based on the bottom-up self-assembly approach. *Angew. Chem.*, **120**, 5834–5838.

18 Altman, M., Shukla, A.D., Zubkov, T., Evmenenko, G., Dutta, P., and van der Boom, M.E. (2006) Controlling structure from the bottom-up: structural and optical properties of layer-by-layer assembled palladium coordination-based multilayers. *J. Am. Chem. Soc.*, **128**, 7374–7382.

19 Bai, F., Wang, D., Huo, Z., Chen, W., Liu, L., Liang, X., Chen, C., Wang, X., Peng, Q., and Li, Y. (2007) A versatile bottom-up assembly approach to colloidal spheres from nanocrystals. *Angew. Chem. Int. Ed.*, **46**, 6650–6653.

20 Tello, M., Garcia, R., Martín-Gago, J.A., Martínez, N.F., Martín-González, M.S., Aballe, L., Baranov, A., and Gregoratti, L. (2005) Bottom–up fabrication of carbon-rich silicon carbide nanowires by manipulation of nanometer-sized ethanol menisci. *Adv. Mater.*, **17**, 1480–1483.

21 Bang, J.H., and Suslick, K.S. (2010) Applications of ultrasound to the synthesis of nanostructured materials. *Adv. Mater.*, **22**, 1039–1059.
22 Liu, B., and Zeng, H.C. (2003) Hydrothermal synthesis of ZnO nanorods in the diameter regime of 50 nm. *J. Am. Chem. Soc.*, **125**, 4430–4431.
23 Miao, Z., Xu, D., Ouyang, J., Guo, G., Zhao, X., and Tang, Y. (2002) Electrochemically induced sol-gel preparation of single-crystalline TiO_2 nanowires. *Nano Lett.*, **2**, 717–720.
24 Ohde, H., Ohde, M., Bailey, F., Kim, H., and Wai, C.M. (2002) Water-in-CO_2 microemulsions as nanoreactors for synthesizing CdS and ZnS nanoparticles in supercritical CO2. *Nano Lett.*, **2**, 721–724.
25 Tsuji, M., Hashimoto, M., Nishizawa, Y., Kubokawa, M., and Tsuji, T. (2005) Microwave-assisted synthesis of metallic nanostructures in solution. *Chem.-Eur. J.*, **11**, 440–452.
26 Wang, F., Zhao, M., and Song, X. (2008) Nano-sized $SnSbCu_x$ alloy anodes prepared by co-precipitation for Li-ion batteries. *J. Power Sources*, **175**, 558–563.
27 Wang, X., and Li, Y.D. (2002) Selected-control hydrothermal synthesis of alpha- and beta-MnO_2 single crystal nanowires. *J. Am. Chem. Soc.*, **124**, 2880–2881.
28 Gou, L.F., and Murphy, C.J. (2003) Solution-phase synthesis of Cu_2O nanocubes. *Nano Lett.*, **3**, 231–234.
29 Sau, T.K., and Murphy, C.J. (2004) Room temperature, high-yield synthesis of multiple shapes of gold nanoparticles in aqueous solution. *J. Am. Chem. Soc.*, **126**, 8648–8649.
30 Sau, T.K., and Murphy, C.J. (2004) Seeded high yield synthesis of short Au nanorods in aqueous solution. *Langmuir*, **20**, 6414–6420.
31 Trentler, T.J., Denler, T.E., Bertone, J.F., Agrawal, A., and Colvin, V.L. (1999) Synthesis of TiO_2 nanocrystals by nonhydrolytic solution-based reactions. *J. Am. Chem. Soc.*, **121**, 1613–1614.
32 Cushing, B.L., Kolesnichenko, V.L., and O'Connor, C.J. (2004) Recent advances in the liquid-phase syntheses of inorganic nanoparticles. *Chem. Rev.*, **104**, 3893–3946.
33 Kuno, M. (2008) An overview of solution-based semiconductor nanowires: synthesis and optical studies. *Phys. Chem. Chem. Phys.*, **10**, 620–639.
34 Narayanan, R., and El-Sayed, M.A. (2005) Catalysis with transition metal nanoparticles in colloidal solution: nanoparticle shape dependence and stability. *J. Phys. Chem. B*, **109**, 12663–12676.
35 Gerbec, J.A., Magana, D., Washington, A., and Strouse, G.F. (2005) Microwave-enhanced reaction rates for nanoparticle synthesis. *J. Am. Chem. Soc.*, **127**, 15791–15800.
36 Hu, X.L., Gong, J.M., Zhang, L.Z., and Yu, J.C. (2008) Continuous size tuning of monodisperse ZnO colloidal nanocrystal clusters by a microwave-polyol process and their application for humidity sensing. *Adv. Mater.*, **20**, 4845–4850.
37 Hu, X.L., and Yu, J.C. (2008) Continuous aspect-ratio tuning and fine shape control of monodisperse alpha-Fe_2O_3 nanocrystals by a programmed microwave-hydrothermal method. *Adv. Funct. Mater.*, **18**, 880–887.
38 Gedye, R., Smith, F., Westaway, K., Ali, H., Baldisera, L., Laberge, L., and Rousell, J. (1986) The use of microwave ovens for rapid organic synthesis. *Tetrahedron Lett.*, **27**, 279–282.
39 Galema, S.A. (1997) Microwave chemistry. *Chem. Soc. Rev.*, **26**, 233–238.
40 Hu, X.L., Zhu, Y.J., and Wing, S.W. (2004) Sonochemical and microwave-assisted synthesis of linked single-crystalline ZnO rods. *Mater. Chem. Phys.*, **88**, 421–426.
41 Komarneni, S. (2003) Nanophase materials by hydrothermal, microwave-hydrothermal and microwave-solvothermal methods. *Curr. Sci. India*, **85**, 1730–1734.
42 Zhu, Y.J., and Hu, X.L. (2003) Microwave-polyol preparation of single-crystalline gold nanorods and nanowires. *Chem. Lett.*, **32**, 1140–1141.
43 Zhu, Y.J., Wang, W.W., Qi, R.J., and Hu, X.L. (2004) Microwave-assisted synthesis of single-crystalline tellurium

nanorods and nanowires in ionic liquids. *Angew. Chem. Int. Ed.*, **43**, 1410–1414.

44 Bruchez, M., Jr., Moronne, M., Gin, P., Weiss, S., and Alivisatos, A.P. (1998) Semiconductor nanocrystals as fluorescent biological labels. *Science*, **281**, 2013–2016.

45 Buehler, G., and Feldmann, C. (2006) Microwave-assisted synthesis of luminescent LaPO4: Ce,Tb nanocrystals in ionic liquids. *Angew. Chem. Int. Ed.*, **45**, 4864–4867.

46 Lu, Q.Y., Gao, F., and Komarneni, S. (2004) Biomolecule-assisted synthesis of highly ordered snowflakelike structures of bismuth sulfide nanorods. *J. Am. Chem. Soc.*, **126**, 54–55.

47 Lu, Q.Y., Gao, F., and Komarneni, S. (2004) Microwave-assisted synthesis of one-dimensional nanostructures. *J. Mater. Res.*, **19**, 1649–1655.

48 Makhluf, S., Dror, R., Nitzan, Y., Abramovich, Y., Jelinek, R., and Gedanken, A. (2005) Microwave-assisted synthesis of nanocrystalline MgO and its use as a bacteriocide. *Adv. Funct. Mater.*, **15**, 1708–1715.

49 Panda, A.B., Glaspell, G., and El-Shall, M.S. (2007) Microwave synthesis and optical properties of uniform nanorods and nanoplates of rare earth oxides. *J. Phys. Chem. C*, **111**, 1861–1864.

50 Panda, A.B., Glaspell, G., and El-Shall, M.S. (2006) Microwave synthesis of highly aligned ultra narrow semiconductor rods and wires. *J. Am. Chem. Soc.*, **128**, 2790–2791.

51 Gabriel, C., Gabriel, S., Grant, E.H., Halstead, B.S.J., and Mingos, D.M.P. (1998) Dielectric parameters relevant to microwave dielectric heating. *Chem. Soc. Rev.*, **27**, 213–223.

52 Kingston, H.M., and Haswell, S.J. (1997) *Microwave-Enhanced Chemistry: Fundamentals, Sample Preparation, and Applications*, American Chemical Society, Washington, DC.

53 Celer, E.B., and Jaroniec, M. (2006) Temperature-programmed microwave-assisted synthesis of SBA-15 ordered mesoporous silica. *J. Am. Chem. Soc.*, **128**, 14408–14414.

54 Hu, X.L., and Yu, J.C. (2006) Microwave-assisted synthesis of a superparamagnetic surface-functionalized porous Fe_3O_4/C nanocomposite. *Chem.-Asian J.*, **1**, 605–610.

55 Hu, X.L., Yu, J.C., Gong, J.M., Li, Q., and Li, G.S. (2007) alpha-Fe_2O_3 nanorings prepared by a microwave-assisted hydrothermal process and their sensing properties. *Adv. Mater.*, **19**, 2324–2329.

56 Leadbeater, N.E. (2005) Fast, easy, clean chemistry by using water as a solvent and microwave heating: the Suzuki coupling as an illustration. *Chem. Commun.*, 2881–2902.

57 Yu, J.C., Hu, X.L., Li, Q., Zheng, Z., and Xu, Y.M. (2005) Synthesis and characterization of core-shell selenium/ carbon colloids and hollow carbon capsules. *Chem.-Eur. J.*, **12**, 548–552.

58 Yu, J.C., Hu, X.L., Quan, L.B., and Zhang, L.Z. (2005) Microwave-assisted synthesis and in-situ self-assembly of coaxial Ag/C nanocables. *Chem. Commun.*, 2704–2706.

59 Pacholski, C., Kornowski, A., and Weller, H. (2002) Self-assembly of ZnO: from nanodots, to nanorods. *Angew. Chem. Int. Ed.*, **41**, 1188–1191.

60 Vayssieres, L. (2003) Growth of arrayed nanorods and nanowires of ZnO from aqueous solutions. *Adv. Mater.*, **15**, 464–466.

61 Wu, J.J., and Liu, S.C. (2002) Low-temperature growth of well-aligned ZnO nanorods by chemical vapor deposition. *Adv. Mater.*, **14**, 215–218.

62 Yang, P.D., Yan, H.Q., Mao, S., Russo, R., Johnson, J., Saykally, R., Morris, N., Pham, J., He, R.R., and Choi, H.J. (2002) Controlled growth of ZnO nanowires and their optical properties. *Adv. Funct. Mater.*, **12**, 323–331.

63 Seelig, E.W., Tang, B., Yamilov, A., Cao, H., and Chang, R.P.H. (2003) Self-assembled 3D photonic crystals from ZnO colloidal spheres. *Mater. Chem. Phys.*, **80**, 257–263.

64 Joo, J., Kwon, S.G., Yu, J.H., and Hyeon, T. (2005) Synthesis of ZnO nanocrystals with cone, hexagonal cone, and rod shapes via non-hydrolytic ester

elimination sol-gel reactions. *Adv. Mater.*, **17**, 1873–1877.
65 Zhong, X.H., Feng, Y.Y., Zhang, Y.L., Lieberwirth, I., and Knoll, W.G. (2007) Nonhydrolytic alcoholysis route to morphology-controlled ZnO nanocrystals. *Small*, **3**, 1194–1199.
66 Liu, B., and Zeng, H.C. (2007) Hollow ZnO microspheres with complex nanobuilding units. *Chem. Mater.*, **19**, 5824–5826.
67 Zhang, Q.F., Chou, T.R., Russo, B., Jenekhe, S.A., and Cao, G.Z. (2008) Aggregation of ZnO nanocrystallites for high conversion efficiency in dye-sensitized solar cells. *Angew. Chem. Int. Ed.*, **47**, 2402–2406.
68 Libert, S., Gorshkov, V., Goia, D., Matijevkic, E., and Privman, V. (2003) Model of controlled synthesis of uniform colloid particles: cadmium sulfide. *Langmuir*, **19**, 10679–10683.
69 Qiu, Y.F., and Yang, S.H. (2007) ZnO nanotetrapods: controlled vapor-phase synthesis and application for humidity sensing. *Adv. Funct. Mater.*, **17**, 1345–1352.
70 Cho, S., Jung, S.H., and Lee, K.H. (2008) Morphology-controlled growth of ZnO nanostructures using microwave irradiation: from basic to complex structures. *J. Phys. Chem. C*, **112**, 12769–12776.
71 Bilecka, I., Elser, P., and Niederberger, M. (2009) Kinetic and thermodynamic aspects in the microwave-assisted synthesis of ZnO nanoparticles in benzyl alcohol. *ACS Nano*, **3**, 467–477.
72 Cao, M.H., Liu, T.F., Gao, S., Sun, G.B., Wu, X.L., Hu, C.W., and Wang, Z.L. (2005) Single-crystal dendritic micro-pines of magnetic alpha-Fe_2O_3: large-scale synthesis, formation mechanism, and properties. *Angew. Chem. Int. Ed.*, **44**, 4197–4201.
73 Chen, J., Xu, L.N., Li, W.Y., and Gou, X.L. (2005) alpha-Fe_2O_3 nanotubes in gas sensor and lithium-ion battery applications. *Adv. Mater.*, **17**, 582–586.
74 Jiao, F., Harrison, A., Jumas, J.C., Chadwick, A.V., Kockelmann, W., and Bruce, P.G. (2006) Ordered mesoporous Fe_2O_3 with crystalline walls. *J. Am. Chem. Soc.*, **128**, 5468–5474.
75 Sun, Z.Y., Yuan, H.Q., Liu, Z.M., Han, B.X., and Zhang, X.R. (2005) A highly efficient chemical sensor material for H2S: alpha-Fe_2O_3 nanotubes fabricated using carbon nanotube templates. *Adv. Mater.*, **17**, 2993–2997.
76 Wen, X.G., Wang, S.H., Ding, Y., Wang, Z.L., and Yang, S.H. (2005) Controlled growth of large-area, uniform, vertically aligned arrays of alpha-Fe_2O_3 nanobelts and nanowires. *J. Phys. Chem. B*, **109**, 215–220.
77 Woo, K., Lee, H.J., Ahn, J.P., and Park, Y.S. (2003) Sol-gel mediated synthesis of Fe_2O_3 nanorods. *Adv. Mater.*, **15**, 1761–1764.
78 Wu, C.Z., Yin, P., Zhu, X., OuYang, C.Z., and Xie, Y. (2006) Synthesis of hematite (alpha-Fe_2O_3) nanorods: diameter-size and shape effects on their applications in magnetism, lithium ion battery, and gas sensors. *J. Phys. Chem. B*, **110**, 17806–17812.
79 Zheng, Y.H., Cheng, Y., Wang, Y.S., Bao, F., Zhou, L.H., Wei, X.F., Zhang, Y.Y., and Zheng, Q. (2006) Quasicubic alpha-Fe_2O_3 nanoparticles with excellent catalytic performance. *J. Phys. Chem. B*, **110**, 3093–3097.
80 Zhong, Z., Ho, J., Teo, J., Shen, S., and Gedanken, A. (2007) Synthesis of porous alpha-Fe_2O_3 nanorods and deposition of very small gold particles in the pores for catalytic oxidation of CO. *Chem. Mater.*, **19**, 4776–4782.
81 Wang, P.C., Ding, H.P., Bark, T., and Chen, C.H. (2007) Nanosized alpha-Fe_2O_3 and Li-Fe composite oxide electrodes for lithium-ion batteries. *Electrochim. Acta*, **52**, 6650–6655.
82 Wu, X.L., Guo, Y.G., Wan, L.J., and Hu, C.W. (2008) alpha-Fe_2O_3 nanostructures: inorganic salt-controlled synthesis and their electrochemical performance toward lithium storage. *J. Phys. Chem. C*, **112**, 16824–16829.
83 Cesar, I., Kay, A., Martinez, J.A.G., and Gratzel, M. (2006) Translucent thin film Fe_2O_3 photoanodes for efficient water splitting by sunlight: nanostructure-directing effect of Si-doping. *J. Am. Chem. Soc.*, **128**, 4582–4583.
84 Cao, S.W., and Zhu, Y.J. (2008) Hierarchically nanostructured

(alpha-Fe$_2$O$_3$ hollow spheres: preparation, growth mechanism, photocatalytic property, and application in water treatment. *J. Phys. Chem. C*, **112**, 6253–6257.

85 Wang, Y.S., Muramatsu, A., and Sugimoto, T. (1998) FTIR analysis of well-defined alpha-Fe$_2$O$_3$ particles. *Colloid Surf. A*, **134**, 281–297.

86 Hayashi, S., and Kanamori, H. (1980) Infrared study of surface phonon modes in α-Fe$_2$O$_3$ microcrystals. *J. Phys. C*, **13**, 1529–1538.

87 Hu, X.L., Yu, J.C., and Gong, J.M. (2007) Fast production of self-assembled hierarchical alpha-Fe$_2$O$_3$ nanoarchitectures. *J. Phys. Chem. C*, **111**, 11180–11185.

88 Brown, K.R., Walter, D.G., and Natan, M.J. (2000) Seeding of colloidal Au nanoparticle solutions. 2. Improved control of particle size and shape. *Chem. Mater.*, **12**, 306–313.

89 Cao, Y.W., Jin, R., and Mirkin, C.A. (2001) DNA-modified core-shell Ag/Au nanoparticles. *J. Am. Chem. Soc.*, **123**, 7961–7962.

90 Lim, B., Jiang, M.J., Camargo, P.H.C., Cho, E.C., Tao, J., Lu, X.M., Zhu, Y.M., and Xia, Y.N. (2009) Pd-Pt bimetallic nanodendrites with high activity for oxygen reduction. *Science*, **324**, 1302–1305.

91 Stamenkovic, V.R., Mun, B.S., Arenz, M., Mayrhofer, K.J.J., Lucas, C.A., Wang, G.F., Ross, P.N., and Markovic, N.M. (2007) Trends in electrocatalysis on extended and nanoscale Pt-bimetallic alloy surfaces. *Nat. Mater.*, **6**, 241–247.

92 Zhao, M.Q., and Crooks, R.M. (1999) Homogeneous hydrogenation catalysis with monodisperse, dendrimer-encapsulated Pd and Pt nanoparticles. *Angew. Chem. Int. Ed.*, **38**, 364–366.

93 Li, C.C., Cai, W.P., Cao, B.Q., Sun, F.Q., Li, Y., Kan, C.X., and Zhang, L.D. (2006) Mass synthesis of large, single-crystal Au nanosheets based on a polyol process. *Adv. Funct. Mater.*, **16**, 83–90.

94 Tsuji, M., Yamaguchi, D., Matsunaga, M., and Alam, M.J. (2010) Epitaxial Growth of Au@Cu Core-Shell Nanocrystals Prepared Using the PVP-Assisted Polyol Reduction Method. *Cryst. Growth Des.*, **10**, 5129–5135.

95 Son, S.U., Jang, Y., Park, J., Na, H.B., Park, H.M., Yun, H.J., Lee, J., and Hyeon, T. (2004) Designed synthesis of atom-economical Pd/Ni bimetallic nanoparticle-based catalysts for sonogashira coupling reactions. *J. Am. Chem. Soc.*, **126**, 5026–5027.

96 Toshima, N., and Yonezawa, T. (1998) Bimetallic nanoparticles – novel materials for chemical and physical applications. *New J. Chem.*, **22**, 1179–1201.

97 Tsuji, M., Miyamae, N., Lim, S., Kimura, K., Zhang, X., Hikino, S., and Nishio, M. (2006) Crystal structures and growth mechanisms of Au@Ag core-shell nanoparticles prepared by the microwave-polyol method. *Cryst. Growth Des.*, **6**, 1801–1807.

98 Liu, S., Lu, F., and Zhu, J.J. (2011) Highly fluorescent Ag nanoclusters: microwave-assisted green synthesis and Cr^{3+} sensing. *Chem. Commun.*, **47**, 2661–2663.

99 Yu, J.C., Hu, X.L., Li, Q., and Zhang, L.Z. (2005) Microwave-assisted synthesis and in-situ self-assembly of coaxial Ag/C nanocables. *Chem. Commun.*, 2704–2706.

100 Anzin, V.B., Eremets, M.I., Kosichkin, Y.V., Nadezhdinskii, A.I., and Shirokov, A.M. (1977) Measurement of energy-gap in tellurium under pressure. *Phys. Status Solidi A*, **42**, 385–390.

101 Araki, K., and Tanaka, T. (1972) Piezoelectric and elastic properties of single crystalline Se-Te alloys. *Jpn. J. Appl. Phys.*, **11**, 472–479.

102 Berger, L.I. (1997) *Semiconductor Materials*, CRC Press, Boca Raton, FL.

103 Cooper, W.C. (1971) *Tellurium*, Van Nostrand Reinhold Co., New York.

104 Lu, Q.Y., Gao, F., and Komarneni, S. (2004) Biomolecule-assisted reduction in the synthesis of single-crystalline tellurium nanowires. *Adv. Mater.*, **16**, 1629–1632.

105 Tangney, P., and Fahy, S. (2002) Density-functional theory approach to ultrafast laser excitation of

semiconductors: application to the A(1) phonon in tellurium. *Phys. Rev. B*, **65**, 054302.

106 Lu, Q.Y., Gao, F., and Komarneni, S. (2005) A green chemical approach to the synthesis of tellurium nanowires. *Langmuir*, **21**, 6002–6005.

107 Mo, M.S., Zeng, J.H., Liu, X.M., Yu, W.C., Zhang, S.Y., and Qian, Y.T. (2002) Controlled hydrothermal synthesis of thin single-crystal tellurium nanobelts and nanotubes. *Adv. Mater.*, **14**, 1658–1662.

108 Liu, Z.P., Hu, Z.K., Xie, Q., Yang, B.J., Wu, J., and Qian, Y.T. (2003) Surfactant-assisted growth of uniform nanorods of crystalline tellurium. *J. Mater. Chem.*, **13**, 159–162.

109 Gautam, U.K., and Rao, C.N.R. (2004) Controlled synthesis of crystalline tellurium nanorods, nanowires, nanobelts and related structures by a self-seeding solution process. *J. Mater. Chem.*, **14**, 2530–2535.

110 Mayers, B., and Xia, Y.N. (2002) Formation of tellurium nanotubes through concentration depletion at the surfaces of seeds. *Adv. Mater.*, **14**, 279–282.

111 Mayers, B., and Xia, Y.N. (2002) One-dimensional nanostructures of trigonal tellurium with various morphologies can be synthesized using a solution-phase approach. *J. Mater. Chem.*, **12**, 1875–1881.

112 Qian, H.S., Yu, S.H., Luo, L.B., Gong, J.Y., Fei, L.F., and Liu, X.M. (2006) Synthesis of uniform Te@Carbon-Rich composite nanocables with photoluminescence properties and Carbonaceous nanofibers by the hydrothermal carbonization of glucose. *Chem. Mater.*, **18**, 2102–2108.

113 Tokuda, H., Hayamizu, K., Ishii, K., Susan, M.A.B., and Watanabe, M. (2004) Physicochemical properties and structures of room temperature ionic liquids. 1. Variation of anionic species. *J. Phys. Chem. B*, **108**, 16593–16600.

114 Liu, J.W., Chen, F., Zhang, M., Qi, H., Zhang, C.L., and Yu, S.H. (2010) Rapid microwave-assisted synthesis of uniform ultralong Te nanowires, optical property, and chemical stability. *Langmuir*, **26**, 11372–11377.

115 Yu, J.C., Hu, X.L., Li, Q., Zheng, Z., and Xu, Y.M. (2006) Synthesis and characterization of core-shell selenium/carbon colloids and hollow carbon capsules. *Chem.-Eur. J.*, **12**, 548–552.

116 Gates, B., Mayers, B., Grossman, A., and Xia, Y.N. (2002) A sonochemical approach to the synthesis of crystalline selenium nanowires in solutions and on solid supports. *Adv. Mater.*, **14**, 1749–1752.

117 Zhu, Y.-J., and Hu, X.-L. (2004) Preparation of powders of selenium nanorods and nanowires by microwave-polyol method. *Mater. Lett.*, **58**, 1234–1236.

118 Chan, W.C.W., and Nie, S. (1998) Quantum dot bioconjugates for ultrasensitive nonisotopic detection. *Science*, **281**, 2016–2018.

119 Schlamp, M., Peng, X., and Alivisatos, A. (1997) Improved efficiencies in light emitting diodes made with CdSe (CdS) core/shell type nanocrystals and a semiconducting polymer. *J. Appl. Phys.*, **82**, 5837–5842.

120 Peng, X.G., Schlamp, M.C., Kadavanich, A.V., and Alivisatos, A.P. (1997) Epitaxial growth of highly luminescent CdSe/CdS core/shell nanocrystals with photostability and electronic accessibility. *J. Am. Chem. Soc.*, **119**, 7019–7029.

121 Peng, Z.A., and Peng, X.G. (2001) Formation of high-quality CdTe, CdSe, and CdS nanocrystals using CdO as precursor. *J. Am. Chem. Soc.*, **123**, 183–184.

122 Yu, W.W., and Peng, X.G. (2002) Formation of high-quality CdS and other II-VI semiconductor nanocrystals in noncoordinating solvents: tunable reactivity of monomers. *Angew. Chem. Int. Ed.*, **41**, 2368–2371.

123 Yu, W.W., Qu, L.H., Guo, W.Z., and Peng, X.G. (2003) Experimental determination of the extinction coefficient of CdTe, CdSe, and CdS nanocrystals. *Chem. Mater.*, **15**, 2854–2860.

124 Washington Ii, A.L., and Strouse, G.F. (2008) Microwave synthesis of CdSe and

CdTe nanocrystals in nonabsorbing alkanes. *J. Am. Chem. Soc.*, **130**, 8916–8922.

125 He, Y., Lu, H.T., Sai, L.M., Lai, W.Y., Fan, Q.L., Wang, L.H., and Huang, W. (2006) Microwave-assisted growth and characterization of water-dispersed CdTe/CdS core-shell nanocrystals with high photoluminescence. *J. Phys. Chem. B*, **110**, 13370–13374.

126 Wada, Y., Kuramoto, H., Anand, J., Kitamura, T., Sakata, T., Mori, H., and Yanagida, S. (2001) Microwave-assisted size control of CdS nanocrystallites. *J. Mater. Chem.*, **11**, 1936–1940.

127 Karan, S., and Mallik, B. (2007) Tunable visible-light emission from CdS nanocrystallites prepared under microwave irradiation. *J. Phys. Chem. C*, **111**, 16734–16741.

128 Zedan, A.F., Sappal, S., Moussa, S., and El-Shall, M.S. (2010) Ligand-controlled microwave synthesis of cubic and hexagonal CdSe nanocrystals supported on graphene. Photoluminescence quenching by graphene. *J. Phys. Chem. C*, **114**, 19920–19927.

129 Li, L., Qian, H., and Ren, J. (2005) Rapid synthesis of highly luminescent CdTe nanocrystals in the aqueous phase by microwave irradiation with controllable temperature. *Chem. Commun.*, 528–530.

130 He, Y., Lu, H.T., Sai, L.M., Lai, W.Y., Fan, Q.L., Wang, L.H., and Huang, W. (2006) Synthesis of CdTe nanocrystals through program process of microwave irradiation. *J. Phys. Chem. B*, **110**, 13352–13356.

131 Ovsyannikov, S.V., and Shchennikov, V.V. (2004) Thermomagnetic and thermoelectric properties of semiconductors (PbTe, PbSe) at ultrahigh pressures. *J. Phys. B*, **344**, 190–194.

132 Cao, H., Gong, Q., Qian, X., Wang, H., Zai, J., and Zhu, Z. (2007) Synthesis of 3-D hierarchical dendrites of lead chalcogenides in large scale via microwave-assistant method. *Cryst. Growth Des.*, **7**, 425–429.

133 Kuang, D., Xu, A., Fang, Y., Liu, H., Frommen, C., and Fenske, D. (2003) Surfactant-assisted growth of novel PbS dendritic nanostructures via facile hydrothermal process. *Adv. Mater.*, **15**, 1747–1750.

134 Levina, L., Sukhovatkin, V., Musikhin, S., Cauchi, S., Nisman, R., Bazett-Jones, D.P., and Sargent, E.H. (2005) Efficient infrared-emitting PbS quantum dots grown on DNA and stable in aqueous solution and blood plasma. *Adv. Mater.*, **17**, 1854–1857.

135 Mcdonald, S.A., Konstantatos, G., Zhang, S., Cyr, P.W., Klem, E.J.D., Levina, L., and Sargent, E.H. (2005) Solution-processed PbS quantum dot infrared photodetectors and photovoltaics. *Nat. Mater.*, **4**, 138–142.

136 Zhu, J., Palchik, O., Chen, S., and Gedanken, A. (2000) Microwave assisted preparation of CdSe, PbSe, and $Cu_{2-x}Se$ nanoparticles. *J. Phys. Chem. B*, **104**, 7344–7347.

137 Sliem, M.A., Chemseddine, A., Bloeck, U., and Fischer, R.A. (2010) PbSe nanocrystal shape development: oriented attachment at mild conditions and microwave assisted growth of nanocubes. *Cryst. Eng. Comm.*, **13**, 483–488.

138 Limaye, M.V., Gokhale, S., Acharya, S., and Kulkarni, S. (2008) Template-free ZnS nanorod synthesis by microwave irradiation. *Nanotechnology*, **19**, 415602.

139 Qian, H., Qiu, X., Li, L., and Ren, J. (2006) Microwave-assisted aqueous synthesis: a rapid approach to prepare highly luminescent ZnSe (S) alloyed quantum dots. *J. Phys. Chem. B*, **110**, 9034–9040.

140 Kim, K.S., Zhao, Y., Jang, H., Lee, S.Y., Kim, J.M., Ahn, J.H., Kim, P., Choi, J.Y., and Hong, B.H. (2009) Large-scale pattern growth of graphene films for stretchable transparent electrodes. *Nature*, **457**, 706–710.

141 Stankovich, S., Dikin, D.A., Dommett, G.H.B., Kohlhaas, K.M., Zimney, E.J., Stach, E.A., Piner, R.D., Nguyen, S.T., and Ruoff, R.S. (2006) Graphene-based composite materials. *Nature*, **442**, 282–286.

142 Lu, C.H., Yang, H.H., Zhu, C.L., Chen, X., and Chen, G.N. (2009) A graphene platform for sensing biomolecules. *Angew. Chem. Int. Ed.*, **48**, 4785–4787.

143 Wang, Y., Shi, Z.Q., Huang, Y., Ma, Y.F., Wang, C.Y., Chen, M.M., and Chen, Y.S. (2009) Supercapacitor devices based on graphene materials. *J. Phys. Chem. C*, **113**, 13103–13107.

144 Yoo, E., Kim, J., Hosono, E., Zhou, H., Kudo, T., and Honma, I. (2008) Large reversible Li storage of graphene nanosheet families for use in rechargeable lithium ion batteries. *Nano Lett.*, **8**, 2277–2282.

145 Zhang, H., Lv, X.J., Li, Y.M., Wang, Y., and Li, J.H. (2010) P25-graphene composite as a high performance photocatalyst. *ACS Nano*, **4**, 380–386.

146 Murugan, A.V., Muraliganth, T., and Manthiram, A. (2009) Rapid, facile microwave-solvothermal synthesis of graphene nanosheets and their polyaniline nanocomposites for energy strorage. *Chem. Mater.*, **21**, 5004–5006.

147 Chen, W., Yan, L., and Bangal, P.R. (2010) Preparation of graphene by the rapid and mild thermal reduction of graphene oxide induced by microwaves. *Carbon*, **48**, 1146–1152.

148 Hassan, H.M.A., Abdelsayed, V., Khder, A.E.R.S., AbouZeid, K.M., Terner, J., El-Shall, M.S., Al-Resayes, S.I., and El-Azhary, A.A. (2009) Microwave synthesis of graphene sheets supporting metal nanocrystals in aqueous and organic media. *J. Mater. Chem.*, **19**, 3832.

149 Kundu, P., Nethravathi, C., Deshpande, P.A., Rajamathi, M., Madras, G., and Ravishankar, N. (2011) Ultrafast microwave-assisted route to surfactant-free ultrafine Pt nanoparticles on graphene: synergistic co-reduction mechanism and high catalytic activity. *Chem. Mater.*, **23**, 2772–2780.

150 Li, B.J., Cao, H.Q., Shao, J., Li, G.Q., Qu, M.Z., and Yin, G. (2011) Co3O4@graphene composites as anode materials for high-performance lithium ion batteries. *Inorg. Chem.*, **50**, 1628–1632.

151 Li, L., Guo, Z.P., Du, A.J., and Liu, H.K. (2012) Rapid microwave-assisted synthesis of Mn3O4-graphene nanocomposite and its lithium storage properties. *J. Mater. Chem.*, **22**, 3600–3605.

152 Lin, Y., Baggett, D.W., Kim, J.W., Siochi, E.J., and Connell, J.W. (2011) Instantaneous formation of metal and metal oxide nanoparticles on carbon nanotubes and graphene via solvent-free microwave heating. *ACS Appl. Mater. Inter.*, **3**, 1652–1664.

153 Liu, X.J., Pan, L.K., Lv, T., Lu, T., Zhu, G., Sun, Z., and Sun, C.Q. (2011) Microwave-assisted synthesis of ZnO-graphene composite for photocatalytic reduction of Cr(VI). *Catal. Sci. Technol.*, **1**, 1189–1193.

154 Liu, X.J., Pan, L.K., Lv, T., Zhu, G., Lu, T., Sun, Z., and Sun, C.Q. (2011) Microwave-assisted synthesis of TiO(2)-reduced graphene oxide composites for the photocatalytic reduction of Cr(VI). *RSC Adv.*, **1**, 1245–1249.

155 Lu, L.Q., and Wang, Y. (2011) Sheet-like and fusiform CuO nanostructures grown on graphene by rapid microwave heating for high Li-ion storage capacities. *J. Mater. Chem.*, **21**, 17916–17921.

156 Zhang, M., Lei, D.N., Yin, X.M., Chen, L.B., Li, Q.H., Wang, Y.G., and Wang, T.H. (2010) Magnetite/graphene composites: microwave irradiation synthesis and enhanced cycling and rate performances for lithium ion batteries. *J. Mater. Chem.*, **20**, 5538–5543.

157 Zhong, C., Wang, J.Z., Chen, Z.X., and Liu, H.K. (2011) SnO$_2$-graphene composite synthesized via an ultrafast and environmentally friendly microwave autoclave method and its use as a superior anode for lithium-ion batteries. *J. Phys. Chem. C*, **115**, 25115–25120.

7
Precisely Controlled Synthesis of Metal Nanoparticles under Microwave Irradiation

Zhi Chen, Dai Mochizuki, and Yuji Wada

7.1
Introduction

7.1.1
General Introduction – Green Chemistry

"Green Chemistry" or "Green and Sustainable Chemistry" have been common phrases when chemists and material scientists talk about new research directions. Green chemistry encourages chemists to keep the following twelve principles in mind [1]:

1) **Prevention** – It is better to prevent waste than to treat or clean up waste after it has been created.

2) **Atom economy** – Synthetic methods should be designed to maximize the incorporation of all materials used in the process into the final product.

3) **Less Hazardous Chemical Syntheses** – Wherever practicable, synthetic methods should be designed to use and generate substances that possess little or no toxicity to human health and the environment.

4) **Designing Safer Chemicals** – Chemical products should be designed to affect their desired function while minimizing their toxicity.

5) **Safer Solvents and Auxiliaries** – The use of auxiliary substances (e.g., solvents, separation agents, etc.) should be made unnecessary wherever possible and innocuous when used.

6) **Design for Energy Efficiency** – Energy requirements of chemical processes should be recognized for their environmental and economic impacts and should be minimized. If possible, synthetic methods should be conducted at ambient temperature and pressure.

Microwaves in Nanoparticle Synthesis, First Edition. Edited by Satoshi Horikoshi and Nick Serpone.
© 2013 Wiley-VCH Verlag GmbH & Co. KGaA. Published 2013 by Wiley-VCH Verlag GmbH & Co. KGaA.

7) **Use of Renewable Feedstocks** – A raw material or feedstock should be renewable rather than depleting whenever technically and economically practicable.

8) **Reduce Derivatives** – Unnecessary derivatization (use of blocking groups, protection/deprotection, and temporary modification of physical/chemical processes) should be minimized or avoided if possible, because such steps require additional reagents and can generate waste.

9) **Catalysis** – Catalytic reagents (as selective as possible) are superior to stoichiometric reagents.

10) **Design for Degradation** – Chemical products should be designed so that at the end of their function they break down into innocuous degradation products and do not persist in the environment.

11) **Real-time Analysis for Pollution Prevention** – Analytical methodologies need to be further developed to allow for real-time, in-process monitoring and control prior to the formation of hazardous substances.

12) **Inherently Safer Chemistry for Accident Prevention** – Substances and the form of a substance used in a chemical process should be chosen to minimize the potential for chemical accidents, including releases, explosions, and fires.

Microwaves can provide some solutions to the above principles of green chemistry. Special effects of microwaves have been demonstrated, as will be described below. Chemical reactions are known to be enhanced under microwaves, resulting in shortening the reaction times. This will give a solution to principle 6. Organic reactions can be carried out without solvents. This would provide a solution to principle 5. Microwaves can provide higher yields and selectivities than conventional heating for organic reactions. This would contribute to a solution of principles 2, 6, and 8. A catalytic reaction has been reported to proceed without the catalyst or with less catalyst under microwaves, corresponding to principle 9.

The ideas embedded in green chemistry should also be applied to the preparation of metal nanoparticles. Accordingly, this chapter describes the preparation of metal nanoparticles using microwaves. In this regard, chemists should appreciate the additional advantages that microwaves can provide in the preparation of metal nanoparticles, namely,

- the precise control of size and size distribution, and
- the precise control of the structure of multi-component metal nanoparticles.

Herein we describe the precise control of size and size distribution of metal nanoparticles in Section 7.2, and then extend the application of microwaves to the precise control of the structures in Section 7.3. The final Section 7.4 is oriented toward industrial applications by demonstrating an example of mass production of Ni nanoparticles by a Japanese company.

7.1.2
Microwave Chemistry for the Preparation of Metal Nanoparticles

When a flask containing a solution of a precursor is immersed in a heated oil bath, the external wall of the flask is heated first, followed by transfer of the heat to the solution through the wall. Therefore, the temperature of the solution in contact with the wall is raised at the beginning, and the heat is further transported to the rest of the solution. This process of heating a solution is referred to as conventional heating. When a metal precursor in solution is heated, it is converted to metal clusters through a reduction process or some other reaction to form metal atoms. Such clusters function as seed nuclei for gathering metal atoms toward the growth of nanoparticles. The metal nuclei are generated at the wall first with conventional heating because the temperature of the wall increases more rapidly than the other parts of the reactor.

By contrast, with microwave irradiaiton the metal nuclei are formed anywhere in the solution because the solution is heated by direct interaction with the microwaves. Therefore, if the distribution of the microwaves in a solution is suitably controlled so as to be homogeneous, the generation of such nuclei will then also be homogeneous in the solution. It is possible to control the generation of nuclei occurring in a short time and then control the particle growth following the nuclei generation. This is very important for the precise control of both the size and size distribution of metal (nano)particles.

Wada *et al.* have realized this idea by demonstrating in the period 1999–2004 that nanosized Ni and Ag metal particles with narrow size distributions can be synthesized using microwaves [2–4]. Figure 7.1 illustrates the homogeneous size distribution of Ag nanoparticles prepared under microwave irradiation. Several studies that deal with the synthesis of nanoparticles have been reported since then.

Wada *et al.* confirmed the above hypothesis by directly observing the formation of silver nanoparticles under microwave irradiation, by means of a Raman spectrometer they constructed in which a probe was inserted into the reaction solutions, as depicted in Figure 7.2 [5].

The time of generation of Ag nanoparticles was determined by the method described below (see Figure 7.3). The intensity of the Raman scattering spectrum of Rhodamine 6G was intensified when the molecule was adsorbed on the surface of Ag nanoparticles (a process referred to commonly as surface-enhanced Raman scattering, or SERS).

Subsequently, the appearance of the Raman spectrum of Rhodamine 6G, amplified by the surface-enhanced effect, was employed to sense the generation of Ag nanoparticles. In the experimental observation of SERS under microwave irradiation, the Raman spectra attributed to the SERS spectrum of Rhodamine 6G appeared simultaneously at the center of the reaction vessel and beside the wall of the reaction tube, as shown in Figure 7.4, when the reaction vessel containing the solution of the precursor of Ag was exposed to microwave radiation.

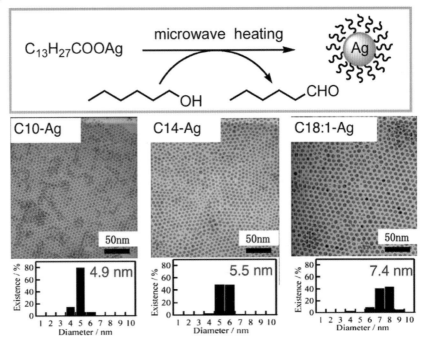

Figure 7.1 Ag nanoparticles prepared by microwave-induced reactions of Ag salts of long chain alkyl carboxylate; C10-Ag; caprate, C14-Ag; myristate, C18:1-Ag; olelate.

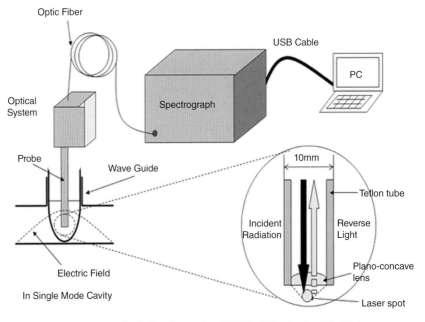

Figure 7.2 Raman system for *in situ* observation of SERS of Rhodamine 6G with Ag nanoparticles prepared under microwave irradiation.

7.1 Introduction | 149

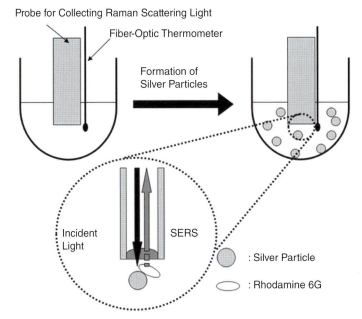

Figure 7.3 Principle for the *in situ* detection of Ag nanoprticles formed under microwave irradiation by means of SERS measurements.

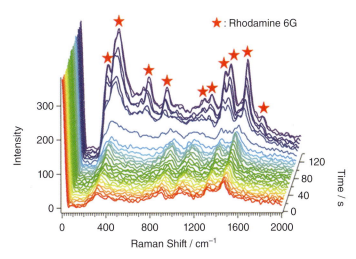

Figure 7.4 *In situ* SERS spectra demonstrating the formation of Ag nanoparticles in solution.

When the reaction vessel was heated in a hot oil bath (see Figure 7.5), however, the Raman spectrum appeared earlier than at the center, as found when the probe was positioned at the wall. The increase in the Raman spectrum coincided completely with the increase in temperature when both microwave heating and oil bath heating were used [5]. This study demonstrated that a reaction solution can

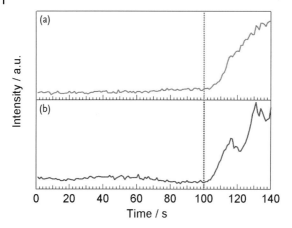

Figure 7.5 Rises of the *in situ* SERS observed at the center (a) and on the side of the wall (b) under microwave irradiation. The peak intensities at 1518 cm^{-1} {n(C–C) vibration} are plotted against MW irradiation time.

be heated in the inner parts of the solution, resulting in the homogeneous reaction conditions that can be achieved under microwave irradiation.

The authors also pointed out two peculiarities of microwave heating [5]. Rapid heating of a solution is an important feature of microwave irradiation. One can heat a solution containing a metal precursor at an extremely high rate such as 100 K min^{-1}. This enables ones to control precisely the structures of multi-component metal particles. Core–shell structures, acorn structures, and so on are prepared by using this feature. As an example, the synthesis of a core (Ag)–shell (Cu) structure achieved by using microwaves is outlined in Figure 7.6 [6]. After Ag nanoparticles are prepared using the microwaves, the solution containing the Ag nanoparticles is then rapidly cooled to room temperature by immersing the reaction vessel in a dry ice–ethanol bath to quench particle growth. Following this, the solution is irradiated by microwaves after addition of the Cu precursor. This method enabled Wada *et al.* to control the Ag core–Cu shell structure of the nanoparticles [6].

The second feature that should be emphasized is the magnetic-field-induced "non-equilibrium local heating" of magnetic nanoparticles. This is induced by the interaction of magnetic nanoparticles with the vibrating magnetic field of the microwaves. When double-component structures having a magnetic metal as a core are synthesized, as demonstrated in Figures 7.7 and 7.8, this feature can be employed to control the structures. The magnetic nanoparticles used as a core can be heated selectively by microwaves and their surface functions as a reaction field for the reduction of the second precursor. This idea has been successfully demonstrated in the synthesis of Ni core–Co shell nanoparticles [7]. Depending on their physical properties, selective heating of substances can be advantageous when the microwaves are separated into their electric and magnetic fields.

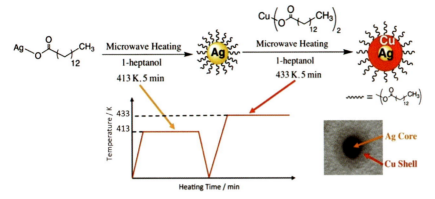

Figure 7.6 Ag core–Cu shell structure obtained by the reduction of a precursor of Cu on a preformed Ag metal core.

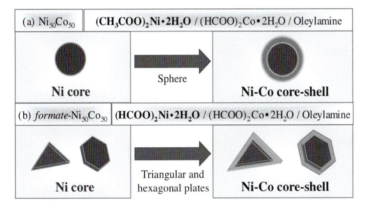

Figure 7.7 Double-component structure prepared by "non-equilibrium local heating" induced by the magnetic field components of the microwaves.

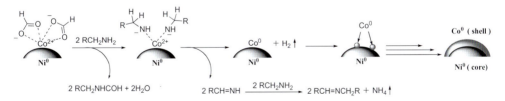

Figure 7.8 Construction of a core–shell structure enabled by non-equilibrium local heating induced by the microwaves. Ni nanoparticles under microwave irradiation experience non-equilibrium local heating, upon which a Co precursor is converted to Co metal.

7.2
Precise Control of Single Component under Microwave Irradiation

7.2.1
Spheres

Metal nanoparticles have attracted much attention in recent years because of the effects that quantum-confinement plays on their electronic, magnetic and other related properties. Microwave heating methods can realize rapid and uniform heating that increases the reaction rates and product yields, while providing a scalable platform for industrial applications. In recent years, microwave dielectric heating has been applied to the rapid synthesis of metallic nanostructures. Water, alcohols, dimethyl formamide (DMF), and ethylene glycol (EG) have high dielectric losses and high reduction ability. Therefore, they are ideal solvents for rapid microwave heating, which has been used in the presence of surfactants to synthesize nanoparticles of various metals (Ni, Ru, Rh, Pd, Ag, Ir, Pt, and Au), metallic compounds, and core–shell structures.

7.2.1.1 Au Nanoparticles
Noble metal nanoparticles are emerging as key materials in catalysis, plasmonic sensing, biological imaging, and in medical therapeutics. The properties of noble metal nanoparticles depend strongly not only on size, shape, crystallinity, and assembly, but also on the local environment of the nanostructures, including substrate, solvent, and capping agent. Gold nanoparticles (Au-NPs) are among the most studied metal nanoparticles with continuously increasing interest from electronics to biological applications due to their promising properties, such as biocompatibility, facile synthesis, and facile surface modification [8].

7.2.1.1.1 Biocompatible Surfactant
Microbial contamination and biofilm formation are serious issues in general health care. With the spread of antibiotic-resistant infections, interest in new antimicrobial agents, such as inorganic materials, has been rising. However, the time required to complete the reaction ranged from 24 to 120 h. Such a long time period is one major drawback of the biological synthesis. In order to form noncytotoxic gold nanoparticles, various synthesis methods have been developed that use biocompatible reducing or stabilizing agents, such as proteins. Although proteins such as bovine serum albumin (BSA) can help to obtain fine Au-NPs, they are characterized by significant drawbacks, such as long reaction time and the need for additional reducing agents, and specific reaction conditions: for example, a certain pH and higher temperature.

Yan *et al.* suggested a new method by combining microwave irradiation and BSA and human serum albumin (HSA) to overcome the problems mentioned above [9]. The reaction time can be shortened from tens of hours to several minutes, thanks to the superheating and non-thermal effects of the microwave radiation. Experimental conditions that include the concentrations of BSA, $HAuCl_4$

and NaOH and the microwave program have been found to be closely related to the quality of the Au nanoparticles.

Geckeler *et al.* have developed a straightforward method for the synthesis of noncytotoxic gold nanoparticles involving the reduction and stabilization by a protein, such as a lysozyme, in conjunction with microwave irradiation [10]. The cell toxicity and the cellular uptake pathways of the as-synthesized gold nanoparticles against mouse embryonic fibroblast NIH-3T3 cells were studied and no significant cytotoxicity was detected. Besides the use of lysozyme as the surfactant, gold nanoparticles with an average diameter of 10.5 nm were also prepared using coconut water through a microwave process within an optimum time of 17 s [11]. Cytotoxicity of the gold nanoparticles was tested on two human cancer cell lines, HeLa (human cervical cancer) and MCF-7 (human breast cancer), and found to be nontoxic. This method exploited an inexpensive and easily available biomaterial that can be used directly for the synthesis of gold nanoparticles; it also has the potential for the synthesis of metallic nanoparticles other than gold. Another direct synthesis of gold nanoparticles was reported by Baruwati *et al.* wherein Au, Ag, Pd and Pt nanoparticles were synthesized in aqueous media using red grape pomace as the solvent, the reducing agent and the capping agent [12].

7.2.1.1.2 Polymer Surfactant

The polyol method is a typical technique to prepare Au nanoparticles of different sizes and shapes by reducing their ionic salts. Generally, spherical nanoparticles are prepared by heating a mixture of reagent and polymer surfactant in EG in an oil bath for several hours. Recently, microwave heating coupled with the polyol method has been used for the rapid preparation of Au nanoparticles.

Polyvinyl alcohol (PVA) has already been considered as a mild reducing agent and an appropriate solvent for microwave-assisted synthesis due to its high loss tangent (tan δ). The polyol process is a popular method for the synthesis of metal nanoparticles and the reaction is carried out in alcohols in which PVA is well-dispersed, even insoluble. A new route based on the microwave-assisted heterogeneous reduction of the metal salts with PVA was reported by Abargues *et al.* for the syntheses of colloidal silver and gold nanoparticles [13]. Size-controlled formation of nanoparticle agglomerates was accomplished with a seed-mediated synthesis; this method is potentially scalable for massive nanoparticle production.

Poly(N-vinylpyrrolidone), PVP, is soluble in water and other polar solvents, and thus has excellent wetting properties and readily forms films, which makes it good as a coating or as an additive to coatings. Morphologies and sizes were controlled by changing various experimental parameters, such as the concentration of metallic salts and surfactant polymers, the chain length of the surfactant polymer, the solvent, and the reaction temperature. Au nanoparticles have been synthesized in EG and glycerol using the microwave technique in the presence of PVP as the stabilizer [14]. The sizes of Au nanostructures depend strongly on the concentrations of $HAuCl_4$ and PVP, and on the length of the PVP chain. Without addition of PVP, only large spherical gold particles with diameters of 100–300 nm were produced. When a short-chain PVP (low molecular weight, $T = 10$ K) is used, small spherical nanoparticles with average diameters of 8 nm are produced, while triangular and hexagonal nanoplates with average edge lengths of 65 and 80 nm are formed when using PVP at 40 and 360 K, respectively. Nanostructures with smaller sizes, narrower size distributions, and a higher degree of crystallinity were obtained with microwave heating than with conventional oil-bath heating, as shown by Tsuji et al. [15]. Kamarudin et al. also prepared various size- and shape-controlled Au nanoparticles for mercury adsorption using a microwave-polyol method in the presence of PVP as a polymer surfactant at different concentrations (1.9, 3.8 and 5.7 mM) [16]. It should be noted that the polygonal particles also increase with increasing PVP concentration. Small spherical particles of 17–59 nm were preferentially produced at the highest PVP concentration (5.7 mM). The as-synthesized Au nanoparticles demonstrated good behavior for mercury adsorption.

7.2.1.1.3 New Reducing Agent

The reduction of gold salt was done within a short time of 60–90 s in CTAB (cetyl trimethyl ammonium bromide) micellar media in the presence of alkaline 2,7-dihydroxynaphthalene (2,7-DHN) as a new reducing agent [17]. Spherical (25 ± 2 nm) Au nanoparticles with other controlled morphologies and sizes were directed by the surfactant/metal ion molar ratios and the reducing agent concentration.

7.2.1.2 Ag Nanoparticles

Silver nanoparticles have been an active research area of nanotechnology because of the importance of silver nanoparticles in catalysis, electronics, photonics, optoelectronics, sensing, and pharmaceuticals [18, 19]. Importantly, silver nanomaterials are most suitable for SERS studies.

7.2.1.2.1 Biocompatible Surfactant

Silver nanoparticles are efficient antimicrobials with size and shape dependence. Antibacterial technology based on silver nanoparticles is already used in commercial products. There are still concerns about their cytotoxicity and environmental impact since nanosized precious metals are conventionally synthesized using toxic reducing agents, such as hydrazine and sodium borohydride. Accordingly, the design of biocompatible and environmentally benign nanomaterials is needed. There is also the need to use a stabilizing and/or capping agent to prevent particle aggregation and to control their growth; synthetic polymers such as PVP are often used.

Faster environmentally friendly methods using microwave heating and starch as protecting/capping agent for the synthesis of silver nanoparticles in water have been reported using D-glucose and the amino acids L-lysine and L-arginine as the reducing agents. Additionally, starch can also work as a reducing agent in autoclave conditions. Synthesis of Ag nanoparticles with size 5 to 22 nm and controlled silver release from cellulosic nanocomposites has been achieved by synthesizing silver nanoparticles under microwave heating in a one-pot, versatile and sustainable process, in which microcrystalline cellulose simultaneously functions as a reducing, stabilizing and supporting agent in water [20]. Using the biopolymer starch as the stabilizing agent and ascorbic acid as the reducing agent, Ag nanoparticles with antibacterial activity were prepared by a green microwave-assisted approach in aqueous medium [21]. Red grape pomace, *bacillus subtilis* or D-glucose have also been introduced for the synthesis of Ag nanoparticles combined with microwave heating; no reducing agent and capping agent were used in these reactions [12, 22, 23]. Recently, Yu *et al.* reported that nearly monodisperse silver nanoparticles of 26.3 ± 2.1 nm were synthesized in large quantities via a microwave-assisted green chemistry method in an aqueous system by using basic amino acids as reducing agents and solvent starch as a protective capping agent [24]. In this reaction, amino acids are indispensable for the uniformity of nanoparticles. Interestingly, self-assembly of starch-capped silver nanoparticles can result in forming multilayered mirror-like films on a glass slide surface. The surface enhancement factor can be efficiently changed by the silver atom concentrations of the films. It was noted that 0.1 g of nearly monodisperse silver nanoparticles can be used for each reaction, and can be potentially applied to large-scale production.

7.2.1.2.2 Polymer Surfactant

The use of microwaves with the reducing agents ethylene glycol, methanol, ethanol, and so on, will increase the kinetics of metal formation. These reducing agents also serve as the solvents. Polymers like PVP have been used widely as protecting agents for producing Ag nanoparticles; size control was realized by varying the reaction conditions, including the metal precursor concentration and the molecular weight of the protecting agent [15, 25–28]. Except for the PVP surfactant, polyvinyl alcohol and polyacrylonitrile (PAN) nanofibers have been used as protecting agents for the controlled synthesis of Ag nanoparticles [13, 29].

Similar to the results for Au nanoparticles, nanostructures with smaller sizes, narrower size distributions, and a higher degree of crystallinity have been obtained with microwave heating relative to those prepared with conventional oil-bath heating.

7.2.1.2.3 New Reducing Agent

Although polymers have been proven to be strong reducing agents for the controlled synthesis of Ag nanoparticles, the removal of the unwanted template or directing agent from the surface of the nanoparticles requires rather harsh conditions and, consequently, increases the difficulty of performing such surface chemistry. Kundu *et al.* reported the reduction of Ag ions for just 60 s in TX-100 (polyoxyethylene isooctylphenyl ether) micellar media in the presence of a new reducing agent, alkaline 2,7-dihydroxynaphthalene (2,7-DHN) [30].

Zhu *et al.* reported the fast microwave-assisted synthesis of polyacrylamide–metal (M = Ag, Pt, Cu) nanocomposites with metal nanoparticles homogeneously dispersed in the polymer matrix, using the corresponding metal salt and the acrylamide monomer in ethylene glycol [31]. This method is based on the single-step simultaneous formation of metal nanoparticles and polymerization of the acrylamide monomer, leading to a homogeneous distribution of metal nanoparticles in the polyacrylamide matrix. No initiator for acrylamide polymerization is needed and neither is a surfactant for stabilization of the metal nanoparticles. This makes it possible to avoid subsequent complicated work-up procedures for the removal of these additives, leading to speed, simplicity, and low-cost in the preparation of polymer–metal nanocomposites.

7.2.1.2.4 Continuous Synthesis

An efficient synthetic process for the large-scale production of monodispersed silver nanoparticles with a narrow size distribution is strongly demanded by many industries due to the wide applications of Ag nanoparticles. Flow-type microwave reactors provide several advantages over batch reactors: stirring is unnecessary and they are more suitable for large-scale production by continuous processing. In this regard, Nishioka *et al.* reported the continuous flow synthesis of silver nanoparticles using an originally designed single-mode microwave cavity reactor that could control the reaction temperature precisely [32]. Monodispersed silver nanoparticles of about 9.8 nm can be prepared using this microwave flow reactor system that can be operated for 5 h. They found that various parameters can influence the reaction rate, product yield, particle size, and uniformity of the size distribution.

7.2.1.3 Pt Nanoparticles

Pt nanoparticles are the subject of substantial research with potential applications in a wide variety of areas, including nanotechnology, medicine, and the synthesis of novel materials with unique properties, due to their antioxidant properties. Such applications are strongly dependent on the size, shape and impurities of the metal nanoparticles.

7.2.1.3.1 Biocompatible Surfactant

As mentioned before, red grape pomace can serve as a solvent, a reducing agent and a capping agent [12]. Wang *et al.* proposed a one-step, rapid and efficient aqueous-phase reaction to produce Pt nanoparticles straightforwardly in high yield without the need for any organic solvent, template or ion replacement. The process was carried out simply by the microwave-assisted heat-treatment of an aqueous solution containing K_2PtCl_4 and 2-[4-(2-hydroxyethyl)-1-piperazinyl]ethanesulfonic acid (HEPES) within 12 s [33]. As-prepared platinum nanoclusters are porous interconnected nanostructures and possess a very high surface area ($41\,m^2\,g^{-1}$), which is advantageous for catalytic applications.

7.2.1.3.2 Polymer Surfactant

The polyol method is a low-temperature process and is environmentally benign because the reactions are carried out under closed system conditions. Generally, alcohols and polyols, such as EG, work as reducing agents as well as solvents, while polymers such as PVP serve as protecting agents. Komarneni *et al.* reported the synthesis process as follows: the starting precursors were mixed with EG and

PVP of different molecular weights as a protective polymer, with and without NaOH in some cases, and treated at 150 °C for 15 min using a microwave digestion system [25, 28, 34, 35]. Platinum nanoparticles in the size range 2–8 nm prepared using EG and microwave dielectric heating were stable in the toluene solution for ≤ 10 months while larger particles were formed at longer microwave heating times, higher metal precursor salt concentrations, and lower pH [34].

Morphology and size were controlled by changing various experimental parameters, such as the concentration of the metallic salts and surfactant polymers, the chain length of the surfactant polymer, the solvent, and the reaction temperature [15, 31, 36]. Uniform Pt nanoparticles with size 25–30 nm have been obtained in a short time (15 s) using glycerol/EG as reducing agent [36]. It should be noted that Pt nanostructures with smaller sizes, narrower size distributions, and a higher degree of crystallinity were obtained with microwave heating relative to those with conventional oil-bath heating, in accord with the results for Au and Ag nanoparticles.

A comparative study was done on the mode of heating and it was found that in the case of microwave heating, nucleation might take place along any plane, whereas in the reflux method it occurs in all planes and axes, resulting in uniform spherical shapes.

7.2.1.4 Pd, Ru, and Rh Nanoparticles

Nanopowders of noble metals are extremely significant materials because of their possible applications in various fields such as catalysis, optical or the electronic industry. Pd nanoparticles are very interesting because of their catalytic properties, which are strongly dependent on the shape, size, and size distribution of the particles.

Ruthenium has been known to show unique and interesting activity as a catalyst, for example, for the ammonia decomposition/synthesis and CO oxidation, and as an electrocatalyst for fuel cell applications. Preparation of Ru nanoparticles has been successfully attempted using methods of colloid chemistry. Rh- and Ru-NPs can be used for olefin hydrogenation.

7.2.1.4.1 Biocompatible Surfactant

Biocompatible agents such as 2-[4-(2-hydroxyethyl)-1- piperazinyl]ethanesulfonic acid (HEPES) and red grape pomace have been successfully used as protecting agents for the synthesis of Pd nanoparticles by the rapid and facile microwave-mediated route [12, 37]. In a HEPES buffer solution, shape and size control could be obtained by a one-step method by varying the capping agents, such as PVP, CTAB, sodium citrate and KBr. Size-dependent catalytic activities of the obtained Pd-NPs for a Suzuki coupling reaction have been examined; results showed that Pd-NPs of less than 10 nm exhibited better catalytic activities.

7.2.1.4.2 Polymer Surfactant

Uniform and stable PVP or PAN nanofibers protected spherical Pd nanoparticles have been synthesized using glycerol as both the solvent and the reducing agent

[28, 29, 36]. Several experimental conditions such as *in situ* irradiation of mixed metal salts and mixing of individual solutions were attempted to understand the mechanism of formation of the nanoparticles. Pd nanoparticles of size 22 to 30 nm have been produced in the microwave-assisted process using glycerol/EG as reducing agent.

The main advantages of microwave-solvothermal methods are: very fast kinetics, phase purity with better yield, and high reproducibility. These methods then seem to be ideally suited for the precise control of the microstructure of nanoparticles. In addition, it is environmentally friendly because the reactions are carried out in closed systems at rather low temperatures for a short time [38, 39]. Uniform and stable Ru nanoparticles with small average diameters (below 2 nm) and narrow size distributions (1–3 nm) were synthesized by reduction of $RuCl_3$ in EG in the presence of PVP using a microwave-assisted solvothermal method. Solvent-stabilized (upon storage under ambient conditions for months) Pt, Rh, and Ru colloidal nanoparticles with narrow size distribution have also been synthesized by a similiar microwave-polyol process [35].

7.2.1.4.3 Other Surfactants

Ionic liquids (ILs) act as a "novel nanosynthetic template" that stabilizes metal nanoparticles on the basis of the ionic charge, high polarity, high dielectric constant, and the supramolecular network of the IL.

Stable Cr, Mo, W, Mn, Re, Ru, Os, Co, Rh, and Ir metal nanoparticles have been obtained reproducibly from $M_x(CO)_y$ in ionic liquids by a facile, rapid (3 min), and energy-saving 10 W microwave irradiation [40]. The microwave-assisted synthesis of the nanoparticles from $M_x(CO)_y$ in ILs, in comparison with the UV-photolytic (1000 W, 15 min) and conventional thermal decomposition (180–250°, 6–12 h) methods, resulted in very small (<5 nm) nanoparticles of uniform size that are prepared without any additional stabilizers or capping molecular agents. Ru, Rh or Ir nanoparticle/IL dispersions are highly active and easily recyclable catalysts for the biphasic liquid–liquid hydrogenation of cyclohexene to cyclohexane.

7.2.1.5 Other Transition Metals

7.2.1.5.1 Cu Nanoparticles

Metallic Cu nanoparticles are used in various fields such as catalysis, electronics and optics [41]. In addition to their low cost compared to Ag nanoparticles, Cu is an important material in the field of electronic devices because of anti-ionic migration.

It is very difficult to obtain uniform and well-dispersed copper nanoparticles at high precursor concentrations by hydrazine or hydrogen reduction. The polyol process is slower and requires refluxing for several hours or even days at a higher temperature. In both the aqueous reaction and the polyol process, a suitable soluble polymer (e.g., PVP) is usually added as a protecting agent to obtain nanoscale copper powder, raising the problem of the difficulty of removing the polymer from the particle surface by simple washing.

Rapid preparations of Cu nanoparticles in environmentally benign stabilizing agents via a green route have been achieved by the well-established microwave-assisted chemistry reduction in aqueous media [21, 42–44]. Cu nanoparticles prepared by a microwave-assisted process in aqueous media, using starch as the stabilizing agent, exhibited surface plasmon absorption resonance (SPR) maxima and interesting antibacterial activity with both gram positive and gram negative bacteria at micromolar concentrations [21]. By using the chelating and reducing power of a polydentate alcohol (e.g., diethyleneglycol) with eco-friendly additives such as ascorbic acid (reducing agent), copper colloidal nanoparticles were synthesized in very high yields and were stable for months in spite of the high metal concentration [42].

Monodispersed Cu nanoparticles with average sizes of 5–6 nm (with the surface plasmon absorption) and 2–3 nm (without surface plasmon absorption) have been prepared using copper(II) octanoate and copper(II) myristate, respectively, as the copper precursors, by reduction with alcohols under microwave-heating at 443 K for 20 min [43]. Zhu et al. reported the synthesis of well-dispersed copper nanoparticles in EG with hydrazine as the reducing agent in the absence of a protective polymer using microwave irradiation for heating [44].

Synthesis conditions do have an effect on particle size, optical properties, and reaction yields. Microwaves provide a rapid uniform heating of reagents and solvent, while accelerating the reduction of metal precursors and the nucleation of metal clusters, resulting in monodispersed nanostructures and large-scale continuous production.

7.2.1.5.2 Ni Nanoparticles

Magnetic nanoparticles, such as Fe, Co, and Ni, have attracted much attention because of their unique properties and applications in various fields. Ni nanoparticles could be used as electrode materials in multilayer ceramic capacitors and catalysts.

Ni nanoparticles with controlled morphology have been obtained with limited success. Binary protecting agent systems were used to control particle shape in the synthesis of Co nanoparticles. Li et al. reported the synthesis of Ni nanoparticles using the microwave-assisted polyol method (M–P process) which was carried out in a binary protecting agent system, including PVP and dodecylamine (DDA). The size and shape of different metals were controlled by using different capping agents with the M–P process [26, 45]. Wada et al. reported the synthesis of uniform Ni nanoparticles with controlled size between 5 and 8 nm by changing the irradiation time and the microwave irradiation power [2].

7.2.1.5.3 Others: (Co, Mn, Zn, Fe)

Co- and Mn-NPs can be used as catalysts for Fischer–Tropsch reactions. Metal carbonyls $M_x(CO)_y$ or other M(0) complexes are interesting precursors for the synthesis of metal nanoparticles (M-NPs) because the metal atoms are already in their final zero-valent oxidation state. Ionic liquids can stabilize metal nanoparticles through their high ionic charge, their polarity and high dielectric constants.

As such, they can therefore function as both stabilizer and solvent for the preparation of small (<5 nm) and (generally) kinetically stabilized M-NPs.

Co-NPs with a diameter of 1.6 ± 0.3 nm and Mn-NPs with a diameter of 4.3 ± 1.0 nm have been prepared by microwave-induced decomposition of the metal carbonyls $Co_2(CO)_8$ and $Mn_2(CO)_{10}$, respectively [46]. The functionalized ionic liquid 1-methyl-3-(3-carboxyethyl)imidazolium tetrafluoroborate [Emim-CO$_2$H][BF$_4$] yielded smaller and better separated particles than the nonfunctionalized ionic liquid 1-n-butyl-3-methylimidazolium tetrafluoroborate [Bmim][BF$_4$]. The particles are stable in the absence of surfactants for more than six months. Stable Cr, Mo, W, Rh, Ru, Os, and Ir metal nanoparticles have also been obtained reproducibly from $M_x(CO)_y$ in the ionic liquids by the microwave irradiation method [40].

7.2.2
Nanorods and Nanowires

One-dimensional (1D) metal nanostructures (rods, wires, and tubes) have received considerable attention from a broad range of researchers because of their wide applications as catalysts, scanning probes, and in various kinds of electronic and photonic nanodevices. Since the catalytic, optical, and electric properties of the 1D nanostructures depend strongly on their shapes and sizes, extensive studies on shape- and size-controlled syntheses of the 1D nanostructures have been carried out.

It has been reported that gold nanorods that have a relatively high aspect ratio are more effective in SERS enhancement than are spherically-shaped particles. A higher enhancement of SERS for a silver nanorod depends on both the direction of laser polarization and the particle orientation; it is largest when the direction of laser polarization coincides with the long axis of the nanorod.

7.2.2.1 Ag Nanorods and Nanowires
The fabrication of rod-like silver nanostructures is a research field that is currently attracting a great deal of interest, as is the study of how their physical phenomena relate to their shapes. The microwave-polyol method is a promising route for the rapid preparation of metallic nanomaterials. Although silver nanorods have been synthesized by reducing AgNO$_3$ with the assistance of Pt seeds in EG solution and in the presence of PVP, it is worthwhile to continue investigating microwave-mediated synthetic methods for the preparation of silver nanorods with such clean surfaces under low-temperature conditions and in the absence of polymers.

Liu *et al.* demonstrated that microwave heating is a feasible method for the synthesis of silver nanorods by reducing silver ions with sodium citrate in the presence of gold seeds [47]. The key requirement for producing the silver nanorods in a high aspect ratio is an adequate microwave heating time.

Tsuji *et al.* conducted a further detailed study on the real role of Pt seeds as nucleation reagents for the preparation of 1D Ag nanostructures by using the microwave-polyol method [48]. The experimental results indictate that pre-formed metallic Pt seeds are probably not responsible for nucleation and subsequent evolution of 1D Ag products, but Cl⁻ ions indeed influence the formation of the 1D Ag nanostructures as well as other perfect crystalline Ag nanoparticles with well-defined crystal structures, including single- and twinned-FCC crystals. It was found that the presence of Cl⁻ ions can accelerate the re-dissolution of formed spherical Ag particles and is favorable to the growth of the 1D Ag and other Ag nanostructures. Ag nanowires or nanorods were produced without the pre-formed seeds; the effects of salt chemistry and concentration on the morphology and yield were also studied [27, 49]. The results highlight that, while the microwave-assisted process is very promising, results cannot be directly anticipated based on the traditional process. Moreover, it appears that the synthesis is highly sensitive to procedural changes.

7.2.2.2 Au, Pt, Ni Nanorods and Nanowires

Gold nanostructures with well-defined dimensions represent a particular class of interesting nanomaterials to synthesize and study because of their wide practical applications. Both the size and the shape of gold nanoparticles were shown to affect optical and electronic properties. Nanowires are applied for connecting and fabrication of microscopic electrodes and quantum devices.

The microwave heating method is an effective approach for the fast production of 1D Au nanostructures [50–53]. Kundu *et al.* reported the synthesis of electronically conductive gold nanowires for the first time that occurred within about 120–180 s by exploiting an electrodeless method [50]. Apparently, the surfactant plays an important role in the formation of 1D Au nanostructures. DNA can serve as a reducing and nonspecific capping agent for the growth of nanowires. The diameter and length of the gold nanorods and nanowires could be controlled by changing the precursor/protecting agent ratio such as $HAuCl_4 \cdot 4H_2O$/PVP. Tetradecylammonium bromide (TOAB) is another candidate surfactant for stabilizing gold nanorods and both TOAB concentrations and reaction temperature programs have an influence on the formation of gold nanorods. Besides concerns on the surfactant process, Zhu *et al.* reported the preparation of single-crystal gold nanorods and nanowires in the liquid phase by using a one-step, seedless and template-free microwave-polyol method [53].

Platinum nanoparticles currently are of intense interest due to their unique catalytic properties. A range of organic chemical reactions can be catalyzed by Pt nanoparticles, such as hydrosilylation, oxidation, and hydrogenation. Furthermore, Pt nanoparticles are catalytically active in room-temperature electro-oxidation reactions for fuel cell applications. Li *et al.* reported the preparation of Pt nanorods by microwave-assisted solvothermal techniques [54]. Size and morphology control of the particles was influenced by specific reaction conditions.

Ni nanoparticles with unique nanostructures have recently been the subject of extensive research because of their many exotic performances, which endow them with great application potentials in various fields and applications, such as magnetic recording, electronic devices for biological labeling, catalysts, sensors, and as electrode materials.

Guo *et al.* reported the microwave-assisted fabrication of nanorod-like nickel powders via a hydrothermal liquid-phase reduction route [55]. Morphology and structure could be easily tuned by adjusting process conditions such as pH and microwave irradiation time. The nanorod-like Ni powders exhibited significantly enhanced ferromagnetic characteristics in comparison with bulk nickel and nickel nanoparticles.

7.2.3
Other Morphologies

7.2.3.1 Au

Among the parameters influencing the properties, shape control has proven to be as effective as size control in fine-tuning the properties and functions of metal nanostructures. Gold is of particular interest because of its fascinating properties and potential applications in catalysis, drug delivery, optoelectronics, and magnetic devices. One- or two-dimensional planar gold nanostructures, such as plates or belts, have received much attention due to their unique properties and potential applications that differ from those of spherical nanoparticles or nanorods.

The microwave-assisted method is an effective approach for the fast production of metal nanoparticles with diversely tunable shapes. Single crystalline planar gold nanostructures, including nanogears, nanobelts, icosahedrons and nanoplates have been prepared by microwave heating in the presence of surfactants [17, 56–58]. Kundu *et al.* synthesized for the first time shape-controlled (spherical, polygonal, rods, and prisms) gold NPs in a reaction time of less than 90 s in the presence of a cationic surfactant [17, 56]. Different shapes and particle sizes were directed by the surfactant/metal ion molar ratios and the reducing agent concentration. Temperature was another important factor for the growth of gold nanoproducts when the reduction of gold salt was done in CTAB (cetyl trimethyl ammonium bromide) micellar media in the presence of alkaline 2, 7-dihydroxynaphthalene (2,7-DHN) as a new reducing agent. Nanobelts were formed with CTAB while nanogears were produced with PVP at 45 °C. Both chloride ion dosage and reaction temperature were important factors when potassium tetrachloroaurate was used as the metal precursor and poly(ethylene oxide)-poly(propylene oxide)-poly(ethylene oxide) (PEO-PPO-PEO) triblock copolymers served as both a reductant and a capping agent under microwave irradiation.

The bulk and shape-controlled synthesis of Au nanostructures with various shapes, such as prisms, cubes, and hexagons has been described; the process occurs via microwave-assisted spontaneous reduction (within ca. 30 to 60 s) of noble metal salts using α-D-glucose, sucrose, and maltose in aqueous solution [59]. Nanostructures of Ag, Pd, and Pt with morphologies of spherical, single crystalline polygonal plates, sheets, rods, wires, tubes, and dendrites can also be generated by the microwave-assisted spontaneous reduction of noble metal salts [15].

7.2.3.2 Ag

SERS is a useful technique with some significant advantages for sensitive chemical analysis and interfacial studies. It is well-known experimentally that the most intense SERS signals are obtained from molecules adsorbed on microscopically rough silver surfaces, such as silver colloids and silver nanoparticles of various shapes. Since it is especially sensitive to the first layer of adsorbates on the roughened surfaces of noble metals, many important applications have been found in the areas of chemical analysis, corrosion, lubrication, heterogeneous catalysis, biological sensors, and molecular electronics, among others. Xia *et al.* reported a novel synthetic method for silver nanostructures with well-defined edges, corners, and sharper surface features under microwave irradiation; such nanostructures are used as sensitive silver substrates for SERS spectroscopy [60]. The SERS peak intensity on silver nanoparticles synthesized under microwave irradiation is about 30 times greater than that on samples synthesized with conventional heating. Nearly monodisperse silver nanoparticles were synthesized in large quantities via a microwave-assisted green chemistry method in an aqueous system by using basic amino acids as reducing agents and soluble starch as a protective capping agent [24]. The self-assembly of starch-capped silver nanoparticles resulted in multilayered mirror-like films forming on a glass slide surface. The surface enhancement factor could be changed efficiently by the silver atom concentrations of the films.

7.2.3.3 Pt, Pd, Ni, and Co

Metal nanoparticles have potential applications in microelectronics, catalysis, photocatalysis, magnetic devices, chemisorptions, aerosols and powder metallurgies. The morphologies of Pd nanoparticles varied from spherical to triangular prism, and star shape when the reaction was changed from conventional reflux conditions to the microwave heating mode [28, 61].

7.2.3.3.1 Nickel

Magnetic nanoparticles have potential use as active components in a number of applications, such as high density magnetic storage devices, contrast enhancement in magnetic resonance imaging, biotechnology and catalysis. Metallic nanostructures such as rods, wires, sheets and plates have been synthesized in an attempt to improve the magnetic properties of nanocrystals. These nanostructures have demonstrated increased magnetic and catalytic properties owing to their high shape anisotropy compared to their spherical counterparts. In this regard, Donegan *et al.* reported on the synthesis of novel, highly faceted, multiple twinned nanocrystals of nickel using microwave irradiation as the heat source [62]. These nickel nanocrystals exhibited enhanced ferromagnetic behavior at room temperature compared to bulk nickel.

7.2.3.3.2 Cobalt

Many efforts have been made to prepare unusual and novel forms of nanostructures, such as nanoplatelets, nanorods, nanowires, nanodisks, nanosheets and flower-like architectures due to their shape-dependent properties. Cobalt nanoparticles are of considerable interest due to their special magnetic and catalytic properties, which can find applications in the fabrication of super alloys, conducting inks, multilayer capacitors, magnetic recording media, magnetic fluids and computer hard disks. Cobalt, like Fe and Ni, demonstrates ferromagnetism at room temperature. In general, most of the reported methods for flower-like structures required long reaction times. Shojaee *et al.* reported the synthesis of flower-like cobalt nanostructures with size distribution and phase control in a solution of water and EG under microwave irradiation without using metal foils or templates for self-assembly [63].

7.3
Precise Control of Multicomponent Structures under Microwave Irradiation

7.3.1
Multicomponent Nanoparticles

In this section, we describe the controlled syntheses and applications of multicomponent metal nanoparticles in colloids. Bimetallic nanoparticles have received intense attention, owing to their different optical, electronic, magnetic, and catalytic properties and improved functions relative to those of the individual metals. Since these properties depend strongly on the composition, shapes, and sizes of nanoparticles, extensive studies have been carried out on the composition-, shape-, and size-controlled syntheses of bimetallic core–shell and alloy nanoparticles.

Among the methods to fabricate bimetallic nanoparticles, the polyol method is a typical technique to prepare them in solution by reducing their ionic salts. The microwave-polyol method coupled with microwave heating is a new rapid preparation method of metallic nanoparticles that can also be applied to the synthesis of bimetallic nanoparticles.

7.3.1.1 Core–Shell Structures
Core–shell metal nanoparticles (MNPs) exhibit superior catalytic, optical and other properties. Much research attention has been paid to the combination of an inexpensive metal (core) and a noble metal (shell) among various combinations of core–shell architectures. The presence of a noble metal (shell) not only protects the core metal but also influences the dramatic modulation of optical, magnetic, and biological responses of the core. Furthermore, such core–shell NPs have received greater significance in catalysis. The existence of a noble metal as the shell significantly enhances the catalytic properties via strain of the core and ligand effects on the supported noble metal.

7.3.1.1.1 Metal–Metal Oxide
The formation of core–shell nanostructures from a metal and its metal oxide is generally achieved by the oxidation of pre-formed metal nanoparticles on the

surface. Chopra *et al.* synthesized size- (6 to 40 nm), shape-, and shell-thickness controlled Ni–NiO core–shell nanoparticles with microwave irradiation [64]. The size of Ni NPs was influenced by the nickel salt to stabilizer ratio and the amount of stabilizer. The shape of the nanoparticles was altered by varying the reaction time. Cu–Cu$_2$O core–shell nanoparticles were formed by the decomposition of copper acetate under microwave heating. Experimental parameters such as the degree of polymerization of EG, the reaction time, the nature of the surfactant and ligands have great effects on the morphology and composition of the products [65].

7.3.1.1.2 Metal–Metal (Ag Core)

Bimetallic nanoparticles having core–shell and alloy nanostructures are desired as new materials because new functions and improvements of functions can be expected in electronic, optical, catalytic, and magnetic properties, which cannot be achieved by individual metals. Therefore, much attention has been paid to the development of novel synthesis methods for bimetallic nanoparticles and their applications.

7.3.1.1.3 Ag Core–Transition Metal Shell

Ag–Ni bimetallic nanoparticles have attracted great attention due to their technologically important catalytic and magnetic properties. Pronounced lattice mismatch, the lower surface energy of Ag and, importantly, the complete immiscibility between Ag and Ni make a phase-segregated core–shell structure the most thermodynamically stable Ag–Ni system.

Tsuji *et al.* reported the fabrication of Ag core–Ni shell nanoparticles in high yield (100%) by reducing AgNO$_3$ and NiSO$_4$·6H$_2$O, NiCl$_2$·6H$_2$O, or Ni(NO$_3$)$_2$·6H$_2$O with EG in the presence of NaOH and PVP under microwave heating for 10 min [66]. The average size of the Ag–Ni nanostructure prepared by a conventional oil-bath heating method for 40 min was 57 ± 17 nm wherein the thickness of the Ni shells was 17 ± 2.3 nm, larger than those obtained under microwave heating by factors of 1.7 and 1.8, respectively.

Ag–Cu or Cu–Ag nanoparticles show new and different functions from either silver or copper nanoparticles with improvements in the catalytic and magnetic functions of the core–shell nanoparticles.

Wada *et al.* reported the rapid preparation of Ag core–Cu shell nanoparticles of size 10–40 nm by a microwave-assisted alcohol reduction process [6]. The core–shell nanoparticles were prepared by successive two-step reduction of a silver precursor and then a copper precursor added after the formation of the silver nanoparticles in 1-heptanol.

The above results further confirm the notion that smaller particles having a narrow size distribution can be prepared in a short time under microwave irradiation owing to the rapid and uniform heating of the reagent solution. Microwave irradiation is an effective method for the preparation of bimetallic nanocomposites.

7.3.1.1.4 Au–Ag

Both Au–Ag alloys and various kinds of Au–Ag core–shell structures can be prepared from Au–Ag reagents using various methods since gold and silver have identical fcc crystal structures, and because the lattice constants of Au (0.4079 nm) and Ag (0.4086 nm) are very similar.

The microwave-polyol method has been successfully introduced for the rapid production of Au–Ag core–shell nanoparticles in the presence of surfactants. Au generally serves as the core and then an Ag shell is formed on the surface since the Au ion is much easier to reduce [67–70]. Tsuji et al. investigated the growth mechanisms of Au–Ag core–shell nanostructures prepared by the microwave-polyol method [67]. Triangular-bipyramidal, cubic, and rod/wire Ag shells having {100} facets were overgrown on the only {111} facets of Au nanocrystalline seeds of triangular twin plate-like, octahedral, and multiple-twinned decahedral geometry; they were prepared in the presence of PVP. Morphology changes between Au cores and Ag shells arise from changes in the adsorption selectivity of PVPs from the {111} facets of Au to the {100} facets of Ag. Total space volumes of Ag shells overgrown on Au cores decreased in the order: cubes > triangular-bipyramidal crystals > rods $\cong$ wires. Liu et al. synthesized core–shell Au–Ag structures from a two-step process wherein $AgNO_3$ was reduced on the preformed gold seeds by sodium citrate in a microwave heating system [70]. The patterned core–shell films on silicon wafers were formed on the surface of a sandwich structure consisting of Au NPs/APTMS/SiO_2 self-assembled by monodisperse Au nanoparticles through specific Au$\cdots NH_2$ interactions onto the silicon oxide substrate surface modified with a pattern of 3-aminopropyltrimethoxysilane (APTMS) groups.

7.3.1.1.5 Ag–Au

Highly dispersed Ag core–Au shell composite nanoparticles with regular spherical morphology were prepared by a high-pressure liquid phase microwave method using Ag clusters as crystal seeds and sodium citrate as the reducing agent [71].

7.3.1.1.6 Au–Pd and Au–Pt

Nanoparticles of Au–Pd and Au–Pt are the most significant because of their special catalytic properties. Au–Pd- and Au–Pt-based catalysts are perfectly suitable for solving the problems of air purification, decreasing the temperature of the combustion of automobile gases, and hydrogen purification for fuel elements.

In order to stabilize the Pd catalyst and overcome the drawbacks such as hysteresis effects due to transformation between Pd oxide and metallic Pd, inclusion of another noble metal, such as Au, Rh, Ru and Pt has been proposed to form hybrid architectures such as core–shell structures depositing a layer of another metal over Pd particles.

Harpeness and Gedanken reported the first microwave synthesis of core–shell gold–palladium bimetallic nanoparticles by the simultaneous reduction of the Au^{III} and Pd^{II} ions [72]. The thickness of the palladium shell was about 3 nm, and the gold core diameter about 9 nm. Belousov et al. reported the reduction of chlorocomplexes of gold(III) from muriatic acid solutions with microwave irradiation of nanocrystalline powders of palladium and platinum at 110 and 130 °C under hydrothermal conditions [73]. The obtained particles of Pd–Au and Pt–Au have a core of the metal reductant covered with a substitutional solid (Au, Pd) solution for the case of palladium, and isolated by a gold layer in the case of platinum. The difference was explained on the basis of the ratio between the rates of aggregation and reduction. Kinetics of the Pd (core)–Au (shell) nanoparticles prepared by the microwave irradiation method can be monitored by the catalyzed transformation of dihydronicotinamide adenine dinucleotide (NADH) to NAD^+ [74].

Clearly, microwave irradiation allows one not only to accelerate the synthesis of particles but also to obtain more homogeneous materials in comparison with conventional heating.

7.3.1.1.7 Au/Au_xAg_{1-x}/Ag Systems

Wu et al. reported that the use of microwave irradiaiton at hydrothermal temperatures promotes shell growth, as well as core–shell alloying of Au–Au_xAg_{1-x}–Ag nanoparticles [75]. The resulting nanoparticle plasmonic behavior is highly sensitive to core–shell or core–alloy–shell morphology, which allows both the precise engineering of the optical properties and the utilization of optical modeling to elucidate the intricacies of the particle ultrastructure.

7.3.1.1.8 Ni–Co

Inexpensive first-row transition-metal (Fe, Co, and Ni) core–shell nanoparticles with ferromagnetic properties have attracted attention due to alternative resources of noble metals and novel physical properties, such as superparamagnetism, giant magneto resistance, and mesoscopic transport effects in magnetic tunnel junctions.

The Ni–Co system is particularly noted for complete liquid and solid solubility, because of the small lattice size mismatch between face-centered cubic (fcc) Ni and fcc Co (3.524 and 3.545 Å), respectively. Wada et al. were the first to propose the preparation of Ni–Co nanoparticles with core–shell structures [7]. Core–shell nanoparticles were fabricated by using both nickel(II) acetate and cobalt(II) formate complexes in the presence of oleylamine in a 1-pot reaction under microwave irradiation. The HAADF-STEM technique showed a nanostructure of a Co-rich shell and a Ni-rich core and and interlayer of mixed Ni–Co alloy in between. The study of the formation mechanism implied that nickel precursors and the crystal shape of the Ni core played a key role in determining the final shape of the Ni–Co system.

7.3.1.1.9 Cu–C and Ni–C Systems

Carbon-coated nanomaterials, especially metals, are of great interest due to their stable nature toward oxidation and degradation; in addition they show potential applications. Jacob et al. developed a simple synthetic route to prepare carbon-coated copper or nickel nanoparticles in an ionic liquid under microwave heating [76]. The carbon acts as a reducing agent for the copper and nickel ions to get pure metal nanoparticles, and also forms a layer structure over the metal nanoparticles.

7.3.1.2 Alloys

Bimetallic nanoalloys have attracted considerable interest due to the additional new properties that may arise from the combination of different compositions of metals on the nanoscale. Alloy nanoparticles take advantage of the diffusion or interdiffusion of atoms, dopants, and defects at the nano-interface which may tailor the morphology and composition to specific properties. Information on the evolution of the electronic structures of bimetallic nanoparticles as a function of size, composition, and shape, and the associated changes in the optical, catalytic, and magnetic properties continues to be a major goal of research in nanostructured materials.

The applications of nanoparticle alloys are thus expected to enhance many fields of advanced materials and relevant technology, particularly in the areas of catalysis, chemical and biological sensors, optoelectronics, drug delivery, and media storage. Alloy interdiffusion is not only predicted to be thermodynamically and kinetically favorable, but nanoalloys are expected to be highly stable, due in large part to the relaxation of interfacial energies.

The requirements of high temperature and inert atmosphere limit the large-scale production of the nanoalloys. For the formation of nanoalloys, the experimental conditions must be chosen to yield binary nucleation events where the initial nuclei contain both metals with compositions that reflect the compositions of the two metal precursors. This requires careful choice of the two metal precursors with almost identical decomposition profiles to ensure the occurrence of binary nucleation events. These conditions are different from those involved in the formation of core–shell nanoparticles. There is a need to develop general and simple synthetic methods operative at near room temperature in air and applicable to a wide range of nanoalloy systems.

The rapid and uniform heating provided by microwaves has potential benefits not only for single metal nanoparticle synthesis but also for binary metal nanoparticles. The binary metal nanoparticles are generally obtained in the presence of a polymer as a protective layer.

7.3.1.2.1 Au-Containing Alloy

Maye *et al.* chose the Au–Pd binary alloy as a model, due in large part to a miscible phase diagram, and differences between Au and Pd SPR signatures, which allow colorimetric observation of shell or alloy growth [77]. They described a layer-by-layer synthesis of Au/Au–Pd core/system alloy shell nanoparticles via microwave-irradiation-based hydrothermal heating of an ionic precursor solution. Alloy shell growth was monitored by the attenuation of the SPR as a function of shell thickness and composition.

El-Shall *et al.* established a microwave irradiaiton approach as a general procedure for the synthesis of a variety of high quality, crystalline bimetallic nanoalloys with controlled size and shape [78]. The controlled synthesis and characterization of bimetallic alloys of Au, Pt, and Pd with Ru, Rh, Ag, Cu, and Ni was investigated by studying the changes in the structural and optical properties of the resulting alloy nanoparticles with respect to the individual metals through a facile and rapid microwave irradiation method.

Under mild conditions using microwave heating, Pal *et al.* prepared Au/Ag alloy nanoparticles of 5–50 nm size in the presence of polyacrylamide as the stabilizing agent and hydrazine hydrate as the reducing agent [79]. The surface plasmon absorption maxima of Au/Ag bimetallic particles appeared in between the peaks corresponding to pure Ag and pure Au; the SPR and change linearly with increasing Au content in the alloy.

7.3.1.2.2 Cu/Ag

There are only a few reports on bimetallic particles of copper and especially with silver due to the large difference in the lattice constants of Ag and Cu (0.409 and 0.361 nm, respectively) making the preparation of their alloy difficult. Also because of differences in redox potential it is difficult to control the simultaneous reduction of Cu and Ag. Further, the instability of copper in an aqueous medium is an added hurdle.

Bimetallic nanoparticles of copper and silver in various proportions were prepared by microwave assisted chemistry reduction in aqueous medium using the biopolymer starch as a stabilizing agent and ascorbic acid as the reducing agent; the SPR for the Cu/Ag alloys appeared in between, depending on the alloy composition [21].

Self-assembled Ll0 Fe–Pt nanoparticle arrays are promising candidates for future generation of magnetic storage applications owing to their large uniaxial magnetocrystalline anisotropy and good chemical stability. Calculations indicate that particles as small as 2.8 nm have sufficient anisotropy energy to be exploited for permanent data storage, leading to significant advances in the hard disk drive area densities over materials currently used.

Evans *et al.* produced monodisperse crystalline fcc particles with controlled size using a simple microwave irradiation method and extended this chemistry to the synthesis of Fe–Pd nanoparticles [80]. The stoichiometrically controlled synthesis of Fe–Pt and Fe–Pd nanoparticles using $Na_2Fe(CO)_4$ and $Pt(acac)_2/Pd(acac)_2$ as the main reactants showed the advantage of the rapid production of monodisperse fcc Fe–Pt nanoparticle metal alloys, which can be converted to the fct phase at lower temperatures (364 °C). Microwave-assisted reactions at high pressure have led to the direct formation of a mixture of fcc and fct phase Fe–Pt nanoparticles. La–Ni–Pt nanoalloy catalysts of the AB5 type showing improved electrocatalytic properties could, by microwave heat treatment, be alloyed at lower temperatures (400–600 °C) within a shorter time period of 120 s without increase in crystallite size and particle size of the samples [81].

7.3.2
Metal Nanoparticles on Supports

7.3.2.1 Metal Oxide Supports

Metal nanoparticles supported on metal oxides, such as CeO_2, TiO_2, SiO_2, and SnO_2, are widely used for various industrially important reactions. The synergistic effect of the metal nanoparticle and metal oxide support results in an improved performance of the catalyst, in addition to the shape and size control of the metal nanoparticles.

Microwave-assisted wet chemical synthesis methods have been widely exploited for the synthesis of organic and inorganic nanocrystals and can also be extended to the preparation of hybrid materials, such as metal nanoparticles supported on metal oxides.

Highly dispersed, uniformly nanosized platinum particles loaded on metal oxides, acting as efficient photocatalysts, have been synthesized by using the rapid microwave-assisted deposition method [78, 82, 83]. Pt nanoparticles deposited quickly on the TiO_2 surface from an aqueous solution of a platinum compound and subsequent reduction with H_2 affords the platinum metal nanoparticles with a narrow size distribution. Smaller platinum nanoparticles were obtained by the microwave heating method than those obtained by conventional heating techniques. Pt/TiO_2 prepared by the microwave heating method exhibited a specifically high H_2 formation activity in the photocatalytic decomposition of aqueous NH_3 in a nearly stoichiometric 3 : 1 (H_2/N_2) molar ratio under inert conditions [82]. Anumol *et al.* described the thermodynamic and kinetic aspects of reduction of metal salts by EG under microwave heating conditions [83]. They identified the temperatures above which the reduction of the metal salt is thermodynamically favorable, and the temperatures above which the rates of homogeneous nucleation of the metal and heterogeneous nucleation of the metal on supports are favored. Different conditions were delineated that favor the heterogeneous nucleation of the metal on the supports over homogeneous nucleation in the solvent medium, based on the dielectronic loss parameters of the solvent and the support and the metal/solvent and metal/support interfacial energies. These Pt/CeO_2 and Pt/TiO_2 hybrids open up possibilities for rational synthesis of high-activity supported catalysts using a fast microwave-based reduction method. Abdelsayed *et al.* reported the controlled synthesis of alloy nanoparticles on ceria nanoparticles and high activity; the thermal stabilities of selected supported nanoalloys as nanocatalysts for CO oxidation were observed to be in the order CuPd > CuRh > AuPd > AuRh > PtRh > PdRh > AuPt [78].

One-dimensional, vertically oriented, self-organized metal oxide nanotube arrays formed by electrochemical anodization are currently of great interest in photocatalysis. The self-assembled thin-wall nanotube array geometry, uprightly oriented from the substrate, promotes ready separation of photogenerated electron–hole pairs and charge transport. Metal nanoparticles, such as Pt, Pd, and Cu, must be loaded onto the semiconductor(s), acting as electron sinks and active sites. Grimes *et al.* presented an easily implemented microwave-assisted solvothermal method for organic ligand-free, rapid, *in situ* and uniform decoration of 1D long nanotube array walls with Pt nanoparticles [84]. The Pt-nanoparticle/ TiO_2 nanotube composite greatly promotes the photocatalytic conversion of CO_2 and H_2O vapor into methane, attributed to the homogeneous distribution of metal co-catalyst nanoparticles over the TiO_2 nanotube array surface.

Catalysts based on ruthenium are important for processes such as ammonia synthesis/decomposition, CO and NO oxidation, N_2O decomposition, and other industrially important reactions. Okal *et al.* prepared Ru catalyst supported on alumina nanoparticles using a microwave-polyol method [85]. The obtained Ru nanocatalyst was free of chlorine contamination, although prepared also from $RuCl_3$, and the catalyst with the 1.6 nm Ru nanoparticles exhibited slightly higher specific activity for complete oxidation of propane than that with the 6 nm Ru nanoparticles. The superior catalytic performance of the Ru nanoparticles could be correlated with a high metallic dispersion and low particle size. The catalyst possessed good stability; the ruthenium phase was not agglomerated.

Glaspell reported a simple microwave-assisted method for preparing Au and Pd nanoparticle catalysts supported on CeO_2, CuO, and ZnO, and compared their activities for CO oxidation [86]. The remarkable enhancement of the activity is directly correlated with the change in the morphology of the catalyst and the efficient dispersion of the active metal on the support achieved by using capping agents during the microwave-assisted synthesis. The significance of the current method lies mainly in its simplicity, flexibility, and the control of the different factors that determine the activity of the nanoparticle catalysts.

Core-shell Ag–SiO_2 nanoparticles were also synthesized by microwave-assisted Ag nanoparticle formation followed by coating with a SiO_2 shell, which was remarkably reduced by microwave heating (70 °C, 1 s) [87].

7.3.2.2 Carbon Material Supports

7.3.2.2.1 Ordinary Carbon

Owing to its stability and low electrical resistance, carbon is one of the most important matrices. Carbon-supported metal catalysts enhance reaction rates significantly and have a number of interesting applications, including deoxygenation of fatty acids, the Heck reaction, formic acid electro-oxidation, and selective oxidation of organic compounds in both small- and large-scale chemical processes.

The microwave-assisted heating method has received great attention as a promising method for the preparation of monodispersed Pt and Pt–Ru catalysts on carbon supports. The microwave-assisted method is superior in terms of its simplicity and rapidity, compared to the conventional heating methods.

Nickel/carbon nanocomposites have been achieved within a few minutes by a microwave-assisted process, starting from nickel salts and a renewable high-content carbon source, tannin. The process involves a simultaneous carbonization of the precursor as well as the reduction of nickel ions to elemental nickel nanoparticles in an ambient atmosphere without requiring the need for hydrogen or inert gas during the transformation [88].

Bimetallic systems allow tuning of reaction activities via composition ratio and structure tailoring (mixed, core–shell, or alloyed NPs). Such systems show a higher catalytic activity and stability than monometallic counterparts. Au-promoted Pd bimetallic catalysts are very promising in potential industrial applications; they display a higher resistance to deactivation. Tang *et al.* reported the preparation of bimetallic Au–Pd/CS composites of Au:Pd atom ratios varying from 0.4 to 4.6 doping near-surface locations via addition of gold to Pd-doped nanoporous CS using the microwave-assisted process [89].

Pt–Ru alloy – Carbon-supported Pt–Ru alloy nanoparticles have been most widely applied in fuel cells as a result of their high electrocatalytic activity accompanied by high protection ability for carbon monoxide poisoning. The preparation of nanoparticles on a carbon support with suitable particle size and good dispersion remains a challenging task. The main problems regarding the preparation of carbon-supported Pt–M alloys are the poor content of non-noble metals in the alloy. The microwave-assisted polyol process has been introduced in the controlled synthesis of supported Pt–Ru nanoparticles in a few minutes [90–93]. Nanocomposites containing 16 wt.% (or 50 wt.%) total metal and alloy nanoclusters of 3.4 nm (or 5.4 nm) average diameter are formed within only 100 s (or 300 s) of total microwave heating [90]. The particle size, composition and catalytic activity of the Pt–Ru/C catalyst are very sensitive to the pH of the reducing solution; the Pt–Ru/C catalyst prepared at pH 8 exhibited a better performance for the electro-oxidation of methanol than the other samples [91]. The size of the nanoparticles decreased as the pH increased; the Pt–Ru mean particle size also decreased from 3.5 to 1.5 nm with increasing pH [92]. Three types of CNFs with very different surface structures, such as platelet, herringbone, and tubular ones, were used as new carbon supports [93]. The dependence of particle sizes and methanol fuel cell activities on the structures of PtRu/CNFs was in the order: platelet > tubular > herringbone.

Pt–Co alloy – Bimetallic catalysts alloying Pt with transition metals such as Co, Ni, Cr, Co–Cr, and Co–Ni were initially developed for the oxygen reduction reaction (ORR) in phosphoric acid fuel cells but have been found to perform well in proton exchange membrane fuel cells (PEMFCs) with fivefold enhancement in activity. Hwang *et al.* utilized microwave heating to lower the alloying temperature and also for good mixing of the alloying components for Pt–Co/C systems [94]. As compared to the Pt/C catalyst, the bimetallic Pt–Co/C sample exhibited an enhancement factor of 3 in mass activity at 0.95 V toward ORR. The alloying extent of Pt is an important parameter by which one can have control over the fine-tuning of the catalytic activities of bimetallic nanoclusters. This activity enhancement may originate from the favorable electronic effects of a well-mixed alloy underneath a thin Pt-rich skin structure of the Pt–Co bimetallic nanoparticles.

7.3.2.2.2 Graphene

The unique structure of graphene from atomically thin 2D aromatic sheets composed of sp^2-bonded carbon atoms results in exceptionally high values of thermal conductivity, mechanical stiffness, and excellent electrical transport properties. These properties give graphene exceptional promise for applications in the fields of super capacitors, sensors, transistors, actuators, reinforcing agents for polymers, and hydrogels, among others.

Metal-decorated graphene has found various applications, such as high-performance catalysts for chemical reactions, electrochemical sensors and substrates for surface-enhanced Raman spectra. The synthesis of metal-decorated graphene is usually achieved by a two-step procedure: synthesis of graphene followed by subsequent decoration of metal particles onto graphene sheets, either by chemical, physical or electrodeposition methods.

Microwave irradiation has also been used for the synthesis of soluble single wall carbon nanotube derivatives and for the exfoliation of graphite intercalation compounds. The main advantage of microwave irradiation over other conventional heating methods is that the reaction mixture is heated uniformly and rapidly. Microwave irradiation allows the simultaneous reduction of graphite oxide (GO) and a variety of metal salts, resulting in the synthesis of metallic and bimetallic nanoparticles supported on the chemically converted graphene (CCG) sheets.

Vadahanambi and Il-Kwon developed an easy and rapid one-pot technique to synthesize metal-decorated graphene using microwave radiation [95]. Under microwave conditions, the intercalated foaming agent between graphite oxide layers played a key role in the rapid and large expansion of the graphene worm along the thickness direction, and in the reduction process of the graphite oxide. All the electrocatalytic performance in sensing organic glucose and inorganic electro-active compounds showed a remarkable increase in electrochemical performance.

Hassan *et al.* have described a facile, convenient and scalable method assisted by microwave irradiation for the synthesis of CCG sheets, as well as metallic and bimetallic nanoparticles supported on the CCG sheets [96]. The rapid simultaneous reduction of graphene oxide (GO) and a variety of metal salts resulted in the dispersion of metallic nanoparticles supported on the large surface area of the thermally stable 2D graphene sheets.

7.3.2.2.3 Carbon Nanotubes

Besides the active metals, the support is another key factor for a highly efficient catalyst, which requires low cost, good chemical stability, and large surface area, as well as high conductivity. Due to their excellent mechanical, electrical, and structural properties, CNTs have attracted a great deal of interest as a potential support for heterogeneous catalysts. CNTs can also be used as a template for the construction of nanostructured materials. The metal NPs were distributed on CNTs to obtain good dispersion and large exposed surfaces. Not only the electronic structure of the support but also the possible interaction with the support, being unavoidable in chemical preparation routes, exhibited strong stabilizing effects on NPs, which are directly associated with the structure of NPs.

Microwave heating methods have also received a great deal of attention in the preparation of metal NPs supported on CNTs, because microwave irradiation allows the uniform heating of substances, resulting in a more homogeneous nucleation and reduction in the time required for crystallization.

Ru/CNT-based catalysts are critical for fuel cells, cellulose conversion into sorbitol, ammonia synthesis and decomposition, hydrogenation, and selective oxidation. In this regard, Zhang et al. reported the rapid formation of Ru NPs on CNTs by microwave-thermolysis without further thermal treatments [97]. Moreover, current synthesis takes advantage of the interactions between metal atoms and localized double bonds on carbon nanostructures with curved basic structural units and the structure of twinned Ru nanoparticles rearranged into Ru single nanocrystals. Nitrogen-doped CNTs (NCNTs) having strong interactions with the metal nanoparticles also provide a good support for Ru and Ag NPs; the metal particles were uniformly dispersed on the surface [98, 99]. The microwave-assisted method allows the synthesis of NPs with a good dispersion, very narrow size distribution and small size to be achieved using shorter reaction times, reduced energy consumption compared to conventional heating. These results provided new insights into the synthesis and application of metal nanoparticle catalysts.

Pt nanoparticles supported on CNTs (Pt/CNTs) have shown higher oxygen reduction activity and durability than Pt/C. The use of CNTs as an electrocatalytic support has shown potential in improving mass transfer. Compared to Pt/C electrodes, Pt/CNTs electrodes provided a 10% increase in voltage, decreased loss of Pt surface area, expanded utilization of Pt by nearly double, and increased performance of the membrane electrode assembly.

Pt, Ru, and Pt–Ru supported on CNTs have displayed good electrocatalytic behavior in fuel cells. Alloying Pt with other non-noble metals could not only reduce the cost, but also enhance the resistance to CO poisoning and improve the electrocatalytic activity for ethanol oxidation. Bimetallic Pt–Sn is featured for its capability for C–C bond scission and shows excellent ethanol electro-oxidation activity.

Chiang et al. prepared Pt nanoparticles on CNTs that had undergone oxidation using a microwave-assisted polyol method [100]. This treatment primarily introduced –OH and –COOH groups to the CNTs, thereby enhancing the reduction of Pt ionic species, resulting in smaller Pt particles with improved dispersion and attachment properties, and also superior durability to those on pristine CNTs or commercially available Pt/C. These improvements have been associated with the percentage of metallic Pt in the particles. Bimetallic Pt alloy nanoparticles have also been prepared by microwave-assisted heating of a polyol process [101–103]. All the as-synthesized metal nanoparticles were uniform in shape and size and well dispersed on the carbon nanotubes. The average metal diameters were about 3.0–3.1 nm for Pt and 2.8 nm for Pt–Ru on HNO_3-treated CNTs [101]. Pt/NCNTs with particle size around 3 nm present a clear increase over Pt/C in activity for alcohol electro-oxidation, due to the improved support, while the bimetallic Pt–Sn/NCNTs had even higher activity owing to the alloying of Pt with Sn [102].

Multiwall Carbon Nanotubes Multiwall carbon nanotubes (MWNTs) are much cheaper to synthesize than SWNTs and the properties are dependent on their chirality, diameters, and structural defects. Support materials play an important role in the performance of electrocatalysts for fuel cells. Multi-walled carbon nanotubes (MWCNTs) have been widely applied as a promising support material in many areas of science and technology due to their specific characteristics, such as large surface area to volume, good adsorption, unique structure, and high thermal and chemical stability.

Deposition of nanoparticles onto MWCNTs assisted by microwave irradiation is a very useful and powerful method of directly controlling the size, morphology, structure, and loading of nanoparticles.

Sakthivel et al. deposited Pt electrocatalysts on MWCNTs with high loading by using a microwave-assisted polyol reduction method and used these for direct methanol fuel cells (DMFC) [104]. Uniform and narrow size distribution of highly dispersed Pt nanoparticles could be achieved by adjusting the weight ratio of surfactant to Pt precursor allowing Pt loadings of up to 60 wt%. The heating time and temperature for the EG oxidation were found to be the key factors for the deposition. The DMFC performance of the surfactant-stabilized cathode catalyst was three times higher than for a commercial catalyst used for comparison, and two times higher than for an unstabilized CNT supported catalyst.

Recently, other efforts have been devoted to using different Pt alloys to reduce the device (sensor/fuel cells) production cost and chances of CO poisoning on the device surface. Pt–Ni alloyed nanoparticles with approximate diameter 3 nm and with different Pt to Ni molar ratios were deposited on the nitrogen-doped multiwall carbon nanotubes (NCNTs) synthesized at 650° from CVD by a microwave-polyol method [105]. Nanoparticles were deposited homogeneously and immobilized at active nitrogen sites. The Pt/NCNTs exhibited the fastest response and recovery and the best sensitivity for conductimetric hydrogen gas sensors.

Ramulifho et al. prepared metal nanoparticles (Pd, Ni, Sn) supported on sulfonated multi-walled carbon nanotubes (SF-MWCNTs) using a very rapid microwave-assisted solvothermal strategy [106]. The mixed Pd-based catalysts obtained by ultrasonic-mixing of the individual Pd and Ni or Sn gave better electrocatalytic activity toward EtOH oxidation in alkaline media than their alloy nanoparticles, while the SF-MWCNT platform gave better electrocatalytic performance compared to the unsulfonated and commercial Vulcan carbons.

Co–Ni alloys have been widely used for decoration, corrosion resistance and magnetic recording devices, and catalysts, among others. The composition and size of the nanoparticles are considered as the key characteristics affecting their magnetic properties. $Co_{1-x}Ni_x$ alloy nanoparticles of quasi-spherical and fcc structure attached on the surface of MWCNTs were prepared by microwave irradiation [107, 108]. The compositon and size of the alloy nanoparticles were controlled by adjusting the atomic ratios of metal Co to Ni in the mixed acetate solution, the microwave power and microwave irradiation time. The coercivity and the saturation magnetization of the $Co_{1-x}Ni_x$ alloy nanoparticles increased with increasing Co concentration from $x = 0.8$ to 0.5, and decreased when Co concentration was increased from $x = 0.5$ to 0.2.

The above results provide some important insights into the electrochemical response of microwave-synthesized bimetallic catalysts for potential applications in direct fuel cell technology.

7.3.2.3 Other Supports

7.3.2.3.1 Single Support

Polyamides (PAs) are among the most common polymers in daily life; their mechanical and thermal properties make their use widespread in several fields, from textile and fabric industries to technological applications. Their noteworthy workability makes them interesting stabilizers of MNPs.

Metal organic frameworks (MOFs) have provided a new class of highly porous crystalline materials with well-defined cavities or channels that have a wide range of applications, including gas storage, separation, sensing, and catalysis.

Alloyed Pt–Ru nanoparticles are the best catalyst materials for polymeric fuel cells. Nanoparticle catalysts dispersed in carbon black have the drawback that part of the active sites remains inaccessible to the methanol, thus reducing the utilization of the active surface area. Platinum-based nanoparticles dispersed in a polymer matrix not only provide access to a larger number of catalytic sites, but also offer the possibility of spent catalyst recovery. Polypyrrole-di(2-ethylhexyl) sulfosuccinate (or PPyDEHS), with good electronic conductivity and solubility in alcohols, provides an excellent protective power when compared to other homopolymers, which offers the possibility of tailoring specific applications and fine-tuning the overall properties of Pt nanoparticles.

Concina *et al.* synthesized PA6-stabilized metal NPs (Pd, Pt, Ag, and Ru) by using a microwave-sustained polyol process, which resulted in a very fast and simple synthetic procedure and allowed complete dispersion of the MNPs into the polymer structure [109]. Bensebaa *et al.* stabilized Pt–Ru nanoparticles within the conductive polymer matrix, PPyDEHS using a scalable and quick two-step microwave heating process [110]. Monodisperse alloyed nanoparticles with an average size of 2.8 nm were obtained; these novel materials may be potential candidates for direct MeOH fuel cell application.

Supported nanoparticles on a porous substrate have also been the subject of many investigations due to their importance and interesting applications in catalysis [111, 112]. Using the highly porous coordination polymer MIL-101 as the support, small (2–3 nm) Pd, Cu, and Pd–Cu nanoparticles were incorporated in the pores, and larger particles (4–6 nm) were supported on the crystal surface by simultaneously activating the MIL-101 pores and rapidly reducing chemically the metal precursors by microwave irradiation in the presence of a reducing agent [111]. Catalytic activity for CO oxidation of Pd nanocatalysts supported by highly porous MIL-101 polymer was significantly higher than any other reported metal clusters supported by metal–organic frameworks; this was attributed to the small metal nanoparticles embedded in the MIL crystal pores. Luque *et al.* prepared self-assembled silica nanotubes coordinated with Cu, Fe, Ni nanoparticles by a simple, low-energy intensive and benign protocol under mild conditions (<100 °C) using microwave irradiation and conventional heating [112]. These nanotubes were found to contain metal nanoparticles that are catalytically active in the microwave-assisted homocoupling of terminal alkynes.

7.3.2.3.2 Mixed Support

Pd-based catalysts were found to show superior activity for formic acid oxidation reaction in the direct formic acid fuel cells (DFAFCs) compared to Pt catalysts. To further improve the catalytic activity and stability of the catalyst in DFAFCs, the carbon support materials were modified with semiconducting oxides. Al_2O_3 appears to be suitable because of its unique physical and chemical properties and high stability in acidic and alkaline solutions. Qu et al. prepared a Pd/Al_2O_3–C catalyst using α-Al_2O_3 and Vulcan XC-72 carbon black as a mixed support for the DFAFC by a microwave-assisted polyol process for the first time [113]. The Pd/Al_2O_3–C catalyst with a mass ratio of α-Al_2O_3 to XC-72 carbon of 1:2 presents the narrowest particle size distribution and exhibits the best activity and stability for formic acid electro-oxidation. The order of activity of the catalysts was: α-Al_2O_3 and Vulcan XC-72 carbon black as a mixed support > carbon black as the support > α-Al_2O_3 as the support, owing to α-Al_2O_3 having poor electrical conductivity.

Pt–Ru alloy nanoparticles supported on multiwalled carbon nanotubes or graphene mixed with other functional compounds were synthesized by the microwave-assisted polyol process [114–116]. Han et al. synthesized polyoxometallate-stabilized ($H_3PMo_{12}O_{40}$, PMo_{12}) Pt–Ru alloy nanoparticles supported on multiwalled carbon nanotubes (Pt–Ru-PMo_{12}-MWNTs) by a microwave-assisted polyol process [114]. Polyoxometallates are known to form self-assembled monolayers on common solid electrodes. The agglomeration of metal nanoparticles and the distribution of CNTs could be prevented effectively by coating the CNTs and metals with a polyoxometallate monolayer which may bring much higher electrocatalytic activity, higher cycle stability, and better tolerance to poisoning species in methanol oxidation. Additionally, microwave reaction time, microwave reaction power, and pH of the reaction solution had effects on the electrocatalytic properties of Pt–Ru-PMo_{12}-MWNTs catalysts. Pt–Ru nanoparticles as electrocatalyst for the fuel cell were deposited onto polypyrrole-MWNT composite under microwave irradiation [115]. Wang et al. developed an effective microwave-assisted one-pot reaction strategy for hybrid metal (Pt–Ru)/metal oxide (SnO_2) nanoparticles on graphene nanocomposites with well-dispersed Pt–Ru nanoparticles [116]. Ethylene glycol acts as solvent and reducing agent for the reduction of Pt–Ru nanoparticles from their precursors and reduction of graphene from graphene oxide; it also gave SnO_2 facilitated by the presence of a small amount of H_2O. SnO_2/graphene hybrid composites have a much higher supercapacitance than pure graphene and Pt–Ru/graphene showed much better electrocatalytic activity for MeOH oxidation compared to the comercial Pt–Ru/C electrocatalysts.

7.4
An Example of Mass Production Oriented to Application

An important question should be answered at the end of this chapter, that is, "Can a chemical process using microwaves be scaled up?". The authors' answer to this question is "Yes, but it is necessary to design a suitable reactor and surrounding apparatus."

The readers of this chapter will be interested in a new material comprised of Ni nanoparticles which has been developed for use in multilayer ceramics capacitors (MLCC, see Figure 7.9). The next generation of MLCC need nickel nanoparticles for use in the inner electrodes in order to increase the charge storage.

Yamauchi *et al.* succeeded in a new method for synthesizing Ni nanoparticles having particle size 500–100 nm, suitable to use in MLCC, under microwave irradiation (Scheme 7.1) [117]. The precursor of nickel is nickel formate which is coordinated with an organic amine. The coordinating organic amine acts as a reducing reagent at the reduction composition of the compound. The particle sizes are controlled by changing the added organic amines as shown in Figure 7.10.

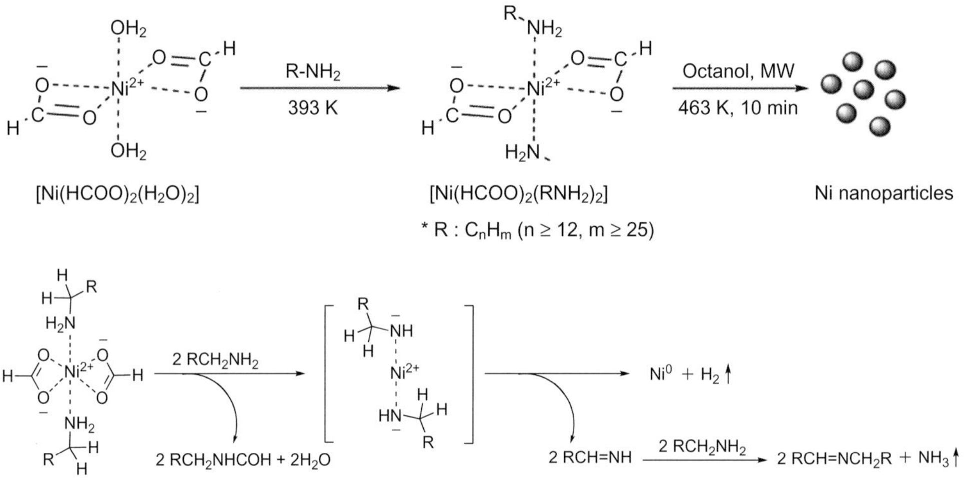

Scheme 7.1 The formation of Ni nanoparticles through the reduction decomposition of nickel formate. Coordination of organic amine leads to size-controlled Ni nanoparticles.

Nippon Steel Chemical Co., Ltd. built a microwave-driven reactor for the mass production of Ni nanoparticles for a commercial process under collaboration with an Osaka University group [118]. The Ni nanoparticles prepared by the process possess an average size of 51 nm with a narrow distribution. This size control can be achieved by applying microwaves to the preparation process.

7.4 *An Example of Mass Production Oriented to Application* | 179

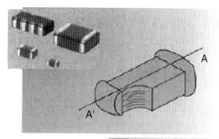

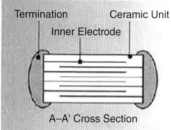

Figure 7.9 Multilayer ceramics capacitors (MLCC). Ni nanoparticles will be used in the inner electrodes.

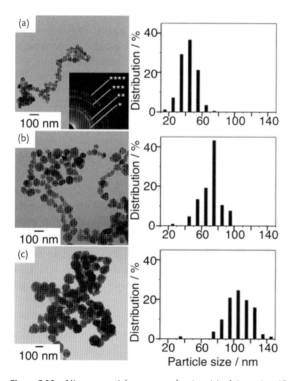

Figure 7.10 Ni nanoparticles prepared using (a) oleic amine ($C_{12}H_{25}NH_2$), (b) myristile amine ($C_{14}H_{29}NH_2$), and (c) laurile amine ($C_{14}H_{27}NH_2$). The particle sizes are controlled by the kinds of coordinating amines.

7.5 Conclusion

The synthesis of nanomaterials is suited to application of microwaves. We can utilize rapid and homogeneous heating, and selective heating by separately applying the electric and magnetic fields. These advantageous features of the microwaves can be usefully employed in other chemical systems containing solid particles, and interfaces. Solid catalysis will hopefully be one application. The surface of solid catalysts is a reaction field, which can be controlled by microwave irradiation. Microwave techniques could provide a broad scope for changing some of the established fields of chemistry.

References

1 Anastas, P.T., and Warner, J.C. (1998) *Green Chemistry: Theory and Practice*, Oxford University Press, New York, p. 30.
2 Wada, Y., Kuramoto, H., Sakata, T., Mori, H., Sumida, T., Kitamura, T., and Yanagida, S. (1999) *Chem. Lett.*, (7), 607–608.
3 Yamamoto, T., Wada, Y., Sakata, T., Mori, H., Goto, M., Hibino, S., and Yanagida, S. (2004) *Chem. Lett.*, 158–159.
4 Yanagida, S., and Wada, Y. (2002) Method for Producing Ultra Microparticles and Ultra Microparticles, US Patent 6387494 B1, Osaka University: USA, May 14; (2001) Japan Patent 3005683.
5 Tsukahara, Y., Nakamura, T., Kobayashi, T., and Wada, Y. (2006) *Chem. Lett.*, **35** (12), 1396–1397.
6 Nakamura, T., Tsukahara, Y., Yamauchi, T., Sakata, T., Mori, H., and Wada, Y. (2007) *Chem. Lett.*, **36** (1), 154–155.
7 Yamauchi, T., Tsukahara, Y., Yamada, K., Sakata, T., and Wada, Y. (2011) *Chem. Mater.*, **23** (1), 75–84.
8 Chen, Z., Della Pina, C., Falletta, E., Lo Faro, M., Pasta, M., Rossi, M., and Santo, N. (2008) *J. Catal.*, **259** (1), 1–4.
9 Yan, L., Cai, Y., Zheng, B., Yuan, H., Guo, Y., Xiao, D., and Choi, M.M.F. (2012) *J. Mater. Chem.*, **22** (3), 1000–1005.
10 Lee, Y., and Geckeler, K.E. (2012) *J. Biomed. Mater. Res. A*, **100A** (4), 848–855.
11 Babu, P.J., Das, R.K., Kumar, A., and Bora, U. (2011) *Int. J. Green Nanotechnol.*, **3** (1), 13–21.
12 Baruwati, B., and Varma, R.S. (2009) *ChemSusChem*, **2** (11), 1041–1044.
13 Abargues, R., Gradess, R., Canet-Ferrer, J., Abderrafi, K., Valdes, J.L., and Martinez-Pastor, J. (2009) *New J. Chem.*, **33** (4), 913–917.
14 Patel, K., Kapoor, S., Dave, D.P., and Mukherjee, T. (2006) *Res. Chem. Intermed.*, **32** (2), 103–113.
15 Tsuji, M., Hashimoto, M., Nishizawa, Y., Kubokawa, M., and Tsuji, T. (2005) *Chemistry–Eur. J.*, **11** (2), 440–452.
16 Kamarudin, K.S.N., Mohamad, M.F., Fathilah, N.N.F.N.M., and Mohamed, M.S. (2010) *J. Appl. Sci.*, **10** (24), 3374–3378.
17 Kundu, S., Peng, L., and Liang, H. (2008) *Inorg. Chem.*, **47** (14), 6344–6352.
18 Chen, Z., Gao, Q., Wu, C., Ruan, M., and Shi, J. (2004) *Chem. Commun.*, (17), 1998–1999.
19 Chen, Z., Gao, Q., Gao, D., Wei, Q., and Ruan, M. (2006) *Mater. Lett.*, **60** (15), 1816–1822.
20 Silva, A.R., and Unali, G. (2011) *Nanotechnology*, **22** (31), 315605.
21 Valodkar, M., Modi, S., Pal, A., and Thakore, S. (2011) *Mater. Res. Bull.*, **46** (3), 384–389.

22 Saifuddin, N., Wong, C.W., and Nur Yasumira, A.A. (2009) *E-J. Chem.*, **6** (1), 61–70.

23 Dong, Z., Richardson, D., Pelham, C., and Islam, M.R. (2008) *Chem. Educ.*, **13** (4), 240–243.

24 Hu, B., Wang, S.-B., Wang, K., Zhang, M., and Yu, S.-H. (2008) *J. Phys. Chem. C*, **112** (30), 11169–11174.

25 Komarneni, S., Li, D., Newalkar, B., Katsuki, H., and Bhalla, A.S. (2002) *Langmuir*, **18** (15), 5959–5962.

26 Komarneni, S., Katsuki, H., Li, D., and Bhalla, A.S. (2004) *J. Phys. Condens. Matter*, **16** (14), S1305–S1312.

27 Li, D., and Komarneni, S. (2010) *J. Nanosci. Nanotechnol.*, **10** (12), 8035–8042.

28 Grace, A.N., and Pandian, K. (2007) *Mater. Chem. Phys.*, **104** (1), 191–198.

29 Chen, J., Li, Z., Chao, D., Zhang, W., and Wang, C. (2008) *Mater. Lett.*, **62** (4–5), 692–694.

30 Kundu, S., Wang, K., and Liang, H. (2009) *J. Phys. Chem. C*, **113** (1), 134–141.

31 Zhu, J.-F., and Zhu, Y.-J. (2006) *J. Phys. Chem. B*, **110** (17), 8593–8597.

32 Nishioka, M., Miyakawa, M., Kataoka, H., Koda, H., Sato, K., and Suzuki, T.M. (2001) *Nanoscale*, **3** (6), 2621–2626.

33 Wang, H., Wang, L., Nemoto, Y., Suzuki, N., and Yamauchi, Y. (2010) *J. Nanosci. Nanotechnol.*, **10** (10), 6489–6494.

34 Ling, X.Y., Liu, Z., and Lee, J.Y. (2005) *J. Metastable Nanocrystal. Mater.*, **23**, 199–202.

35 He, B., Chen, Y., Liu, H., and Liu, Y. (2005) *J. Nanosci. Nanotechnol.*, **5** (2), 266–270.

36 Patel, K., Kapoor, S., Dave, D.P., and Mukherjee, T. (2005) *J. Chem. Sci. (Bangalore, India)*, **117** (4), 311–316.

37 Zhang, W., Wang, Q., Qin, F., Zhou, H., Lu, Z., and Chen, R. (2011) *J. Nanosci. Nanotechnol.*, **11** (9), 7794–7801.

38 Zawadzki, M. (2007) *J. Alloys Compd.*, **439** (1–2), 312–320.

39 Zawadzki, M., and Okal, J. (2008) *Mater. Res. Bull.*, **43** (11), 3111–3121.

40 Vollmer, C., Redel, E., Abu-Shandi, K., Thomann, R., Manyar, H., Hardacre, C., and Janiak, C. (2010) *Chemistry – Eur. J.*, **16** (12), 3849–3858.

41 Chen, Z., Della Pina, C., Falletta, E., and Rossi, M. (2009) *J. Catal.*, **267** (2), 93–96.

42 Blosi, M., Albonetti, S., Dondi, M., Martelli, C., and Baldi, G. (2011) *J. Nanopart. Res.*, **13** (1), 127–138.

43 Nakamura, T., Tsukahara, Y., Sakata, T., Mori, H., Kanbe, Y., Bessho, H., and Wada, Y. (2007) *Bull. Chem. Soc. Jpn.*, **80** (1), 224–232.

44 Zhu, H., Zhang, C., and Yin, Y. (2005) *Nanotechnology*, **16** (12), 3079–3083.

45 Li, D., and Komarneni, S. (2006) *J. Am. Ceram. Soc.*, **89** (5), 1510–1517.

46 Marquardt, D., Xie, Z., Taubert, A., Thomann, R., and Janiak, C. (2011) *Dalton Trans.*, **40** (33), 8290–8293.

47 Liu, F.-K., Huang, P.-W., Chang, Y.-C., Ko, C.-J., Ko, F.-H., and Chu, T.-C. (2005) *J. Cryst. Growth*, **273** (3–4), 439–445.

48 Tsuji, M., Matsumoto, K., Jiang, P., Matsuo, R., Tang, X.-L., and Kamarudin, K.S.N. (2008) *Colloids Surf. A Physicochem. Eng. Aspects*, **316** (1–3), 266–277.

49 Nandikonda, S., and Davis, E.W. (2011) *ISRN Nanotechnol.*, **104086**, 7.

50 Kundu, S., and Liang, H. (2008) *Langmuir*, **24** (17), 9668–9674.

51 Tsuji, M., Hashimoto, M., Nishizawa, Y., and Tsuji, T. (2004) *Mater. Lett.*, **58** (17–18), 2326–2330.

52 Liu, F.-K., Chang, Y.-C., Ko, F.-H., and Chu, T.-C. (2003) *Mater. Lett.*, **58** (3–4), 373–377.

53 Zhu, Y.-J., and Hu, X.-L. (2003) *Chem. Lett.*, **32** (12), 1140–1141.

54 Li, D., and Komarneni, S. (2006) *Z. Naturforsch. B Chem. Sci.*, **61** (12), 1566–1572.

55 Guo, Y., Wang, G., Wang, Y., Huang, Y., and Wang, F. (2012) *Mater. Res. Bull.*, **47** (1), 6–11.

56 Kundu, S., and Liang, H. (2010) *J. Nanosci. Nanotechnol.*, **10** (2), 746–754.

57 Pei, M.S., Li, J., Wang, L.Y., Wu, Z.Y., Wu, X.Z., and Tao, X.T. (2010) *Mater. Res. Innov.*, **14** (2), 127–132.

58 Liang, W., and Harris, A.T. (2011) *Mater. Lett.*, **65** (14), 2307–2310.

59 Mallikarjuna, N.N., and Varma, R.S. (2007) *Cryst. Growth Des.*, **7** (4), 686–690.
60 Xia, L., Wang, H., Wang, J., Gong, K., Jia, Y., Zhang, H., and Sun, M. (2008) *J. Chem. Phys.*, **129** (13), 134703/1–134703/7.
61 Pal, A., Shah, S., Chakraborty, D., and Devi, S. (2008) *Aust. J. Chem.*, **61** (11), 833–836.
62 Donegan, K.P., Godsell, J.F., Tobin, J.M., O'Byrne, J.P., Otway, D.J., Morris, M.A., Roy, S., and Holmes, J.D. (2011) *CrystEngComm*, **13** (6), 2023–2028.
63 Shojaee, K., Edrissi, M., and Izadi, H. (2010) *J. Nanopart. Res.*, **12** (4), 1439–1447.
64 Chopra, N., Claypoole, L., and Bachas, L.G. (2010) *J. Nanopart. Res.*, **12** (8), 2883–2893.
65 Guo, Q., Miao, J.-J., Geng, J., and Zhu, J.-J. (2010) *Yingyong Huaxue*, **27** (12), 1438–1443.
66 Tsuji, M., Hikino, S., Matsunaga, M., Sano, Y., Hashizume, T., and Kawazumi, H. (2010) *Mater. Lett.*, **64** (16), 1793–1797.
67 Tsuji, M., Miyamae, N., Lim, S., Kimura, K., Zhang, X., Hikino, S., and Nishio, M. (2006) *Cryst. Growth Des.*, **6** (8), 1801–1807.
68 Tsuji, M., Miyamae, N., Matsumoto, K., Hikino, S., and Tsuji, T. (2005) *Chem. Lett.*, **34** (11), 1518–1519.
69 Ling, S.-M., Shen, W.-W., and Long, J.-Q. (2003) *Jingxi Huagong*, **20** (4), 201–204.
70 Liu, F.-K., Huang, P.-W., Chang, Y.-C., Ko, F.-H., and Chu, T.-C. (2005) *Langmuir*, **21** (6), 2519–2525.
71 Ling, S., Shen, W., and Long, J. (2003) *Huaxue Shijie*, **44** (12), 622–625.
72 Harpeness, R., and Gedanken, A. (2004) *Langmuir*, **20** (8), 3431–3434.
73 Belousov, O.V., Belousova, N.V., Sirotina, A.V., Solovyov, L.A., Zhyzhaev, A.M., Zharkov, S.M., and Mikhlin, Y.L. (2011) *Langmuir*, **27** (18), 11697–11703.
74 Gopalan, A., Ragupathy, D., Kim, H.-T., Manesh, K.M., and Lee, K.-P. (2009) *Spectrochim. Acta A Mol. Biomol. Spectrosc.*, **74A** (3), 678–684.
75 Wu, W., Njoki, P.N., Han, H., Zhao, H., Schiff, E.A., Lutz, P.S., Solomon, L., Matthews, S., and Maye, M.M. (2011) *J. Phys. Chem. C*, **115** (20), 9933–9942.
76 Jacob, D.S., Genish, I., Klein, L., and Gedanken, A. (2006) *J. Phys. Chem. B*, **110** (36), 17711–17714.
77 Njoki, P.N., Solomon, L.V., Wu, W., Alam, R., and Maye, M.M. (2011) *Chem. Commun.*, **47** (36), 10079–10081.
78 Abdelsayed, V., Aljarash, A., El-Shall, M.S., Al Othman, Z.A., and Alghamdi, A.H. (2009) *Chem. Mater.*, **21** (13), 2825–2834.
79 Pal, A., Shah, S., and Devi, S. (2007) *Colloids Surf. A Physicochem. Eng. Aspects*, **302** (1–3), 51–57.
80 Nguyen, H.L., Howard, L.E.M., Giblin, S.R., Tanner, B.K., Terry, I., Hughes, A.K., Ross, I.M., Serres, A., Buerckstuemmer, H., and Evans, J.S.O. (2005) *J. Mater. Chem.*, **15** (48), 5136–5143.
81 Yang, S.-T., Yang, W.-G., Yin, Y.-H., Yang, J.-X., and Yue, H.-Y. (2006) *Cailiao Rechuli Xuebao*, **27** (5), 22–25.
82 Fuku, K., Kamegawa, T., Mori, K., and Yamashita, H. (2012) *Chemistry–Asian J.*, **7** (6), 1366–1371.
83 Anumol, E.A., Kundu, P., Deshpande, P.A., Madras, G., and Ravishankar, N. (2011) *ACS Nano*, **5** (10), 8049–8061.
84 Feng, X., Sloppy, J.D., LaTempa, T.J., Paulose, M., Komarneni, S., Bao, N., and Grimes, C.A. (2011) *J. Mater. Chem.*, **21** (35), 13429–13433.
85 Okal, J., Zawadzki, M., and Tylus, W. (2011) *Appl. Catal. B Environ.*, **101** (3–4), 548–559.
86 Glaspell, G., Fuoco, L., and El-Shall, M.S. (2005) *J. Phys. Chem. B*, **109** (37), 17350–17355.
87 Nishioka, M., Miyakawa, M., Kataoka, H., Koda, H., Sato, K., and Suzuki, T.M. (2011) *Chem. Lett.*, **40** (10), 1204–1206.
88 Gunawan, G., Bourdo, S., Saini, V., Biris, A.S., and Viswanathan, T. (2011) *J. Wood Chem. Technol.*, **31** (4), 345–356.
89 Tang, S., Vongehr, S., He, G., Chen, L., and Meng, X. (2012) *J. Colloid Interface Sci.*, **375** (1), 125–133.
90 Boxall, D.L., Deluga, G.A., Kenik, E.A., King, W.D., and Lukehart, C.M. (2001) *Chem. Mater.*, **13** (3), 891–900.

91 Chu, Y.-Y., Wang, Z.-B., Jiang, Z.-Z., Gu, D.-M., and Yin, G.-P. (2010) *Fuel Cells*, **10** (6), 914–919.
92 Liang, Y., and Liao, D.-W. (2008) *Wuli Huaxue Xuebao*, **24** (2), 317–322.
93 Tsuji, M., Kubokawa, M., Yano, R., Miyamae, N., Tsuji, T., Jun, M.-S., Hong, S., Lim, S., Yoon, S.-H., and Mochida, I. (2007) *Langmuir*, **23** (2), 387–390.
94 Hwang, B.J., Kumar, S.M.S., Chen, C.-H.M., Cheng, M.-Y., Liu, D.-G., and Lee, J.-F. (2007) *J. Phys. Chem. C*, **111** (42), 15267–15276.
95 Vadahanambi, S., Jung, J.-H., and Oh, I.-K. (2011) *Carbon*, **49** (13), 4449–4457.
96 Hassan, H.M.A., Abdelsayed, V., Khder, A.R.S., AbouZeid, K.M., Terner, J., El-Shall, M.S., Al-Resayes, S.I., and El-Azhary, A.A. (2009) *J. Mater. Chem.*, **19** (23), 3832–3837.
97 Zhang, B., Ni, X., Zhang, W., Shao, L., Zhang, Q., Girgsdies, F., Liang, C., Schloegl, R., and Su, D.S. (2011) *Chem. Commun.*, **47** (38), 10716–10718.
98 Maben, L.F., Ray, S.S., and Coville, N.J. (2011) *Ceram. Eng. Sci. Proc.*, **32** (7), 33–42.
99 Mphahlele, K., Ray, S.S., Onyango, M.S., and Mhlanga, S.D. (2011) *Ceram. Eng. Sci. Proc.*, **32** (7), 113–124.
100 Chiang, Y.-C., and Ciou, J.-R. (2011) *Int. J. Hydrogen Energy*, **36** (11), 6826–6831.
101 Chen, W.-X., Lee, J.Y., and Liu, Z. (2004) *Mater. Lett.*, **58** (25), 3166–3169.
102 Wang, X., Xue, H., Yang, L., Wang, H., Zang, P., Qin, X., Wang, Y., Ma, Y., Wu, Q., and Hu, Z. (2011) *Nanotechnology*, **22** (39), 395401/1–395401/6.
103 Han, X.-F., Chen, W.-X., Zhao, J., and Nie, Q.-L. (2005) *Zhejiang Daxue Xuebao, Gongxueban*, **39** (12), 1871–1874.
104 Sakthivel, M., Schlange, A., Kunz, U., and Turek, T. (2010) *J. Power Sources*, **195** (20), 7083–7089.
105 Sadek, A.Z., Zhang, C., Hu, Z., Partridge, J.G., McCulloch, D.G., Wlodarski, W., and Kalantar-Zadeh, K. (2010) *J. Phys. Chem. C*, **114** (1), 238–242.
106 Ramulifho, T., Ozoemena, K.I., Modibedi, R.M., Jafta, C.J., and Mathe, M.K. (2012) *Electrochim. Acta*, **59**, 310–320.
107 Wu, H., Cao, P., Zhang, N., Mao, L., and Li, M. (2012) *Mater. Res. Bull.*, **47** (1), 1–5.
108 Wu, H., Cao, P., Li, W., Ni, N., Zhu, L., and Zhang, X. (2010) *J. Alloys Compd.*, **509** (4), 1261–1265.
109 Concina, I., and Zecca, M. (2011) *J. Nanopart. Res.*, **13** (3), 1289–1300.
110 Bensebaa, F., Farah, A.A., Wang, D., Bock, C., Du, X., Kung, J., and Le Page, Y. (2005) *J. Phys. Chem. B*, **109** (32), 15339–15344.
111 El-Shall, M.S., Abdelsayed, V., Khder, A.R.S., Hassan, H.M.A., El-Kaderi, H.M., and Reich, T.E. (2009) *J. Mater. Chem.*, **19** (41), 7625–7631.
112 Gonzalez-Arellano, C., Balu, A.M., Luque, R., and MacQuarrie, D.J. (2010) *Green Chem.*, **12** (11), 1995–2002.
113 Qu, W.-L., Wang, Z.-B., Jiang, Z.-Z., Gu, D.-M., and Yin, G.-P. (2012) *RSC Adv.*, **2** (1), 344–350.
114 Han, D.M., Guo, Z.P., Zhao, Z.W., Zeng, R., Meng, Y.Z., Shu, D., and Liu, H.K. (2008) *J. Power Sources*, **184** (2), 361–369.
115 Bae, H.-B., Oh, S.-H., Woo, J.-C., and Choi, S.-H. (2010) *J. Nanosci. Nanotechnol.*, **10** (10), 6901–6906.
116 Wang, S., Jiang, S.P., and Wang, X. (2011) *Electrochim. Acta*, **56** (9), 3338–3344.
117 Yamauchi, T., Tsukahara, Y., Sakamoto, T., Kono, T., Yasuda, M., Baba, A., and Wada, Y. (2009) *Bull. Chem. Soc. Jpn.*, **82**, 1044–1051.
118 Kono, T. (2011) KurosakiHarima Technical Report, pp. 9–12, No. 159.

8
Microwave-Assisted Nonaqueous Routes to Metal Oxide Nanoparticles and Nanostructures

Markus Niederberger

8.1
Introduction

Athough the first report on the use of microwave heating was published for the preparation of an inorganic material, namely titania xerogel spheres [1], microwave chemistry became more popular in organic rather than in inorganic synthesis [2, 3]. However, microwave-assisted synthesis routes to inorganic materials, especially nanomaterials, are well established. The preparation of inorganic nanoparticles using solution routes in combination with microwave heating experienced a major boost in the last few years [4, 5]. The immense reduction in the synthesis time is one of the main advantages contributing to the growing popularity of this non-conventional heating technique. Synthesis times of the order of just a few minutes are not unusual [6]. From a more scientific point of view, the volumetric heating offered by microwave irradiation represents another major benefit. Microwave irradiation produces efficient internal heating, increasing the temperature of the whole reaction volume simultaneously and homogeneously (although in relatively small volumes due to the restricted penetration depth). Considering the pronounced sensitivity of all the characteristics of inorganic nanoparticles (e.g., crystallinity, size and size distribution, shape, defect concentrations and surface chemistry) on the synthesis conditions, and especially on the temperature, it is obvious that these parameters have to be particularly carefully controlled. Slight temperature inhomogeneities in the reaction solution disturb the fragile balance of nucleation and growth, and thus lower the quality of the final nanoparticles.

Almost 30 years after the pioneering work of Komarneni and Roy on titania xerogels [1], the library of inorganic nanomaterials produced by microwave-assisted solution routes has grown dramatically, including (but not restricted to) metals and alloys, metal oxides and hydroxides, metal chalcogenides, metal phosphides/phosphates and metal halides in different sizes, shapes and combinations [4, 5]. Similarly broad as the nanoparticle compositions are the reaction systems

Microwaves in Nanoparticle Synthesis, First Edition. Edited by Satoshi Horikoshi and Nick Serpone.
© 2013 Wiley-VCH Verlag GmbH & Co. KGaA. Published 2013 by Wiley-VCH Verlag GmbH & Co. KGaA.

(i.e., precursors, solvents and possibly additives) and the synthesis techniques used for the preparation of these materials. This is a remarkable observation, because it underlines the flexibility of the microwave approach. It can basically be combined with the full spectrum of available liquid-phase syntheses, ranging from coprecipitation, hydrolytic and non-hydrolytic sol–gel processes, hydrothermal and solvothermal processing, to sonochemistry, template and biomimetic approaches. However, in this chapter the discussion will be strongly focused on the microwave-assisted nonaqueous sol–gel routes to metal oxide nanoparticles.

The nonaqueous sol-gel process can be defined as the chemical conversion of a molecular precursor dissolved in an organic solvent into a metal oxide framework, typically without any addition of water. However, in some cases nonaqueous processes are not completely anhydrous. Traces of water can initially be present in the reaction mixture, either intentionally added, or introduced with hydrated precursors or solvents, rendering these systems hydrolytic. Here, nonaqueous processes are only regarded as non-hydrolytic, if no hydroxy-containing intermediates are formed at any stage of the reaction.

A unique feature of nonaqueous processes is that organic reactions provide the "monomers" for the nucleation of the inorganic nanoparticles. The strong relation between these organic reaction pathways and the formation of inorganic particles not only offers a versatile tool to control the morphological characteristics of the particles by varying the organic species in the reaction mixture, but also makes it possible to study particle formation mechanisms through monitoring the organic side reactions [7].

The fundamental role of organic reactions in nonaqueous sol–gel routes and the fact that organic reactions are very sensitive to microwave irradiation make it obvious to combine both techniques, potentially providing a way to control the formation of inorganic nanoparticles by affecting (e.g., accelerating) the organic reaction pathways by microwave irradiation. The possibility to subtly control the irradiation time, the temperature and the output power makes microwave-assisted nonaqueous chemistry a powerful tool for the efficient synthesis of a large variety of metal oxide nanoparticles.

8.2
Nonaqueous Sol–Gel Chemistry

The sol–gel process can roughly be defined as the conversion of a precursor solution into a metal oxide by specific chemical reactions. In aqueous sol–gel chemistry metal alkoxides constitute the most widely used class of precursors. Upon hydrolysis and condensation, the metal alkoxide is transformed into a sol (dispersions of colloidal particles in a liquid), which reacts further to a gel, an interconnected, rigid and porous inorganic network enclosing a continuous liquid phase [8]. In aqueous sol–gel processes, the oxygen for the formation of the oxidic compound is supplied by the water molecules. In nonaqueous systems, where basically no water is present, the oxygen required for the formation of the metal oxide nanoparticles is provided either by the organic solvent (ethers, alcohols, ketones or aldehydes) or by the organic constituent of the precursor (alkoxides, acetates or acetylacetonates) [7, 9, 10]. The most frequently found condensation steps for the formation of a metal–oxygen–metal bond as the basic structural unit in metal oxides are summarized in Scheme 8.1 [11, 12]:

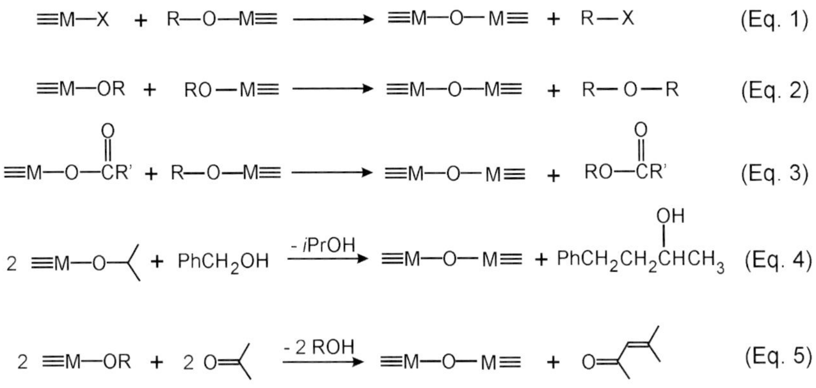

$$\equiv M-X + R-O-M\equiv \longrightarrow \equiv M-O-M\equiv + R-X \quad \text{(Eq. 1)}$$

$$\equiv M-OR + RO-M\equiv \longrightarrow \equiv M-O-M\equiv + R-O-R \quad \text{(Eq. 2)}$$

$$\equiv M-O-\underset{\underset{O}{\parallel}}{C}R' + R-O-M\equiv \longrightarrow \equiv M-O-M\equiv + RO-\underset{\underset{O}{\parallel}}{C}R' \quad \text{(Eq. 3)}$$

$$2 \equiv M-O-\hspace{-4pt}\diagup\hspace{-4pt} + PhCH_2OH \xrightarrow{-iPrOH} \equiv M-O-M\equiv + PhCH_2CH_2\underset{\underset{OH}{|}}{C}HCH_3 \quad \text{(Eq. 4)}$$

$$2 \equiv M-OR + 2\, O\hspace{-4pt}=\hspace{-4pt}\diagup\hspace{-4pt} \xrightarrow{-2\,ROH} \equiv M-O-M\equiv + O\hspace{-4pt}=\hspace{-4pt}\diagup\hspace{-4pt} \quad \text{(Eq. 5)}$$

Scheme 8.1 Condensation reactions in nonaqueous sol–gel processes leading to the formation of metal–oxygen–metal bonds: Alkyl halide elimination (eqn. 1), ether elimination (eqn. 2), ester elimination (eqn. 3), C–C bond formation between benzylic alcohols (Ph = phenyl) and alkoxides (eqn. 4) and aldol-like condensation reactions (eqn. 5).

In nonaqueous sol–gel routes, metal oxide precursors, such as metal halides, alkoxides, acetates or acetylacetonates are chemically transformed into metal oxide nanoparticles by reacting them with different organic solvents, such as alcohols, amines, or ketones, in the temperature range from about 50 to 250 °C. Heating techniques include simple oil baths, autoclaves or microwave reactors. Figure 8.1 provides an overview of nonaqueous sol–gel processes to metal oxide nanoparticles with different functionalities.

In comparison to aqueous sol–gel chemistry, nonaqueous processes offer several advantages that justify the use of organic solvents. On the one hand, they give access to a great variety of metal oxide nanoparticles with different sizes, shapes and compositions, including binary, ternary and doped compounds [10, 12–14]. Especially, multi-metal oxides are difficult to prepare in an aqueous medium due to the different chemical reactivities of the precursors toward water. On the other hand, the synthesis protocols are simple, robust (i.e., less sensitive to slight variations in the synthesis conditions), and the products are characterized by high crystallinity, which is in contrast to aqueous systems that often produce amorphous nanoparticles. However, the most important advantage is the possibility to correlate the organic reactions with the formation of the inorganic nanoparticles [15], offering detailed information about chemical mechanisms involved in nanoparticle formation. Knowledge of particle formation mechanisms is essential on

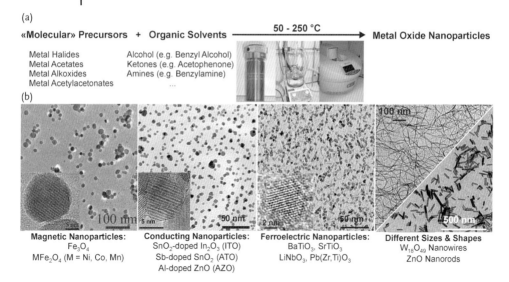

Figure 8.1 (a) Overview of nonaqueous liquid-phase routes to metal oxide nanoparticles with precursors, solvents, typical temperature range, heating tools (autoclave, oil bath, microwave reactor). (b) Transmission electron microscopy (TEM) images of selected metal oxide nanoparticles with composition and physical properties.

the way to improving the synthesis routes to nanoparticles and to introduce some rationality to this endeavor that is typically based on trial-and-error experiments.

Nonaqueous sol–gel approaches can be divided into two families: surfactant-assisted and surfactant-free (or solvent-directed) routes. In the first case, surfactants are present in the reaction mixture, either as solvent or as additive. Surfactants play multiple roles, which are often not completely understood and hardly experimentally verified on a molecular level. On the one hand they contribute to the size and shape control, and on the other hand they cap the surface of the nanoparticles and thus determine the assembly and dispersibility properties. However, great care has to be taken when interpreting the function of surfactants in such reaction systems. Less obvious effects might come into play. For example, counter ions of charged surfactants can influence the size and the shape [16], or the surfactants can be involved in chemical reactions [17, 18], producing new organic species that might influence the whole reaction system. To simplify things, one can leave the surfactants out and just work with a precursor and a solvent. Such two-component systems are particularly suitable to study nanoparticle crystallization and formation mechanisms. General disadvantages with respect to the quality of the nanoparticles include a broader particle size distribution, less defined or developed shapes, and a stronger tendency to agglomerate [13]. Irrespective of whether surfactants are used or not, nonaqueous sol–gel approaches offer the perfect chemistry to be combined with microwave irradiation. However, whereas microwave heating became quite popular for the surfactant-assisted nonaqueous synthesis of III–V (InGaP, InP) and II–VI (CdS, CdSe) quantum dots [19–21], it is much less studied for metal oxides. Indeed, in comparison to the large number of papers dedicated to surfactant-assisted routes to metal oxide nanoparticles involving conventional heating [22, 23], only very few reports focused on microwave heating in organic solvents containing surfactants. Nevertheless, the great success of surfactant-mediated non-hydrolytic/nonaqueous synthesis approaches makes it obvious that the combination with microwave heating opens up great opportunities which, however, remain to be explored. In this chapter, the discussion will be focused on surfactant-free nonaqueous routes due to the larger number of illustrative examples.

8.3
Polyol Route

The polyol method involving the reaction of metal salts with polyalcohols was originally developed by Fiévet *et al.* for the preparation of metal particles [24] and later extended by the same group to metal oxides [25]. However, in most of these cases, a defined amount of water is added to transform the precursors into the corresponding metal oxides, rendering these nonaqueous processes hydrolytic. Nowadays, the polyol route is a well-established synthesis methodology for a wide variety of metal oxide nanoparticles [26, 27]. Because polyalcohols like ethylene glycol, 1,3-propanediol, 1,4-butanediol or glycerol have high loss tangents [28], they are particularly susceptible to microwave heating.

A major problem of sol–gel chemistry is that the products obtained are often characterized by low crystallinity. Although this problem is more pronounced in aqueous systems, nonaqueous processes also struggle with this issue. Multi-metal oxides containing two or more metals are particularly difficult to access with high crystallinity and homogeneity. Gedanken *et al.*, for example, reported a microwave-assisted soft-chemical route for $BaTiO_3$, $Ba_6Ti_{17}O_{40}$, $BaZrO_3$ and $PbTiO_3$ in ethylene glycol under atmospheric pressure [29]. The products were nearly amorphous and only annealing at 700 °C led to highly crystalline particles. In addition to ethylene glycol as solvent, the authors used hydrated precursors and solid KOH to adjust the pH, which means that the reaction system is nonaqueous, but not non-hydrolytic. Analogously, YAG:Eu^{3+} nanoparticles were prepared from hydrated precursors in different diols by microwave irradiation [30]. Here also, calcination is required to produce crystalline materials.

Another interesting example of the microwave-assisted polyol method is the synthesis of monodisperse ZnO colloidal nanocrystal clusters [31]. More details related to this example are presented in Chapter 6.

Microwave heating seems to be particularly suitable for the synthesis of doped metal oxides (see also Section 8.4). One example along these lines is the preparation of Co-doped ZnO nanoparticles from the corresponding metal acetates and ethylene glycol [32]. Doping levels of up to 15 mol% were achieved although the product was not completely phase pure. One year later, the same approach was extended to other dopants such as Mn^{2+}, Ni^{2+}, and Cr^{3+} [33].

As mentioned before, most of the conventionally heated polyol routes rely on the addition of water to the organic solvent, and this is also true for microwave-assisted methods. Feldmann *et al.*, for example, mixed an excess of diethylene glycol with small amounts of water to produce indium tin oxide nanocrystals that could be used for the preparation of transparent conductive layers [34]. This approach was later extended by the same group to other types of transparent conducting oxide materials such as In:ZnO (IZO), Al:ZnO (AZO) and $F:SnO_2$ [35, 36]. As a proof of concept, thin layers were deposited on glass substrates from IZO and AZO suspensions by spin-coating (Figure 8.2a and 8.2b). Specific resistivities of $2.1 \times 10^{-1} \Omega cm$ and $2.6 \times 10^{-1} \Omega cm$, respectively, were obtained after annealing of the films.

The amount of water and the chain length of the alkanediols are important parameters in determining the crystal size in the microwave-assisted polyol synthesis of metal oxide nanoparticles. Yanagida *et al.* reported that the crystal size of anatase nanoparticles can be tailored in the range of about 4.5 to 10 nm, either by adding more water or by increasing the chain lengths from ethanediol to propanediol, butanediol, pentanediol and hexanediol [39]. The importance of the type of polyol, and especially of the water concentration, was already recognized by Fievet *et al.* many years ago [25]. Independent of the polyol used, water had to be present to obtain zinc oxide from zinc acetate as precursor.

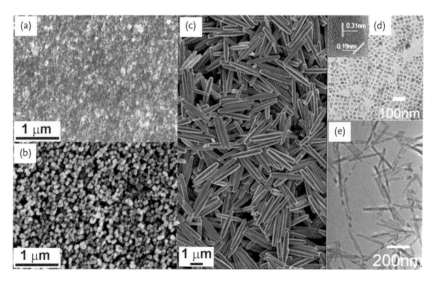

Figure 8.2 Scanning electron microscopy (SEM) images of (a) IZO and (b) AZO films on glass substrates. Reproduced from ref [35] with permission of Elsevier Inc. (c) SEM images of ZnO microrods. Reproduced from ref [37] with permission of Wiley-VCH. TEM images of cerium oxide (d) nanocubes and (e) nanorods. Reproduced from ref [38] with permission of Elsevier B.V.

In addition to well-accepted specific microwave effects, such as solvent superheating, minimization of wall effects, and selective heating of strongly microwave-absorbing species, there is still some speculation about non-thermal microwave effects. In this context, Orel and Kappe *et al.* presented an in-depth and critical examination of the specific role of microwave dielectric heating during the synthesis of ZnO microrods (Figure 8.2c) in a mixture of water and ethylene glycol [37]. Carefully performed control experiments clearly demonstrated that in that particular reaction system no differences between conventional and microwave heating existed. The ZnO particles had the same crystal structure, primary crystallite size, shape and size distribution, independent of the heating technique [37].

In order to increase the control over the nanoparticle shape, the polyol solvent can be combined with a surfactant. Tao *et al.* reported the synthesis of cerium dioxide nanoparticles in ethylene glycol, oleic acid and *tert*-butylamine [38]. The morphology could be tuned from spherical nanoparticles to nanocubes and nanorods (Figure 8.2d and 8.2e, respectively) simply by changing the irradiation time.

8.4
Benzyl Alcohol Route

Benzyl alcohol plays a unique role as reaction medium for the synthesis of metal oxide nanoparticles and nanostructures as well as oxide-based organic–inorganic hybrid materials [13, 40–43]. There is probably no other organic solvent that gives access to such a large variety of metal oxide nanomaterials [12, 13]. With a dielectric loss tangent of 0.667, benzyl alcohol belongs to the family of strongly microwave absorbing solvents [28]. Although it is a relatively bulky molecule, its relaxation time in the microwave field is not much larger than that for ethanol, which is probably due to a localized rotation of the CH_2OH group rather than of the whole molecule [28]. Accordingly, the combination of benzyl alcohol and microwave chemistry is expected to provide a powerful and versatile reaction system for nanoparticle synthesis.

In 2008 Niederberger *et al.* reported the microwave-assisted synthesis of different types of binary and ternary metal oxide nanoparticles in benzyl alcohol [6]. Starting from different metal organic precursors, CoO, ZnO, Fe_3O_4, MnO, Mn_3O_4 and $BaTiO_3$ nanoparticles with good crystallinity were produced after just a few minutes of microwave irradiation (Figure 8.3). These results gave a first indication that the benzyl alcohol route can easily be extended to microwave-mediated approaches. However, some empirical differences that are of practical importance were observed between conventionally and microwave heated systems. For example, it is difficult to control the crystal size of the nanoparticles in conventionally heated benzyl alcohol by variation of the reaction time or the precursor concentration. In the case of microwave heating, these parameters are more influential and it is possible to control the crystal size of the nanoparticles simply by adjusting the irradiation time [6, 44].

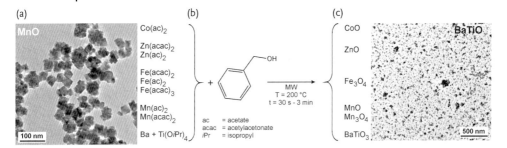

Figure 8.3 Reaction scheme for the microwave-assisted synthesis of metal oxide nanoparticles in benzyl alcohol (b), TEM images of selected samples ((a) MnO, and (c) BaTiO₃).

Magnetite nanoparticles are particularly easy to obtain from the microwave-assisted benzyl alcohol route [6, 45–47]. Depending on the surface properties, they can be dispersed in apolar organic solvents or in water. Catechol-derived stabilizers were attached to the surface of iron oxide nanoparticles in a postsynthetic step, leading to superparamagnetic nanoparticles stable under physiological conditions, making them interesting candidates for biomedical applications [48]. If the benzyl alcohol route is combined with surfactants, then dispersibility in apolar solvents can be achieved. Roig *et al.* combined benzyl alcohol with oleic acid to produce iron oxide nanoparticles (γ-Fe_2O_3/Fe_3O_4) with a crystal size of 6.3 ± 1.5 nm (according to TEM) and which can easily be dispersed in hexane [47].

The formation of the inorganic nanoparticles in organic solvents under anhydrous conditions is, in most cases, the result of an organic reaction between the metal oxide precursor and the solvent, which means that not only the precursor, but also the solvent molecules undergo a chemical reaction. The fact that organic reactions are usually easier to monitor than inorganic processes opens up the unique possibility to correlate organic with inorganic chemistry. For example, the acetate ligand of a metal acetate reacts with an alcohol to give the corresponding organic ester and, formally, water, which provides the oxygen for the metal oxide formation (cf. Scheme 8.1, eqn. 3). Accordingly, characterization of the organic product, in this case an ester, makes it possible to elaborate a chemical reaction pathway responsible for nanoparticle formation. In addition, time-dependent quantification of the organic product provides valuable information about the reaction progress. How such studies can look was shown for the formation of zinc oxide nanoparticles in benzyl alcohol [44, 49]. Further details are given in Chapter 6.

One way to subtly tailor the chemical and physical properties of nanoparticles is by incorporating dopants into the host lattice. Careful selection of the dopant element and its concentration enables one to engineer the electronic, optical, photochemical and magnetic properties of nanomaterials toward specific applications [50]. However, due to the small volume of the nanoparticles, homogeneous distribution of the dopants in substitutional sites without segregation and formation of clusters is challenging. From a chemical point of view, it is therefore essential to match the reactivity of the dopant precursor with the precursor of the host material, to ensure the formation of a solid solution rather than separated phases. Unfortunately, there is basically no information available about the reactivity of different precursors toward specific solvents, which renders the doping of nanoparticles a rather random process.

Although examples are still scarce and the reasons are not yet well understood, it seems that microwave irradiation is a powerful tool for the doping of metal oxide nanoparticles. For example, doping of zinc oxide nanoparticles with transition metal ions has proven to be much more difficult in a solvothermal process [51] than in a microwave reactor [52]. Indeed, the microwave-assisted synthesis of ZnO nanoparticles in benzyl alcohol gave access not only to a broad family of different dopants (Figure 8.4), but also to exceptionally high "doping" levels in the range of 20–30 atom% in the case of Co and Fe [52].

Doping not only allows one to change the magnetic properties of ZnO, but also the electrical conductivity. Materials such as Al:ZnO, In:ZnO or Ga:ZnO are studied as transparent conducting oxides (TCO) that might be able to replace the increasingly expensive indium tin oxide (ITO), which represents the state-of-the-art. Industrially, most of the TCO layers are produced by physical deposition methods, like magnetron sputtering. The quality of such coatings, typically on glass, in terms of transparency and electrical conductivity has reached a very high level. On the other hand, the process is expensive due to the high vacuum needed

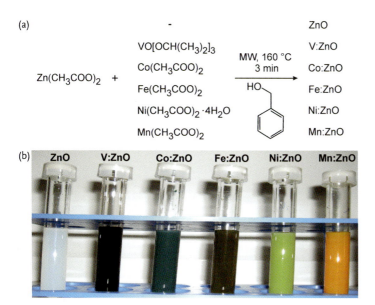

Figure 8.4 (a) Reaction scheme with precursors, solvent, temperature and irradiation time (MW = microwave irradiation). (b) Photographs of nanoparticle dispersions in benzyl alcohol with different colors depending on the dopant element. Adapted from ref [52] with permission of the American Chemical Society.

during the deposition of the films, the costly targets, and the frequently required procedures to clean the deposition chambers, which usually leads to a halt in production for several days. For all these reasons, wet chemical deposition techniques strongly gained attractiveness. Deposition of colloidal TCO nano-inks promises to be a simple, non-vacuum-based process, offering the possibility to coat two- and three-dimensional substrates at low temperature. The moderate deposition conditions, enabling the use of flexible and thermally labile substrates, are an integral aspect on the way to fabricating flexible electronics. Similar to the self-assembly of nanoparticles into hierarchically structured materials with new properties, nano-inks also provide nearly endless possibilities to combine different nanoparticles into films with defined architectures. However, one big challenge remains: the quality of the as-deposited films is not good enough to use them directly as TCO. The large amount of impurities (solvent, stabilizers/surfactants) and the low connectivity of the grains make it inevitable to heat the films under a controlled atmosphere.

In this respect, microwave irradiation could offer some interesting solutions. Selective heating effects might allow the deposition of nanoparticles on appropriate substrates and local overheating of the films might be an interesting way to increase the interparticle conductivity. In spite of these promising prospects, the direct microwave-assisted deposition of nanoparticles on substrates in organic solvents has hardly been explored.

One example, not related to TCO, is the deposition of ferrite MFe_2O_4 (M = Fe, Co, Mn and Ni) nanocrystal films on glass substrates [45, 46]. Whereas the microwave irradiation of the corresponding metal acetylacetonates in benzyl alcohol for 12 min led to normal nanoparticle dispersions (Figure 8.5a), immersion of a glass slide in the initial reaction solution made it possible to deposit some of the newly formed nanoparticles directly on the substrate (Figure 8.5b). The overview SEM image of $CoFe_2O_4$ as a representative example indicates that the films were porous, consisting of a homogeneous arrangement of uniformly sized nanoparticles (Figure 8.5c). By adjusting the precursor concentration, the reaction time and temperature, the film thickness can be varied from 20 to 80 nm and the crystal

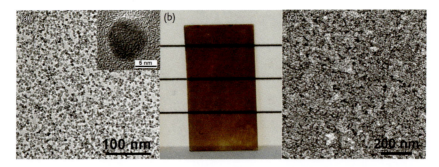

Figure 8.5 (a) TEM image of $CoFe_2O_4$ nanocrystals (inset: HRTEM image of one crystal), (b) photograph of a $CoFe_2O_4$ film on a glass slide, (c) SEM overview image of a $CoFe_2O_4$ film.

size of the nanoparticles can be changed from 4 to 8 nm. Due to the small crystal size, all these ferrite nanoparticles were superparamagnetic.

Interestingly, this approach can easily be extended to $BaTiO_3$ layers [45], but not to ZnO-based TCO films. Whereas the $BaTiO_3$ nanocrystals form a homogeneous porous film comparable to those of the ferrites, the direct microwave-deposition of ZnO nanoparticles leads to the growth of isolated islands on the substrate. One possible reason for the inhomogeneous deposition of ZnO nanoparticles might be their significantly larger size (around 20 nm) in comparison to the ferrites and $BaTiO_3$ (4–5 nm). However, the uniformity of the films can be greatly improved by applying a combination of dip-coating of nano-inks and a post-deposition microwave treatment [53]. In a first step, Sn:ZnO nanoparticles were prepared by irradiating a solution of tin *tert*-butoxide and zinc acetate in benzyl alcohol with microwaves. The wet precipitate was redispersed in a mixture of cyclohexane and oleic acid/oleylamine and a porous Sn:ZnO film on fused silica was obtained by dip-coating from such a nano-ink. After annealing, these films were used as seed layers for the next steps (Figure 8.6a). The glass slide with the porous film was immersed in a fresh reaction solution containing the precursors dissolved in benzyl alcohol. Microwave irradiation resulted in the formation of another set of doped ZnO nanoparticles, however, this time preferentially in the pores of the film, leading to a significant densification of the film. During this densification step, a different type of material compared to the seed layer can, in principle, be nucleated so that a multicomponent film is formed (Figure 8.6a, lower pathway). For example, it is possible to use Sn:ZnO nanoparticles as a seed layer, followed by the deposition of Al:ZnO nanoparticles inside the pores, resulting in a homogeneous Sn:ZnO/Al:ZnO film [53]. Figure 8.6b shows a SEM top view image of a Sn:ZnO film, which was densified with the same material, whereas Figure 8.6d

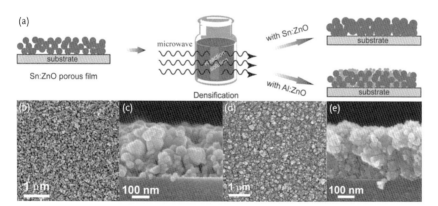

Figure 8.6 (a) Schematic illustration of the microwave-assisted densification procedure starting from a preformed seed layer, which was deposited from nanoparticle dispersions by dip-coating. SEM images of Sn:ZnO films densified with Sn:ZnO (top view (b) and cross-section (c)) and with Al:ZnO (top view (d) and cross-section (e). Both films were annealed again after microwave densification in air and nitrogen.

presents a Sn:ZnO/Al:ZnO film. The cross-section images clearly prove that the multicomponent film (Figure 8.6e) consists of smaller grains than the single component Sn:ZnO film (Figure 8.6c). Obviously, the Sn:ZnO nanoparticles during the densification step tend to nucleate on the surface of the seed layer particles, whereas the Al:ZnO nanoparticles mainly grow inside the pores. With a resistivity of 0.6 Ωcm, the multicomponent film additionally offered the highest electrical conductivity after annealing in air and nitrogen [53].

A seeded growth process, although for powders rather than films, was proposed for the preparation of flower-like mixed-phase titania [54]. In this case, a commercially available titania powder (P25) was coated with flower-like titania nanostructures grown from $TiCl_4$ in benzyl alcohol by microwave heating. Of course, the seeded growth method can be extended to multicomponent particles in such a way that the seed and the shell have different compositions. Such a core–shell configuration makes it possible to combine two materials with different properties within the same particle. For example, coating magnetite nanoparticles with silica provides a platform that can be used as a support for peptide synthesis [55]. Whereas the magnetic properties are important with respect to the removal of the colloidal supports from the reaction solution by an external magnetic field, the silica shell offers the appropriate chemistry to modify the surface in such a way that it can be used for peptide synthesis. The magnetite seed particles were produced by a nonaqueous microwave-assisted approach in benzyl alcohol, whereas the silica shell was obtained from a microwave-assisted process based on aqueous sol–gel chemistry [55]. In comparison to the solid-phase supported peptide synthesis, colloidal supports in "quasi-solution" offer larger surface areas and better access to the surfaces, where the growth of the peptides occurs. As a matter of fact, the $Fe_3O_4/SiO_2/NH$-$(\beta Ala)_2$-Rink core–shell particles were applicable as colloidal supports for the synthesis of a tetrapeptide model and a potentially bioactive undecapeptide, both exhibiting excellent purities as isolated crude products [55].

It seems that the use of microwave irradiation might be particularly suitable for the synthesis of supported nanoparticles due to potential selective heating of the surface of the substrate, which would facilitate the local deposition of nanoparticles. First indications along these lines were recently proposed for the microwave-assisted synthesis of SnO_2- and Fe_3O_4-based graphene nanocomposites in benzyl alcohol [56]. The authors proposed that the graphene oxide acted as the principal microwave absorber and thus was selectively heated in such a way that the metal oxide nanoparticles nucleated on its surface. Proof came from the fact that the synthesis of tin oxide and iron oxide in the absence of graphene oxide required a much larger microwave power to reach the same temperature. In addition, syntheses performed by conventional heating resulted in composites with much lower particle coverage. Figure 8.7 shows SEM, TEM and HRTEM images of graphene nanosheets, prepared under microwave irradiation, that are fully coated with the metal oxides without any isolated nanoparticles. The composites showed good lithium intercalation–deintercalation performances at high rates and good cycling stability compared to the pure metal oxides [56].

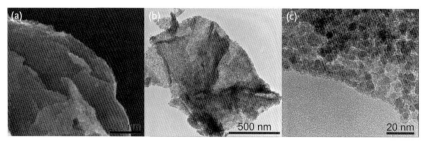

Figure 8.7 (a) SEM, (b) TEM and (c) HRTEM images of magnetite/graphene nanocomposites. Reproduced from ref [56] with permission of the Royal Society of Chemistry.

Another system that has been studied using the microwave-assisted benzyl alcohol route was dedicated to tungsten oxide nanoparticles [57]. Depending on the synthesis temperature and the WCl_6 concentration, the structure can be varied from tungstite to cubic WO_3. Additionally, the particle size can be controlled by adjusting the heating rates. The authors compared some features of the products obtained by microwave with those obtained by conventional heating, and they found an unexpected difference. The organic residues attached to the surface of the nanoparticles can be removed at lower temperature in the case of the microwave product. The presence of organic impurities is a serious limitation on the way to apply metal oxide nanoparticles prepared in nonaqueous media, and therefore the possibility to lower the calcination temperature is a significant progress.

8.5
Other Mono-Alcohols

In this short section, the microwave-assisted synthesis of metal oxide nanoparticles in other mono-alcohols than benzyl alcohol will be briefly discussed. Schneider et al. studied the formation of ZnO nanoparticles from zinc acetylacetonate or zinc oximate in different alkoxyethanols, such as methoxyethanol, ethoxyethanol and n-butoxyethanol [58]. They found that the particle morphology, aggregation behavior and defect chemistry were strongly dependent on the precursor type. Especially, the defect chemistry is of great importance for potential applications in electronic devices. Thin films obtained by spin-coating of ZnO nanoparticle dispersions were tested for their field-effect transistor device properties. However, the films had to be thermally processed or treated with UV irradiation to improve the electronic characteristics [58].

Nanostructured titanium dioxide with high photocatalytic activity was produced from $Ti(O-^iPr)_4$ and ethanol in the presence of poly(vinyl pyrrolidone) [59]. According to XRD and Raman measurements, the sample consisted of phase-pure anatase nanoparticles and TEM investigations showed spherical agglomerates in the size range 40–100 nm built up of primary crystallites of 7–10 nm. In another example, the titania polymorph can be varied systematically from anatase to rutile and brookite by reacting $TiCl_4$ or $TiCl_3$ with various alcohols like ethanol, propanol, butanol and octanol under microwave irradiation [60]. However, in this case the precursors are dissolved in an aqueous HCl solution, which renders this approach nonaqueous, but hydrolytic. Whereas the synthesis in ethanol, propanol and butanol resulted in phase-pure anatase nanocrystals when mixed with $TiCl_4$ as precursor, the reaction between $TiCl_3$ and butanol yielded a mixture of brookite as the main phase and a minor amount of rutile. The $TiCl_4$–octanol system produced rutile nanoparticles. For all the combinations, the primary crystallite size was in the range 7–8 nm, however, with varying tendencies to form agglomerates with different morphologies. The authors proposed that the strong dependence of the crystal structure and the morphology of the nanomaterials on the precursor and solvent is related to the complex role of HCl as catalyst, the change in chemical reactivity of the solvent towards the precursors with increasing chain length, and the increase in steric hindrance with increasing number of carbon atoms in the alcohols.

8.6
Ionic Liquids

Such properties as high thermal stability, low vapor pressure and the possibility to dissolve many organic, inorganic and organometallic compounds make ionic liquids very attractive solvents for the synthesis of inorganic materials [61, 62]. Together with their high ionic conductivity and polarizability, they can effectively be combined with microwave-assisted methods [63, 64]. In addition, ionic liquids can be used as co-solvents to increase the absorbance level of nonpolar solvents. Shi et al. added a small amount of [bmim][BF_4] to dibenzyl ether to be able to heat the reaction mixture to 250 °C by microwave irradiation to obtain monodisperse magnetite nanocrystals [65].

Similar to the polyol-route, the synthesis of metal oxide nanostructures in ionic liquids usually occurs in the presence of small amounts of water. For example, Wang and Zhu used a mixture of an imidazolium salt and water to produce ZnO nanostructures with different morphologies, including needles and flowers [66]. On the other hand, it is possible to use the crystal water of the precursor for metal oxide formation in an ionic liquid. Irradiation of iron nitrate nonahydrate in 1-butyl-3-methylimidazolium tetrafluoroborate, [bmim]$^+$[BF_4]$^-$, with microwaves resulted in the formation of Fe_2O_3 nanocrystals [67]. Ionic liquids are rather difficult to obtain in completely dry form and, therefore, metal oxides can be produced even by starting from anhydrous precursors. Ding et al. synthesized [bmim]$^+$[BF_4]$^-$ and reacted it with titanium isopropoxide [68]. Thanks to the 2.5 wt% water in the ionic liquid the precursor was hydrolyzed under microwave irradiation and crystalline anatase nanoparticles with a bipyramid shape formed (Figure 8.8a).

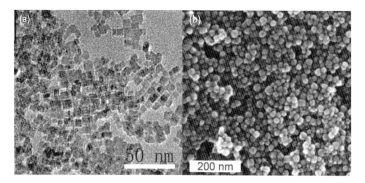

Figure 8.8 (a) TEM image of anatase nanocrystals. Reproduced from ref [68] with permission of the American Chemical Society. (b) SEM image of indium tin oxide nanoparticles. Reproduced from ref [69] with permission of Wiley-VCH.

A major problem of nanoparticle synthesis in organic solvents is the necessity to remove the organic residues from the product by a thermal post-treatment. Because ionic liquids are known for their non-coordinating properties, they should be advantageous in view of a facile and complete removal of any by-products from the surface of the nanoparticles. The resulting problem of low stability of colloidal crystals at elevated temperatures can be minimized by a fast and short-time microwave heating. This strategy was successfully accomplished by Feldmann et al., who prepared highly conductive indium tin oxide nanocrystals in a mixture of $[N(CH_3)(C_4H_9)_3][N(SO_2CF_3)_2]$ as the ionic liquid, dimethyl formamide, ethanol and $[N(CH_3)_4]OH$ as the base [69]. According to SEM analysis, the particles were 25 nm in size with a uniform spherical shape (Figure 8.8b). Ink-jet printing from alcoholic dispersions of as-prepared indium tin oxide nanoparticles on standard overhead projector transparencies resulted in films, whose electrical conductivity and optical transparency could be improved by densification through simple mechanical pressing [69]. The final transparency was 99% and a resistivity of $6.8 \times 10^{-2} \Omega$ cm at a layer thickness of 500 nm could be achieved.

Other examples of microwave-assisted syntheses in ionic liquids (and water) include CuO [70], $ZnFe_2O_4$ [71], and Fe_3O_4 [72]. It was reported for TiO_2 nanoparticles that the crystallization of the anatase phase was strongly dependent on the initial water concentration, the irradiation time and the concentration of the ionic liquids [73].

8.7
Nonaqueous Microwave Chemistry beyond Metal Oxides

Although the nonaqueous sol–gel routes were mainly designed for the preparation of metal oxide nanomaterials, the microwave-assisted polyol and the benzyl alcohol routes have been extended to many materials beyond metal oxides. However, only a small selection of non-oxide materials is given herein. For a more detailed overview of nanomaterials prepared by microwave irradiation, including aqueous and nonaqueous processes, the reader is referred to a recent review by Cozzoli and Kappe et al. [5].

Ethylene glycol is a reducing agent and therefore it can be applied with great success to the preparation of metal and alloy nanoparticles such as PbTe, PbSe [74], Ni [75], Ag, Au, Pt, Au–Pd and Au/Ag [76–79]. If metal salts are heated in ethylene glycol in the presence of thiourea, then metal sulfides are accessible. Based on this idea, Yu et al. presented the synthesis of hierarchically porous $ZnIn_2S_4$ submicrospheres via a microwave-solvothermal process [80].

In addition, polyols have proven to be excellent solvents for the synthesis of cathode materials for lithium ion batteries. Manthiram et al. used tetraethyleneglycol as solvent to produce $LiFePO_4$ in a microwave reactor [81, 82]. In comparison to water as solvent, the particles were much smaller, which is important for the electrochemical performance. After coating the nanorods with carbon at 700 °C, these nanomaterials showed excellent cyclability and high rate capability [82]. The same approach was applied to the synthesis of Li_2MSiO_4 (M = Mn and Fe) [83]. The high reaction temperatures of 300 °C made it necessary to work with a microwave reactor that can stand high pressures.

Similar to polyols, benzyl alcohol can successfully be used as a reaction medium for the microwave-assisted synthesis of $LiMPO_4$ with M = Fe and Mn [84–86]. The morphology is strongly dependent on the concentration of the reactants and can be varied from platelets (Figure 8.9a) [84] to strongly anisotropic sticks (Figure 8.9b) [86]. Interestingly, the platelets are composed of nanocrystalline building blocks, which are, however, crystallographically aligned to form a nearly perfect single crystal (Figure 8.9a). The sticks look like an anisotropic assembly of rhombohedral particles (Figure 8.9b), which points to an unusual growth process. Therefore, the formation of these particles was investigated in detail [86] and it was found that, in a first step, lithium phosphate rods formed with iron-containing particles attached to their surface. The surface-adsorbed particles start to react with the lithium phosphate, resulting in an intermediate structure of sticks with a lithium rich core and an iron rich shell (Figure 8.9c). Once

Figure 8.9 SEM images of (a) $LiFePO_4$ platelets and (b) $LiFePO_4$ sticks. (c) Dark-field scanning TEM image of an intermediate of $LiFePO_4$ with the Li_3PO_4 core surrounded by Fe-rich particles.

the diffusion of the iron ions into the stick core was finished, phase-pure LiFePO$_4$ was formed.

As mentioned above for the ZnO synthesis in benzyl alcohol, it seems that the use of microwave irradiation is particularly suitable for the doping of materials. In agreement with this statement, the LiFePO$_4$ particles can easily be doped with a large variety of divalent (Mn, Ni, Zn), trivalent (Al) and tetravalent (Ti) metal ions in different concentrations [85]. Upon doping, the electrochemical performance was significantly improved in comparison to the undoped material. The short treatment in the microwave reactor of just a few minutes makes this approach a very time-efficient method.

If the OH function of benzyl alcohol is replaced by a SH function, then it is possible to extend the nonhydrolytic sol–gel method in such a way that metal sulfides (ZnS, SnS$_2$) rather than metal oxides are obtained [87]. Taking advantage of the excellent microwave-absorbing properties, MoS$_2$ was selectively deposited on graphene oxide sheets based on the same approach [88]. However, benzyl mercaptan was replaced by butyl mercaptan as the sulfur source and dimethyl formamide was used as the dispersion agent for graphene oxide and as the solvent. Microwave irradiation was performed for 1 h, resulting in a temperature of 215 °C. Similar to Pinna's work [56], the authors also found that it was possible to reach a higher temperature with the same microwave power with graphene oxide in the reaction mixture rather than without. Once more, this observation underlines the high microwave absorption properties of graphene oxide. It is obvious that the microwave-assisted nonaqueous particle synthesis in the presence of strongly microwave-absorbing carbon compounds like graphene bears great potential to prepare nanocomposites in a selective way.

8.8
Summary and Outlook

Although the field of microwave-assisted nanoparticle synthesis is relatively young, there is no doubt that the use of microwave irradiation opens up unique opportunities that cannot be provided by any other heating technique. Up to now, the great potential of microwave chemistry has been explored mainly for organic syntheses, but it is clear that in addition to the many advantages already observed for these organic processes, new effects and features can be expected arising specifically from heterogeneous systems. Although the rate enhancement as a consequence of the efficient and fast heating process of a reaction mixture by microwaves represents a major benefit, the greatest potential of microwave-assisted syntheses lies in the possibility to selectively heat compounds with different microwave-absorbing properties. The preparation of core–shell particles and hetero-nanostructures, the growth of nanoparticles inside another (host) material, or deposition of nanoparticles as films on suitable substrates are just a few ideas along these lines. Examples in the literature are still scarce and, therefore, unprecedented, compositionally complex multicomponent materials can be expected to be synthesized in the next few years by using the unique features of microwave chemistry.

On the other hand, in spite of some recent progress, microwave reactors still resemble a kind of black box, which is not yet fully understood. For example, a crucial parameter in microwave-assisted approaches is the temperature. However, it is rather difficult to get information about local temperatures (e.g., on the surface of nanoparticles) and temperature distributions (e.g., between solid and solvent), and questions about the influence of microwaves on nucleation and growth and on the surface chemistry (i.e., attachment of monomers, solvent molecules or surfactants) of the nanoparticles are still open.

The only way to learn more about the processes inside a microwave reactor is to connect it to *in situ* probes. However, due to the field containment and the interaction with metallic objects in the field [89], the combination of microwave chemistry with *in situ* reaction monitoring is challenging. Nevertheless, in the last few years several combinations of microwave reactors with characterization tools, such as X-ray techniques [89–91], neutron scattering [92, 93], and Raman spectroscopy [89, 94–96] have been proposed to study organic as well as inorganic reactions under the respective experimental conditions.

Another important task is the scale-up of microwave routes to industrial quantities under consideration of all the necessary safety issues. Combination with flow reactor technology might represent a promising strategy.

Microwave chemistry is a somewhat controversially discussed, but nevertheless highly fascinating research topic. It offers unique opportunities that no other synthesis technique can provide. Although quite empirical – like most of the synthesis techniques – it will unfold its full potential once applied in a more rational way. However, already today microwave chemistry is a prime source of inspiration for the development of new materials, giving important impulses for the whole field of nanoscience and nanotechnology.

References

1 Komarneni, S., and Roy, R. (1985) *Mater. Lett.*, **3**, 165–167.
2 Kappe, C.O., Dallinger, D., and Murphree, S.S. (2008) *Practical Microwave Synthesis for Organic Chemists*, Wiley-VCH Verlag GmbH, Weinheim, Germany.
3 Kappe, C.O., and Dallinger, D. (2009) *Mol. Divers.*, **13**, 71–193.
4 Bilecka, I., and Niederberger, M. (2010) *Nanoscale*, **2**, 1358.
5 Baghbanzadeh, M., Carbone, L., Cozzoli, P.D., and Kappe, C.O. (2011) *Angew. Chem. Int. Ed.*, **50**, 11312–11359.
6 Bilecka, I., Djerdj, I., and Niederberger, M. (2008) *Chem. Commun.*, 886–888.
7 Niederberger, M., and Garnweitner, G. (2006) *Chem. Eur. J.*, **12**, 7282–7302.
8 Brinker, C.J., and Scherer, G.W. (1990) *Sol-Gel Science: The Physics and Chemistry of Sol-Gel Processing*, Academic Press, Inc., San Diego, CA, USA.
9 Vioux, A. (1997) *Chem. Mater.*, **9**, 2292–2299.
10 Mutin, P.H., and Vioux, A. (2009) *Chem. Mater.*, **21**, 582–596.
11 Niederberger, M., and Antonietti, M. (2007) in *Nanomaterials Chemistry: Recent Developments and New Directions* (eds C.N.R. Rao, A. Müller, and A.K.

Cheetham), Wiley-VCH Verlag GmbH & Co. KGaA, Weinheim, Germany, pp. 119–137.

12 Niederberger, M., and Pinna, N. (2009) *Metal Oxide Nanoparticles in Organic Solvents: Synthesis, Formation, Assembly and Application*, Springer-Verlag, London, UK.

13 Pinna, N., and Niederberger, M. (2008) *Angew. Chem. Int. Ed.*, **47**, 5292–5304.

14 Debecker, D.P., and Mutin, P.H. (2012) *Chem. Soc. Rev.*, **41**, 3624–3650.

15 Garnweitner, G., and Niederberger, M. (2008) *J. Mater. Chem.*, **18**, 1171–1182.

16 Kovalenko, M.V., Bodnarchuk, M.I., Lechner, R.T., Hesser, G., Schäffler, F., and Heiss, W. (2007) *J. Am. Chem. Soc.*, **129**, 6352–6353.

17 Steckel, J.S., Yen, B.K.H., Oertel, D.C., and Bawendi, M.G. (2006) *J. Am. Chem. Soc.*, **128**, 13032–13033.

18 Liu, H., Owen, J.S., and Alivisatos, A.P. (2007) *J. Am. Chem. Soc.*, **129**, 305–312.

19 Zhu, J.J., Palchik, O., Chen, S.G., and Gedanken, A. (2000) *J. Phys. Chem. B*, **104**, 7344–7347.

20 Murugan, A.V., Sonawane, R.S., Kale, B.B., Apte, S.K., and Kulkarni, A.V. (2001) *Mater. Chem. Phys.*, **71**, 98–102.

21 Gerbec, J.A., Magana, D., Washington, A., and Strouse, G.F. (2005) *J. Am. Chem. Soc.*, **127**, 15791–15800.

22 Niederberger, M. (2007) *Acc. Chem. Res.*, **40**, 793–800.

23 Jun, Y.W., Choi, J.S., and Cheon, J. (2006) *Angew. Chem. Int. Ed.*, **45**, 3414–3439.

24 Fievet, F., Lagier, J.P., Blin, B., Beaudoin, B., and Figlarz, M. (1989) *Solid State Ionics*, **32–3**, 198–205.

25 Jezequel, D., Guenot, J., Jouini, N., and Fievet, F. (1995) *J. Mater. Res.*, **10**, 77–83.

26 Feldmann, C. (2005) *Adv. Funct. Mater.*, **13**, 101–107.

27 Feldmann, C. (2005) *Solid State Sci.*, **7**, 868–873.

28 Gabriel, C., Gabriel, S., Grant, E.H., Halstead, B.S.J., and Mingos, D.M.P. (1998) *Chem. Soc. Rev.*, **27**, 213–223.

29 Palchik, O., Zhu, J.J., and Gedanken, A. (2000) *J. Mater. Chem.*, **10**, 1251–1254.

30 Nakamura, T., Yanagida, S., and Wada, Y. (2006) *Res. Chem. Intermed.*, **32**, 331–339.

31 Hu, X.L., Gong, J.M., Zhang, L.Z., and Yu, J.C. (2008) *Adv. Mater.*, **20**, 4845–4850.

32 Fidelus, J., Piticescu, R.R., Piticescu, R.M., Lojkowski, W., and Giurgiu, L. (2008) *Z. Naturforsch. B Chem. Sci.*, **63**, 725–729.

33 Lojkowski, W., Gedanken, A., Grzanka, E., Opalinska, A., Strachowski, T., Pielaszek, R., Tomaszewska-Grzeda, A., Yatsunenko, S., Godlewski, M., Matysiak, H., and Kurzydowski, K.J. (2009) *J. Nanopart. Res.*, **11**, 1991–2002.

34 Hammarberg, E., Prodi-Schwab, A., and Feldmann, C. (2008) *Thin Solid Films*, **516**, 7437–7442.

35 Hammarberg, E., Prodi-Schwab, A., and Feldmann, C. (2009) *J. Colloid Interface Sci.*, **334**, 29–36.

36 Avadhut, Y.S., Weber, J., Hammarberg, E., Feldmann, C., Schellenberg, I., Pottgen, R., and der Gunne, J.S.A. (2011) *Chem. Mater.*, **23**, 1526–1538.

37 Baghbanzadeh, M., Skapin, S.D., Orel, Z.C., and Kappe, C.O. (2012) *Chem. Eur. J.*, **18**, 5724–5731.

38 Tao, Y., Wang, H., Xia, Y.P., Zhang, G.Q., Wu, H.P., and Tao, G.L. (2010) *Mater. Chem. Phys.*, **124**, 541–546.

39 Yamamoto, T., Wada, Y., Yin, H.B., Sakata, T., Mori, H., and Yanagida, S. (2002) *Chem. Lett.*, 964–965.

40 Niederberger, M., Bartl, M.H., and Stucky, G.D. (2002) *J. Am. Chem. Soc.*, **124**, 13642–13643.

41 Pinna, N. (2007) *J. Mater. Chem.*, **17**, 2769–2774.

42 Pucci, A., and Pinna, N. (2010) *Z. Naturforsch. B Chem. Sci.*, **65**, 1015–1023.

43 Pinna, N., Karmaoui, M., and Willinger, M.G. (2011) *J. Sol-Gel Sci. Technol.*, **57**, 323–329.

44 Bilecka, I., Elser, P., and Niederberger, M. (2009) *ACS Nano*, **3**, 467.

45 Kubli, M., Luo, L., Bilecka, I., and Niederberger, M. (2009) *Chimia*, **64**, 170–172.

46 Bilecka, I., Kubli, M., Amstad, E., and Niederberger, M. (2011) *J. Sol-Gel Sci. Technol.*, **57**, 313–322.

47 Pascu, O., Carenza, E., Gich, M., Estrade, S., Peiro, F., Herranz, G., and Roig, A. (2012) *J. Phys. Chem. C.*, **116**, 15108–15116.

48 Amstad, E., Gillich, T., Bilecka, I., Textor, M., and Reimhult, E. (2009) *Nano Lett.*, **9**, 4042–4048.

49 Ludi, B., Süess, M.J., Werner, I.A., and Niederberger, M. (2012) *Nanoscale*, **4**, 1982–1995.

50 Pearton, S.J. (2010) *Nanoscale*, **2**, 1057.

51 Djerdj, I., Garnweitner, G., Arcon, D., Pregelj, M., Jaglicic, Z., and Niederberger, M. (2008) *J. Mater. Chem.*, **18**, 5208–5217.

52 Bilecka, I., Luo, L., Djerdj, I., Rossell, M.D., Jagodic, M., Jaglicic, Z., Masubuchi, Y., Kikkawa, S., and Niederberger, M. (2011) *J. Phys. Chem. C*, **115**, 1484–1495.

53 Luo, L., Häfliger, K., Xie, D., and Niederberger, M. (2013) *J. Sol-Gel Sci. Technol.* doi:10.1007/s10971-012-2709-7

54 Hsu, Y.C., Lin, H.C., Chen, C.H., Liao, Y.T., and Yang, C.M. (2010) *J. Solid State Chem.*, **183**, 1917–1924.

55 Stutz, C., Bilecka, I., Thünemann, A.F., Niederberger, M., and Börner, H.G. (2012) *Chem. Commun.*, **48**, 7176–7178.

56 Baek, S., Yu, S.H., Park, S.K., Pucci, A., Marichy, C., Lee, D.C., Sung, Y.E., Piao, Y., and Pinna, N. (2011) *RSC Adv.*, **1**, 1687–1690.

57 Le Houx, N., Pourroy, G., Camerel, F., Comet, M., and Spitzer, D. (2010) *J. Phys. Chem. C*, **114**, 155–161.

58 Schneider, J.J., Hoffmann, R.C., Engstler, J., Klyszcz, A., Erdem, E., Jakes, P., Eichel, R.A., Pitta-Bauermann, L., and Bill, J. (2010) *Chem. Mater.*, **22**, 2203–2212.

59 Jena, A., Vinu, R., Shivashankar, S.A., and Madras, G. (2010) *Ind. Eng. Chem. Res.*, **49**, 9636–9643.

60 Yoon, S., Lee, E.S., and Manthiram, A. (2012) *Inorg. Chem.*, **51**, 3505–3512.

61 Ma, Z., Yu, J.H., and Dai, S. (2010) *Adv. Mater.*, **22**, 261–285.

62 Ahmed, E., Breternitz, J., Groh, M.F., and Ruck, M. (2012) *Cryst. Eng. Comm.*, **14**, 4874–4885.

63 Bühler, G., Zharkouskaya, A., and Feldmann, C. (2008) *Solid State Sci.*, **10**, 461–465.

64 Martinez-Palou, R. (2010) *Mol. Divers.*, **14**, 3–25.

65 Hu, H.Y., Yang, H., Huang, P., Cui, D.X., Peng, Y.Q., Zhang, J.C., Lu, F.Y., Lian, J., and Shi, D.L. (2010) *Chem. Commun.*, **46**, 3866–3868.

66 Wang, W.W., and Zhu, Y.J. (2004) *Inorg. Chem. Commun.*, **7**, 1003–1005.

67 Jacob, D.S., Bitton, L., Grinblat, J., Felner, I., Koltypin, Y., and Gedanken, A. (2006) *Chem. Mater.*, **18**, 3162–3168.

68 Ding, K.L., Miao, Z.J., Liu, Z.M., Zhang, Z.F., Han, B.X., An, G.M., Miao, S.D., and Xie, Y. (2007) *J. Am. Chem. Soc.*, **129**, 6362–6363.

69 Bühler, G., Thölmann, D., and Feldmann, C. (2007) *Adv. Mater.*, **19**, 2224–2227.

70 Xu, X.D., Zhang, M., Feng, J., and Zhang, M.L. (2008) *Mater. Lett.*, **62**, 2787–2790.

71 Cao, S.W., Zhu, Y.J., Cheng, G.F., and Huang, Y.H. (2009) *J. Hazard. Mater.*, **171**, 431–435.

72 Yin, S., Luo, Z.J., Xia, J.X., and Li, H.M. (2010) *J. Phys. Chem. Solids*, **71**, 1785–1788.

73 Liu, Y.H., Liu, P.I., Chung, L.C., Shao, H., Huang, M.S., Horng, R.Y., Yu, S.W., Yang, A.C.M., and Chang, M.C. (2011) *J. Mater. Sci.*, **46**, 4826–4831.

74 Kerner, R., Palchik, O., and Gedanken, A. (2001) *Chem. Mater.*, **13**, 1413–1419.

75 Li, D.S., and Komarneni, S. (2006) *J. Am. Ceram. Soc.*, **89**, 1510–1517.

76 Tsuji, M., Miyamae, N., Matsumoto, K., Hikino, S., and Tsuji, T. (2005) *Chem. Lett.*, **34**, 1518–1519.

77 Tsuji, M., Hashimoto, M., Nishizawa, Y., Kubokawa, M., and Tsuji, T. (2005) *Chem. Eur. J.*, **11**, 440–452.

78 Patel, K., Kapoor, S., Dave, D.P., and Mukherjee, T. (2006) *Res. Chem. Intermed.*, **32**, 103–113.

79 Tsuji, M., Miyamae, N., Lim, S., Kimura, K., Zhang, X., Hikino, S., and Nishio, M. (2006) *Cryst. Growth Des.*, **6**, 1801–1807.

80 Hu, X.L., Yu, J.C., Gong, J.M., and Li, Q. (2007) *Cryst. Growth Des.*, **7**, 2444–2448.

81 Murugan, A.V., Muraliganth, T., and Manthiram, A. (2008) *Electrochem. Commun.*, **10**, 903–906.

82 Murugan, A.V., Muraliganth, T., and Manthiram, A. (2008) *J. Phys. Chem. C*, **112**, 14665–14671.

83 Muraliganth, T., Stroukoff, K.R., and Manthiram, A. (2010) *Chem. Mater.*, **22**, 5754–5761.

84 Bilecka, I., Hintennach, A., Djerdj, I., Novak, P., and Niederberger, M. (2009) *J. Mater. Chem.*, **19**, 5125–5128.

85 Bilecka, I., Hintennach, A., Rossell, M.D., Xie, D., Novak, P., and Niederberger, M. (2011) *J. Mater. Chem.*, **21**, 5881–5890.

86 Carriazo, D., Rossell, M.D., Zeng, G., Bilecka, I., Erni, R., and Niederberger, M. (2012) *Small*, **8**, 2231–2238.

87 Ludi, B., Olliges-Stadler, I., Rossell, M.D., and Niederberger, M. (2011) *Chem. Commun.*, **47**, 5280–5282.

88 Firmiano, E.G.S., Cordeiro, M.A.L., Rabelo, A.C., Dalmaschio, C.J., Pinheiro, A.N., Pereira, E.C., and Leite, E.R. (2012) *Chem. Commun.*, **48**, 7687–7689.

89 Tompsett, G.A., Panzarella, B., Conner, W.C., Yngvesson, K.S., Lu, F., Suib, S.L., Jones, K.W., and Bennett, S. (2006) *Rev. Sci. Instrum.*, **77**, 124101.

90 Robb, G.R., Harrison, A., and Whittaker, A.G. (2002) *PhysChemComm*, **5**, 135–137.

91 Panzarella, B., Tompsett, G., Conner, W.C., and Jones, K. (2007) *Chemphyschem*, **8**, 357–369.

92 Whittaker, A.G., Harrison, A., Oakley, G.S., Youngson, I.D., Heenan, R.K., and King, S.M. (2001) *Rev. Sci. Instrum.*, **72**, 173–176.

93 Harrison, A., Ibberson, R., Robb, G., Whittaker, G., Wilson, C., and Youngson, D. (2002) *Faraday Discuss.*, **122**, 363–379.

94 Leadbeater, N.E., and Schmink, J.R. (2008) *Nat. Protoc.*, **3**, 1–7.

95 Schmink, J.R., Holcomb, J.L., and Leadbeaterr, N.E. (2008) *Chem. Eur. J.*, **14**, 9943–9950.

96 Schmink, J.R., Holcomb, J.L., and Leadbeater, N.E. (2009) *Org. Lett.*, **11**, 365–368.

9
Input of Microwaves for Nanocrystal Synthesis and Surface Functionalization Focus on Iron Oxide Nanoparticles

Irena Milosevic, Erwann Guenin, Yoann Lalatonne, Farah Benyettou, Caroline de Montferrand, Frederic Geinguenaud, and Laurence Motte

9.1
Introduction

Nanomaterials have a significant potential application in the fields of physics, chemistry, biology, medicine and material science because of their unique electronic, optical, thermal and catalytic properties. It has also been generally acknowledged that the physicochemical properties of nanomaterials strongly depend on their shape and size besides their inherent chemical constitutions. To improve the dispersibility of nanoparticles in suitable solvents and to translate intrinsic properties of nanoparticles to these extended applications, the nanoparticle surface functionalization itself is a key factor. Hence, a major current research direction is to develop new synthetic methodologies for the tunability of the particle size, shape, composition, and surface properties.

An outstanding class of functional materials with potential applications in almost all fields of technology consist of metal oxide nanoparticles, and especially iron oxide nanoparticles [1]. Iron oxide nanocrystals have been of much scientific and technological interest over the past decades, in particular for various magnetic applications, such as in magnetic recording devices, ferrofluids, magnetic refrigeration systems, and catalysis [2–5]. The recent focus on magnetic particles is in view of a large range of biomedical applications, such as magnetic separation systems for biomolecules and cells, magnetic resonance imaging contrast enhancement, therapy like hyperthermia, and drug delivery, as well as for multidetection systems based on biosensors [6–11].

For these applications, the chemical composition, morphology, size and surface controls of magnetic nanoparticles appear to be crucial. These characteristics are strongly dependent on the method of preparation. Many approaches have been employed to prepare iron oxide nanoparticles, including: co-precipitation, sol–gel methods, thermal decomposition, microemulsion and hydrothermal methods. In comparison to these well-established approaches, the use of microwave irradiation

Microwaves in Nanoparticle Synthesis, First Edition. Edited by Satoshi Horikoshi and Nick Serpone.
© 2013 Wiley-VCH Verlag GmbH & Co. KGaA. Published 2013 by Wiley-VCH Verlag GmbH & Co. KGaA.

represents a relatively new strategy in nanoparticle synthesis and surface functionalization.

In organic chemistry, microwave heating has become a standard tool for the efficient execution of organic reactions, often with excellent yields being obtained in reaction times considerably shorter than when conventional heating is used. Excellent reviews have been published on various aspects of microwave-assisted organic chemistry [12–17].

The intention of the present review is to highlight the input of microwaves for nanocrystal synthesis and surface functionalization. We focus on water dispersible iron oxide nanoparticles.

In addition to the introduction and conclusion, the review is divided into four parts. In Section 9.2 we briefly discuss the potential medical applications of these nanoparticles. Section 9.3 reports the conventional synthetic methodologies inducing either hydrophilic or hydrophobic iron oxide nanoparticles. Section 9.4 presents the specific conjugation methods that are used currently to functionalize nanoparticle surfaces. Section 9.5 discusses the microwave-assisted preparation routes that have been developed in recent years to synthesize and surface functionalize iron oxide nanoparticles.

9.2
Biomedical Applications of Iron Oxide Nanoparticles

"Nanomedicine" according to the National Institutes of Health refers to the applications of nanotechnology for treatment, diagnosis, monitoring, and control of biological systems [18]. Nanoparticles have a high ratio of surface area to volume as well as tunable optical, electronic, magnetic, and biologic properties, and they can be engineered to have different sizes, shapes, chemical compositions, surface chemical characteristics, and hollow or solid structures [19, 20]. These properties are being incorporated into new generations of drug-delivery vehicles, biological labeling, contrast agents and diagnostic devices, some of which are currently undergoing clinical investigation or have been approved for use in humans [21]. Within the different chemical compositions of nanomaterials, superparamagnetic iron oxide nanoparticles (SPION) are certainly the most promising material for medical applications, such as immunoassays, imaging, nanovehicles and drug carriers, hyperthermia treatment, and so on Figure 9.1.

The immunoassay is one of the most important methods in the field of clinical and environmental analysis. Immunoassay is based on the exceptional specificity that an antibody has for its own target antigen. The highly specific affinity coupled with a very low limit of detection has made immunoassay a widely accepted analytical method in various fields. Immunoassays using a radioisotopic technique have been heavily utilized since the 1960s [22] because radioisotopic labels have shown superior performance in both high selectivity and low detection limits. The use of radioisotopes, however, has many problems related to handling and

Figure 9.1 Examples of nanotechnology biomedical applications.

disposal; therefore, immunoassays using a nonisotopic technique have increased dramatically and have gained a certain status as a common analytical technique in recent years. The use of magnetic particles in immunological assays has grown considerably, as the particles' magnetic properties permit their easy separation and/or concentration in large volumes, allowing faster assays and, in some cases, improved sensitivity over currently available commercial methods [23]. This binding reaction is measured by detection of a magnetic signal. Recently, a new detection and characterization method was developed to take advantage of the non-linearities of superparamagnetic materials for magnetic immunoassays [24, 25]. The non-linearity measurement has the advantage of being both very sensitive and non-competitive with the biological environment. This allows the detection of nanograms of SPION within biological systems [26], Figure 9.2.

SPION are currently used as a contrast agent for magnetic resonance imaging (MRI). In nanoparticle-enhanced MRI, a contrast can be observed between tissues with and without SPION, owing to a difference in the precession frequency of the protons. Indeed the magnetism of SPION enhances the static and applied magnetic fields. Then the stronger local magnetic field shortens the relaxation times of protons, resulting in a stronger local MR signal. This property is currently used for MRI contrast enhancement, especially with T2 relaxation time alteration [27]. SPION are currently used to detect liver lesions, since these particles are taken up by macrophages [28]. SPION are also attractive as sensitive contrast agents for cancer imaging. The nanoparticles remained in the tumors for 24 h after injection,

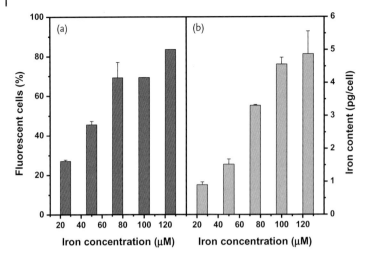

Figure 9.2 Quantification of the internalization of γFe_2O_3 nanoparticles by SW 480 colon carcinoma cells. (a) Percentage of transfected cells versus the iron concentration in the culture media (data obtained by flow cytometry). (b) Average weight of iron load per cell versus the iron concentration in the culture media (data measured via non-linear magnetization) [26].

as compared with 1 h for gadolinium-chelate contrast agents [29]. The reason for this difference is that the smaller nanoparticles are more easily taken up by tumor cells and diffuse out of the tumor more slowly [30].

Most nanoparticles used for drug targeting are hollow structures, such as liposomes [31]. Nevertheless, inorganic structures present a special interest for theranostic applications, that is, diagnostic and therapeutic combined systems. For example, antitumor magnetic nanoparticles have been designed by coating iron oxide nanoparticles with a clinically relevant antitumor agent: alendronate [32]. The antitumor agent anchors onto the nanoparticle surface through phosphonate groups. These nanoparticles have both MRI contrast agent and anti-cancer properties. This nanoparticle functionalization spectacularly improved cell penetration by alendronate and its antitumor effect. In particular, this behavior was enhanced in the presence of an appropriate magnetic field. *In vivo* experiments confirmed the therapeutic efficacy of γFe_2O_3@alendronate nanocrystals in the presence of a magnetic field [32].

Hyperthermia, also called thermal therapy or thermotherapy, is a type of cancer treatment in which body tissue is exposed to high temperatures (above 45 °C). Since SPION are excitable within a magnetic field these particles present a specific interest to deliver local heating (hyperthermia). Several methods of hyperthermia are currently under study, including local, regional, and whole-body hyperthermia [33–36].

Table 9.1 Reaction time and temperature of different conventional methods for synthesis of iron oxide nanoparticles.

Method	Reaction time	Reaction temperature
Coprecipitation	Less than 1 h	Room temperature to 90 °C
Forced hydrolysis	Hours to days	50 °C to 200 °C
Hydrothermal method	Up to 24 h	160 °C to 400 °C
Microemulsion	2 h	Above Kraft temperature
Sol–gel	Hours to days	Up to 400 °C
Polyol	30 min to hours	Polyol boiling temperature ($150 < T < 250$ °C)
Thermal decomposition	30 min to hours	Near solvent boiling temperature

9.3 Nanoparticle Synthesis

Many chemical methods have been widely used to produce nanostructured materials, Table 9.1. Particles from nanometers to micrometers are produced by controlling particle size during synthesis and by using competition between nucleation and growth. The preparation of colloidal particles of iron oxide has been the object of attention in many research fields. The synthesis of particles in a solution occurs by chemical reactions forming stable nuclei with subsequent particle growth. First, the solution reaches a critical supersaturation of the particle-forming species, then only one burst of nuclei occurs. A continuous diffusion of solutes onto the existing nuclei leads to the growth of the particles.

9.3.1 Synthesis in Aqueous Solution

9.3.1.1 Coprecipitation Method

The coprecipitation method is a synthesis that leads to the formation of ferrite nanoparticles. Ferrites $M^{II}Fe_2O_4$ (M = Fe, Co, Ni, Mn, Zn, Cd...) are ferric oxides of divalent metal. They crystallize in a spinel cubic structure. This method involves alkalizing an aqueous mixture of Fe^{3+} and M^{2+} ions. The size and shape of particles obtained are highly dependent on synthesis conditions and in particular: (i) the type of salt used (e.g., chlorides, sulfates, nitrates, etc.) the ratio of metal ions, (ii) the reaction temperature varying from room temperature to 90 °C, (iii) the pH, (iv) the ionic strength, (v) the presence of molecules that can be coordinated to the particle surface and (vi) the other reaction parameters (e.g., stirring rate, dropping speed of basic solution). The control of morphology and size polydispersity of the nanoparticles has been a scientific challenge since the early 1980 [37–39].

9.3.1.2 Forced Hydrolysis

The easiest way to prepare uniform colloidal metal (hydrous) oxides is based on forced hydrolysis of metal salt solutions [40]. The precipitation of iron oxides by hydrolysis of Fe^{3+} ions in aqueous media depends on the type of Fe(III)-salt and its concentration, pH, the presence of a complexing agent, temperature and time. Matijevic and Scheiner [41] demonstrate that small changes in reaction conditions produce particles with different morphologies. Under forced hydrolysis at elevated temperature, rods, spheres, cubes, double ellipsoid, star and cigar-shaped particles precipitate [41, 42]. For example, by using a complexing agent from the family of catecholamine, named dopamine (DOPA or H_3L^+), it is possible to form rod-like shapes (2 h aging, 90 °C) or cigar shapes (4 h aging, 90 °C) akaganeite particles. The DOPA is a natural organic molecule composed of a catechol part which has a very strong affinity to iron(III) [43].

9.3.1.3 Hydrothermal Method

The hydrothermal method includes the various techniques of crystallizing substances from high-temperature aqueous solutions at high vapor pressures. For iron oxide synthesis, the hydrothermal method leads to various shapes with interesting magnetic properties [44–49]. For example, unusual polyhedral structures of cubic Fe_3O_4 were fabricated by Duan et al. [50] in high yield via a facile hydrothermal method in the presence of a surfactant cetyltrimethyl-ammonium bromide. Hexagonal, dodecahedral, truncated octahedral, and octahedral shapes can be prepared by changing the concentration of the CTAB in L-arginine solutions.

9.3.1.4 Aqueous Sol–Gel Method

The aqueous sol-gel process can briefly be defined as the conversion of a precursor solution into an inorganic solid via inorganic polymerization reactions induced by water. This process involves the evolution of inorganic networks through formation of a colloidal suspension [22] and gelation of the sol to form a network in a continuous liquid phase (gel). Three reactions generally describe the sol–gel process: hydrolysis, alcohol condensation and water condensation. Many parameters can be tuned, such as pH, temperature and time of reaction, reagent concentrations, catalyst nature and concentration, H_2O/Si molar ratio, aging temperature and time, and drying. These parameters can affect the rate of hydrolysis and the condensation reaction. The precursors used for this synthesis are metal or metalloid elements surrounded by various reactive ligands [51]. Among other preparative techniques the sol–gel process has proved to be a convenient method for the synthesis of nanocomposites [52].

9.3.1.5 Direct Micelles Microemulsion Method

A microemulsion is a thermodynamically stable isotropic dispersion of two immiscible phases (water and oil) using surfactant molecules. Self-assembled structures of surfactants can be formed, ranging from spherical and cylindrical micelles to lamellar phases. The use of microemulsions is one of the most promising methods for obtaining shape- and size-controlled iron oxide particles. The principle is based on the decomposition by a weak base (for example, dimethylamine [24, 53]) of micelles formed by iron dodecyl sulfate molecules, Figure 9.3.

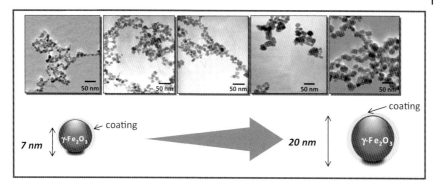

Figure 9.3 Transmission electron microscopy images of iron oxide nanoparticles differing by their sizes synthesized using direct micelle procedure [24].

9.3.2
Synthesis in Non-Aqueous Solvent

9.3.2.1 Reverse Micelle Microemulsion Method
The principle is the same as the direct micelles microemulsion method but here we have a water-in-oil emulsion, so the continuous phase is a non-aqueous solvent [54].

9.3.2.2 Non-Aqueous Sol–Gel Method
In non-aqueous sol–gel chemistry the transformation of the precursor takes place in an organic solvent under exclusion of water. The list of potential precursors is longer than is the case for the aqueous sol–gel method and includes, in addition to inorganic metal salts and metal alkoxides, also metal acetates and metal acetylacetonates.

Pinna et al. described a simple one-pot reaction involving iron(III) acetylacetonate dissolved in benzyl alcohol and treated in an autoclave between 175 and 200 °C. This approach leads to monocrystalline magnetite particles with sizes ranging from 12 to 25 nm [55].

9.3.2.3 Polyol Synthesis
The principle is based on the reduction of metallic salts in a dialcohol solution. The nano-objects are collected from the dialcohol solution. The polyol process has been developed by Fiévet et al. [56] and consists in dispersing metal precursors (hydroxide, acetate, oxide, etc.) in a polyol solution (usually α-diol) and then in warming without exceeding the polyol boiling point. Polyols are used as solvents because they can act as reducing polar solvents and also because of their high boiling temperature which allows solubilization of a large number of metal salts and reduction of the latter over a large temperature range. The main advantage of the polyol process, except that it easily provides several grams of nano-objects, comes from the richness of the parameters which control the reaction. This allows

one to obtain a large variety of nanoparticles shapes, dimensions and composition [57–61].

9.3.2.4 Thermal Decomposition

Thermolysis of the organic precursor leads to the formation of germs. The number of germs formed depends on the stability of the precursors used. Thus, the phases of nucleation and growth can be controlled. Organic ligands present in the medium control the size dispersion of particles by a more or less strong interaction with the particle surface. The key factors of this method are (i) the ratio of reagents, (ii) the synthesis temperature and (iii) the reaction time. If the precursor is used in the zerovalent state, thermolysis leads to the formation of metallic particles which will then be oxidized by refluxing in air or by addition of an oxidant. Different precursors can be employed for the production of iron oxide nanostructures: $Fe(CO)_5$ [62–64], iron acetylacetonate [65, 66] and iron oleate complexes [67–69].

9.4
Nanoparticle Surface Functionalization

Basically a biomedical mNP consists of an inorganic nanoparticle core and a biocompatible surface coating that provides stability under physiological conditions and may allow further modifications to provide specific properties. As discussed above, such nanohybrid materials are prepared using two different approaches:

- Water compatible iron oxide nanocrystals (coprecipitation or using normal micelles)
- Nonpolar iron oxide nanocrystals (Precipitation in reverse microemulsions or thermal decomposition of organometallic coumpounds in the presence of hydrophobic ligands)

Both nanocrystals should be coated to be suitable for biological applications; then they may interact with systems such as ligands, receptors, or drugs. For nonpolar nanocrystals, the hydrophobic surfactant coating needs to be replaced by a hydrophilic one by using ligand exchange and/or surface silanization and polymer or micelle coating [70].

Water-soluble nanocrystals may be formulated directly with their final properties in a one step strategy. Indeed, the surface presents hydronium or hydroxy groups (depending on pH). These groups will interact strongly with a high coordination capacity functional group (carboxylic acid, phosphonate, etc.), Figure 9.4. Another subsidiary strategy is to coat the surface with a bifunctional ligand interacting strongly with the surface and simultaneously expose headgroup functionality such as amine, carboxyl, and aldehyde groups. This indirect functionalization is by far the most used strategy as it allows a precise molecular control on the surface.

Surface engineering of the particles is a key step for the bioactive mNP concept. Both strategies allow decoration of the magnetic nanoplatform surface with a large

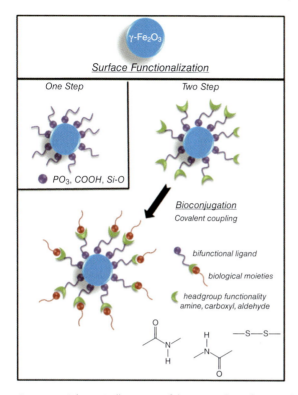

Figure 9.4 Schematic illustration of direct or indirect functionalization of nanoparticle surface with biomolecules.

range of biological properties (fluorescence, targeting, drug delivery, imaging, etc.) [71].

9.4.1
Hydrophobic Nanocrystals

High-temperature organometallic synthetic routes produce nanoparticles with hydrophobic surface ligands such as oleic acid (OA), trioctylphosphine oxide (TOPO) and oleyl amine (OLA). There are four main routes to modify hydrophobic nanoparticles and render them soluble in aqueous solution, as illustrated in Figure 9.5 for native oleic acid-capped nanoparticles.

9.4.1.1 Ligand Exchange
In the first route, called "ligand-exchange", the native monolayer of hydrophobic surface ligands is exchanged with ligands containing head groups that bind on one end of the magnetic nanoparticle surface and hydrophilic tails that interact with aqueous solvent with the other end (bifunctional molecule). The complexing strength of the surface anchoring group must be high enough to ensure maximum

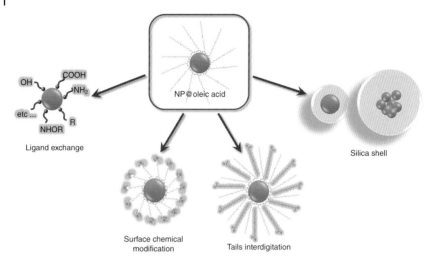

Figure 9.5 From a hydrophobic oleic acid-capped nanoparticle to a water-soluble nanoprobe.

surface coverage. Many different hydrophilic ligands can be grafted onto the NPs, such as catechols [72], polypeptides [73] bisphosphonates [74, 75], betaine hydrochloride [76], silanes [77], polymers [78, 79]. The transfer conditions are very varied: some of the ligand exchanges are carried out in very easy conditions, in a few minutes with sonication [75, 80]. The transfer agent acts as a ligand intermediate to help the water solubilization of the nanoparticle. Tetramethylammonium hydroxide (TMAOH) is the most usual transfer agent: once around the particle, this quaternary ammonium salt brings a hydrophilic surface charge, which is then removed by the final ligand.

9.4.1.2 Surface Chemical Modification

Another approach is the chemical modification of the oleic acid tail so as to have a hydrophilic terminal function on the surface of the NPs. The oleic acid molecule has a double link carbon–carbon in the middle of its hydrocarbon chain; the ozonolysis of this double bond via potassium permanganate ($KMnO_4$) transforms oleic acid into azaleic acid and yield a second carboxylic acid functionality [76, 81]. The azaleic acid coating is water-soluble and the outer carboxylic acid function allows the direct biofunctionalization of the NPs, Figure 9.6.

9.4.1.3 Tails Interdigitation

A third route for transferring NPs in aqueous media consists in keeping the first ligand coating and adding a second one composed of equivalent-structured molecules. Interdigitation between hydrocarbon tails of acid oleic and hydrophilic molecules takes place [82, 83]. This "swift" transfer is used as an intermediate step for silica [84] or polymer [85] shell formation. The tails interdigitation is softer than ligand place-exchange. The surface of the NPs is not modified, but the bilayer

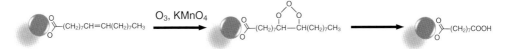

Figure 9.6 Transformation of oleic acid into azaleic acid via ozonolysis.

coating can be sensitive to pH variation or salts addition and is thus less stable than a grafted hydrophilic coating.

9.4.1.4 Silica or Polymer Shell

The last way to obtain water-soluble NPs is by encapsulation inside a silica shell. An intermediate step can be bilayer formation, followed by the addition of tetraethyl orthosilicate (TEOS). In that way single NP [86] or several NPs [87] can be embedded in a mesoporous silica shell.

9.4.2
Water Soluble Nanocrystals

9.4.2.1 Direct Surface Functionalization

The direct functionalization of a nanoparticle is a reaction that allows the binding of a molecule, for example, a bioactive molecule, onto the surface of a particle without any intermediate or additional reactive species. Once the molecule is bound, the particle can be used for its own therapeutic properties or as an intermediate for further reactions.

Direct functionalization involves adsorption mechanisms between the nanoparticle surface (adsorbent) and a molecule of interest (adsorbate). Adsorption occurs at the nanocrystal's surface where many lattice defects are observed and where adsorbates can be attracted to the surface via electrostatic forces. Three types of adsorption mechanism are generally described: physisorption, chemisorption and electrostatic attraction.

For physical adsorption the binding of the adsorbate is mainly guided by Van der Waals dipole–dipole and dipole–quadrupole interactions. Owing to the nature of the forces applied, which do not involve significant changes in the electronic structure of the adsorbed species, physisorption implies generally low binding energies and, therefore, is a fully reversible phenomenon, which can lead to desorption of the molecules.

In chemisorption, forces involved lead to drastic changes in the electronic state of the adsorbate with binding energies characteristic of a covalent bond. The modifications can be easily detected by physical methods such as infrared or UV–visible absorption spectroscopy, Figure 9.7. In order to avoid a strong desorption phenomenon; chemisorption via carboxylate, sulfate, phosphate groups or derivatives will be preferred for the anchoring of a small molecule on a particle.

The third adsorption mechanism is by electrostatic repulsion. Metallic oxide nanoparticles can be stabilized in a colloidal solution by simply adjusting the pH

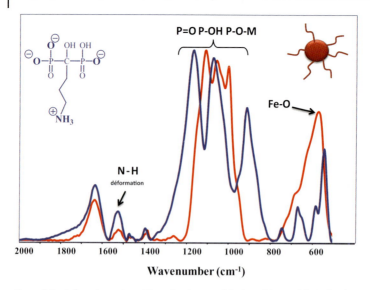

Figure 9.7 Infrared spectra of free alendronate (blue) and iron oxide@alendronate particles (red) [32].

of the solution and thus without any additional coating molecules. Indeed, at the solid/liquid interface the coordination of the metallic cation is conserved via chemisorption of a water molecule, leading to the hydroxylation of the metal. This confers to the nanoparticle an acid–base behavior in Brönstedt terms [88]. When the pH is below the isoelectric point (IEP) particles are positively charged (M-OH^{2+}, where M is a metal atom) while beyond this point, as the pH is increasing, the surface charge becomes negative (M-O$^-$), as confirmed by zeta potential values. We can take advantage of this surface property to drive the adsorption of molecules having an opposite charge onto a nanoparticle.

For example, bisphosphonates (BPs) show high affinity for divalent cations [89]. BPs can also be anchored to the surface of a metal oxide through the formation of strong M–O–P bonds via a heterocondensation reaction and coordination of the phosphoryl oxygen. Coordination of one phosphonic acid can lead to the formation of three M–O–P bonds with the metal oxide surface, thus conferring to the BPs a great binding affinity to the iron oxide nanoparticles [90]. Motte et al. applied this technique with success for the surface passivation of maghemite nanocrystals using different bisphosphonates [10, 32, 89, 91, 92]. The FTIR spectroscopic data, Figure 9.7, are indicative of large modifications in the P–O stretching region confirming thus the chemical adsorption of phosphonic acids onto the maghemite surface.

The direct functionalization approach is smart, and simple. Most of the syntheses are carried out in water at ambient temperature. It is an easy way to obtain bioactive nanoparticles. Direct functionalization, by decreasing synthesis intermediates and derivatives, by reducing the utilization of organic solvent, and by

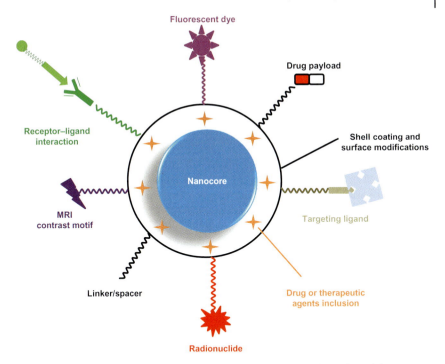

Figure 9.8 Optionally functionalized and devised nanoparticles could be achieved for individualized diagnosis and treatments.

reducing the temperature of reaction, could be considered as alternative techniques for achieving green chemistry goals in nanotechnology.

9.4.2.2 Two-Step Surface Functionalization

To insert biomolecules on the particles surface with control of their architecture and surface density, several chemical methodologies are described. The two more widespread strategies are the classical technique which leads to the formation of amide, ester, carbamate bonds, and the modern technique of "click chemistry". A third methodology not using covalent linkage but strong biological interaction is also described, biotin–streptavidin linkage being the best known example.

9.4.2.2.1 Covalent Coupling

Covalent linkages are strong and stable bonds, which can be specifically formed between functional groups such as amino, carboxylic acid, and thiol groups and other corresponding groups on the NP surface. It allows introduction on the nanoparticle surface of many types of biomolecules such as fluorescent molecules, drugs, targeting molecules and so on [9], Figure 9.8.

The conjugation of ligands to nanoparticles requires various chemistries. As reported in Veiseh's review [93], NP surfaces functionalized with amine, sulfhydryl, aldehyde, carboxylic acid and active hydrogen groups can be targeted. These strategies are particularly suitable for small molecule conjugation but were extended to peptides, proteins and antibodies. Most commonly, amines or carboxylic acids are present on the nanoparticle surface and carbodiimide-coupling chemistry is used for biomolecule conjugation. Usually, these functional groups are added to the NP surface via its first surface functionalization step (see Figure 9.4).

The carbodiimide chemistry, Figure 9.9, induces the formation of an amide bond between a carboxylic acid and an amine, using as coupling agent 1-ethyl-3-(dimethylaminopropyl) carbodiimide hydrochloride (EDC). This derivative is completely soluble in water. N-hydroxysuccinimide (NHS) or sulfo-NHS are usually used to enhance the coupling with EDC (Figure 9.9). The reaction involves the formation of an active intermediate NHS ester, which then reacts with the amine to give the amide bond. The coupling rate of this reaction could also be greatly enhanced by the used of microwave energy [91].

In our group, we used bisphosphonate molecules binding to the surface and exposing a functional group (amino, carboxylic acid) [10, 89, 91]. We have shown that the nanoparticle surface presents 1200 reactive amino groups per NP. These amine functionalities were successfully covalently conjugated with the activated carboxylic function of Rhodamine molecules, leading to the elaboration of magneto-fluorescent iron oxide nanoparticles with both therapeutic and dual imageries (fluorescence and MRI) properties.

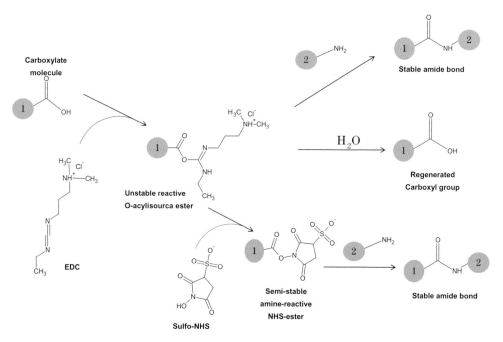

Figure 9.9 Different activation mechanisms for covalent coupling on carboxylic surface functionalized nanoparticles.

Click chemistry is the synthetic approach to linking two molecules covalently that has impacted modern chemistry and biological chemistry more this last decade. Click chemistry has been used in various domains, such as medicinal chemistry, material and polymer chemistry, imaging, and so on [94–97]. It has been widely used for nanoparticle functionalization recently. The concept of click chemistry was introduced by K. Barry Sharpless in 2001 [98]. This term describes chemical reactions which can covalently bond two different species rapidly and efficiently (ideally quantitative reactions without by-product formation). These reactions must be specific, chemo-selective, and should be realized at ambient temperatures, in aqueous or hydro-alcoholic media, and in the presence of air or moisture. Few chemical reactions have been described that have been accepted as click chemistry reactions: dipolar [4+2] cycloaddition (Diels–Alder reaction), oxime or hydrazone formation, thiol-ene and thiol-yne reactions and the 1,3 dipolar cycloaddition between azide and alkyne catalyzed by copper (I). This latter reaction, originally described by Huisgen in 1963 [99] and re-introduced by Sharpless, is by far the most common example of a click chemistry reaction [100, 101], Figure 9.10.

Several examples may be found in the literature. Hayshi *et al* [102, 103]. Developed folic acid-functionalized Fe_3O_4 nanoparticles from Huisgen 1,3 dipolar cycloaddition. They showed that this functionalization enhanced the cellular affinity of SPIONs to glioma cells and allowed a better entry into the cells. Von Maltzahn *et al.* [104] covalently clicked alkyne-modified cyclic LyP-1 targeting peptides to the nanoparticles in order to direct their binding to p32-expressing tumor cells. This targeting was assessed *in vitro* and *in vivo* on mice bearing human MDA-MB-435 cancer xenografts and injected intravenously via the tail vein. Click chemistry was also used by Devaraj *et al.* [105] to 18F labeled nanoparticles for *in vivo* PET-CT Imaging. Monocrystalline iron oxide nanocompounds coated with dextran were cross-linked with epichlorin hydrin, aminated, and labeled with near-infrared fluorochrome. The remaining amine functions at the surface of the nanoparticles were linked to an azido PEG which was reacted with a 18F labeled PEG alkyne. The obtained nanosystem showed very high signal-to-noise in *in vivo* dynamic PET imaging and pharmacokinetic analysis.

The most challenging task in nanoparticles functionalization by click chemistry is the coupling of bio-macromolecules such as antibodies, enzymes or DNA. Cutler *et al.* [106] used the copper-catalyzed azide-alkyne reaction to form a dense monolayer of oligonucleotides on a superparamagnetic nanoparticle core. They coupled oligonucleotide modified with a terminal alkyne to azide-functionalized

Figure 9.10 1,3 Dipolar cycloaddition between azide and alkyne catalyzed by copper (I).

iron oxide particles. A last example showing the interest of click chemistry assisted bioconjugation was given by Thorek et al. [107]. They carried out a comparative analysis of nanoparticle–antibody conjugations between carbodiimide methodology and click chemistry. They evaluated the efficiency of conjugations between antibodies and superparamagnetic iron oxide nanoparticles coated with dextran and bearing azide or carboxylic functions. The conjugation was done using click chemistry with alkynated antibodies or by formation of carbodiimide cross-linking to native antibodies. The click reaction allowed a higher extent and efficiency of labeling compared with carbodiimide, thus requiring less antibody. Further, conjugates prepared via the click reaction exhibited improved binding to target receptors.

Though click chemistry coupling via 1,3 cycloaddition on nanoparticles has many unique advantages there is a limitation to its use. The Cu catalyst needed for completion of the reaction must be removed before use in several applications. First, it is known that use of Cu *in vivo* could be linked to several disorders [94]. Moreover, other click chemistry reactions not involving metal catalysis are described which may certainly be used in the future for inorganic nanoparticles biofunctionalization.

9.4.2.2.2 Non-Covalent Interactions

Another route to surface functionalized inorganic nanoparticles is to create a non-covalent but very strong and specific interaction, such as found in nature, for example protein–ligand interactions. Functionalization of nanoparticles with biomolecules by specific affinity interactions uses this long known methodology developed initially for the functionalization of macrosize supports. Among all the techniques described, the leading one is still the streptavidin–biotin conjugation method. Several examples of utilization of the biotin–streptavidin interaction to functionalized nanoparticles with biomolecules can be found in the literature [7, 108–110].

9.5
Microwave-Assisted Chemistry

Since the first use of microwave energy in organic chemistry [111], this field, now known as MAOS (microwave-assisted organic synthesis) is well established in most modern laboratories. In almost all areas of organic chemistry there are reports of the increased yield, the enhanced product purities and the drastic decrease in reaction time using microwave "flash heating" [12, 16, 112–114]. These advantages make MAOS an eco-friendly synthesis approach, termed as green chemistry.

Microwave-enhanced chemistry is based on the more efficient heating of materials compared to conventional heating procedure (oil bath, heating mantle, etc.). This effect is due to the so-called "microwave dielectric heating" effects. This phenomenon is the conversion of the absorbed microwave energy by the media into heat. Two major mechanisms are responsible for this dielectric heating: the dipolar polarization mechanism which is due to the effect of microwaves on dipolar moment creating friction and collisions between molecules and thus generating heat, and the conduction mechanism which is due to the interaction of microwaves with ions in solution, generating movement and thus releasing heat [115]. Since the first uses of microwaves some authors have suggested the existence of possible "non-thermal microwave effects". This highly controversial effect is put forward when obtained enhanced results cannot be explained by thermal or kinetic effects. Some answers to this debate could be found in recent work on the rational design of microwave-actuated organic reactions [116].

Microwave-assisted synthesis is becoming increasingly used in all parts of chemistry, for example microwave-assisted drug discovery [117–119], microwave-assisted polymer chemistry [120, 121] and microwave-assisted solid phase peptide synthesis [122] have been reviewd recently. In the area of nanotechnology only recently have reviews been published on the use of microwaves [123, 124], though assisted microwave synthesis is really well suited to this area of chemistry using for example high temperature, highly polar reactants, heterogeneous media, and so on. Here we will focus on the use of microwave energy for the synthesis and functionalization of inorganic nanoparticles and, more precisely, on iron oxide nanoparticles.

9.5.1
Microwave-Assisted Synthesis of Nanoparticles

The use of microwaves in organic chemistry has become widely popular due to its advantages, such as rapid volumetric heating, higher reaction rate, reduced reaction time, and increasing yield of products compared with conventional heating methods. A few works exist on iron oxide nanoparticles synthesis but its almost instantaneous "in core" heating of materials in a homogeneous and selective manner, different from the classical methods is very attractive in this field. Microwave energy can heat the entire object to the crystallization temperature rapidly and uniformly. Satisfactory control over the size of nanoparticles can be achieved if there is a quick nucleation event leading to the solution supersaturation. This results in homogeneous nucleation, a subsequent growth of the initially formed embryos and a shorter crystallization time when compared to the conventional method. Mono-dispersive particles can be prepared by carefully controlling the kinetics of the reaction. microwave-assisted hydrothermal (MW-HT) and solvothermal (MW-ST) methods are exploited for nanomaterials synthesis as they offer a clean, low-cost approach to the synthesis of nanocrystals within a very short reaction time (minutes), Table 9.2. This section aims to describe the advantages of microwaves in iron oxide synthesis and the progress made in the last decade.

9.5.1.1 Microwave-Assisted Hydrothermal Method
The first report dates back to 1998 and concerns the formation of acicular goethite/ magnetite particles by decomposition of ferric and ferrous salts in aqueous

Table 9.2 Conventional iron oxide synthesis compared to similar synthesis implemented into the microwave device.

Method	Conventional		Microwave-assisted	
	T (°C)	Time	T(°C)	Time
Coprecipitation	<90	<1 h	–	–
Forced hydrolysis	<200	>hours	200	<30 min
Hydrothermal	<400	~24 h	150–220	<25 min
Microemulsion	>T (kraft)	2 h	>T (kraft)	2 min
Sol-gel	<400	>hours	200	<30 min
Polyol	<250	<hours	200	<15 min
Thermal decomposition	T (boiling)	<hours	/	6 min

solution under microwave radiation with a power ranging from 60 to 410 W at 80 °C for 6 h. Wang et al. demonstrated that microwave radiation can provide an efficient energy source in driving the hydrolysis of an aqueous solution of ferric and ferrous ions to produce either goethite or magnetite, depending on the reaction conditions [125]. After that, a microwave one-step "flash" thermohydrolysis of nitrate iron (III) solutions for hematite nanoparticles synthesis was published in 1999 by Rigneau et al. [126]. Then, microwave-assisted synthesis of iron oxide particles became increasingly popular. For instance, many articles show that the synthesis of hematite (α-Fe_2O_3) can be achieved by microwave-assisted hydrothermal reactions [126–129] or by microwave heating in an open vessel [130, 131].

In 2007, Sreeja and Joy [132] reported the synthesis of maghemite nanoparticles with average particle size of 10 nm, at 150 °C, in a short time duration of 25 min, by an efficient microwave hydrothermal method. Their work showed that lower temperature and less reaction time are required to obtain comparable results by using microwave heating.

Komanerni et al. also compare the conventional hydrothermal method to MW-HT conditions from room temperature to 150 °C. Magnetite nanoparticles were synthesized from $FeCl_2$ using different alkaline sources. The crystallinity of magnetite increased with increasing temperature under microwave-assisted conditions. Moreover, the use of microwaves significantly increased the size of the magnetite particles [133].

A high-yield route to monodisperse cubic-phase spinel maghemite (γ-Fe_2O_3) nanoparticles in immiscible H_2O/toluene media was also presented by Baruwati et al. [134]. An aqueous solution of nitrate or chloride metal salts was heated at 160 °C for 1 h under MW irradiation. Oleic acid molecules were dissolved in toluene and act as both stabilizer and transferring agent at the water/toluene interface. The authors reported a production in grams of uniform spherical particles in the 4–9 nm size range.

The direct micelle synthesis was also implemented in the microwave device by our group. We show in Figure 9.11a and b that using microwave irradiation, the

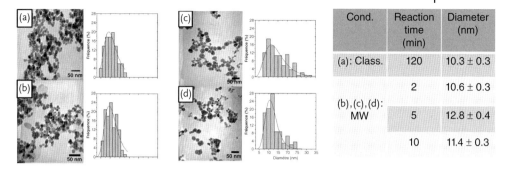

Figure 9.11 (a) Classical direct micelle synthesis compared to (b, c, d) MW-assisted direct micelle synthesis at different irradiation times.

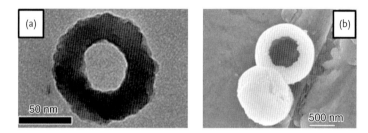

Figure 9.12 (a) TEM image of a nanoring [135] and (b) SEM image of hollow α-FeOOH [137].

reaction time is reduced from 120 to 2 min in order to obtain spherical nanoparticles of 10 nm in size. By increasing the irradiation time, we obtain larger particles sizes (Figure 9.11b, c and d) (article in progress).

A large variety of differently shaped particles can be produced using the microwave-assisted method. In the literature, hematite particles produced under microwave irradiation are in the form of nanorings[135], spindles [136], spheres [127, 137], ellipsoid [138], snowflake [139], and so on.

The hydrolysis of an iron precursor ($FeCl_3$) under microwave irradiation at 220 °C for 25 min promotes nucleation in the presence of $NH_4H_2PO_4$ and aggregation, as well as the "coordination-assisted dissolution" by the ligand of phosphate ions. Due to the superheating and nonthermal effects induced by the MW-HT process, a hole is generated gradually at the center of each primarily formed hematite nanodisk, resulting in the formation of nanorings [135, 136], Figure 9.12a. This synthesis is based on a non-microwave process at 220 °C for 48 h that leads to single crystalline nanotubes of hematite [48].

Hu *et al.* show that the MW-HT process can lead to various morphologies of α-Fe_2O_3, such as nanoparticles, dendritic crystals with different sizes, plates and self-organized nanorods, by adjusting experimental conditions [139]. They reduce the synthesis time from days to minutes using microwaves.

Hollow spheres [137] of α-Fe$_2$O$_3$, Figure 9.12b, can be obtained through the microwave preparation of α-FeOOH at 160 °C during 30 min irradiation of Fe(NO$_3$)$_3$·9H$_2$O, urea and 1-n-butyl-3-methyl imidazolium tetrafluoroborate ([BMIM][BF$_4$]) in water.

For magnetite or maghemite phases, microwave-assisted synthesis permits one to obtain particles in the form of roses [140], wires [141], nanoplates [142], and so on.

Zhou et al. [142] successfully prepared hexagonal magnetite (Fe$_3$O$_4$) nanoplates with an average edge length of 80 nm in large quantities by a facile microwave-assisted route at 240 W for 10 min using NaH$_2$PO$_2$ as the shape-controlling agent.

Conventionnally, Fe$_3$O$_4$ nanorods and nanowires are obtained by means of long (>10 h) hydrothermal conditions in a stainless steel autoclave at high temperature (>100 C) in presence or not of a magnetic field [40, 143–146]. A coprecipitation process in the presence of PVP and sonochemical oxidation has also been reported [147, 148].

Using FeSO$_4$ as precursor, Muraliganth et al. [141] obtained 20–50 nm thick Fe$_3$O$_4$ nanowires under MW-HT conditions with polyethylene glycol (PEG-400) acting as a soft template in water at temperatures as low as 150 °C in just 15 min, Figure 9.13.

In our group we use microwave irradiation for the flash phase transformation of an iron oxide phase to obtain maghemite nanorods [43]. In this procedure, the γ-Fe$_2$O$_3$ nanorods are produced through microwave-assisted reduction using β-FeOOH nanorods obtained by forced hydrolysis and hydrazine as precursors and redactor, respectively, Figure 9.14. This reaction takes less than 2 min under microwave irradiation.

First qualitative observations tend to prove the modification of the structure and magnetic properties. After 4 MW cycles of 30 s at 100 °C, the color of the colloidal suspension changes from orange to black and the final particles are easily attracted by a permanent neodymium magnet, Figure 9.14b and 9.14c. These two qualitative modifications suggest the formation of Fe$_3$O$_4$ or γ-Fe$_2$O$_3$ in our sample.

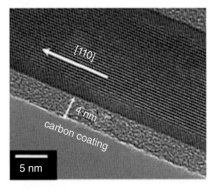

Figure 9.13 TEM image of Fe$_3$O$_4$ nanowires synthesized under MW-HT conditions [142].

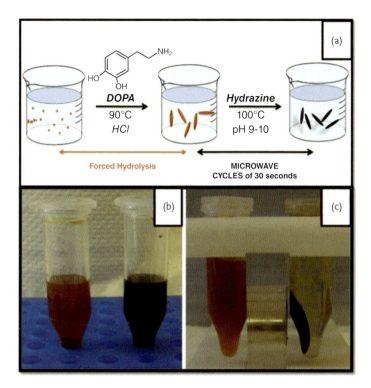

Figure 9.14 (a) Schematic representation of the two steps synthetic route. (b, c) Initial iron hydroxide (orange) and final iron oxide (black) colloid. (b) Nanoparticles are well dispersed and stable at pH 2. (c) The solutions are at pH 7 and final iron oxide nanoparticles (in an aggregated state) are attracted by a neodymium magnet [43].

The shape of the final particles was found to be strongly dependent on the reaction conditions. After 4 MW cycles, the transformation reaction did not affect the shape of the particles. A typical TEM image of the as-prepared sample is shown in Figure 9.15a. Comparing with the β-FeOOH precursor, the aspect ratio of the final nanorod decreases. This corresponds to an increase in the nanorod width with concurrent decrease in average length. This result suggests that the MW reduction process occurs via a dissolution–recrytallization mechanism [149]. Moreover, on increasing the MW cycles from 4 to 6 or 8, the morphology of the nanorods changes to hexagon-like particles (Figure 9.15c and 9.15e).

9.5.1.2 Microwave-Assisted Solvothermal Method

Even if the microwave hydrothermal method is by far the most reported in iron oxide microwave-assisted synthesis, other non-aqueous synthesis methods have been recently implemented in the microwave device. The mainly described MW-ST process is the polyol one. This method is based on exploiting differential dielectric constants to induce preferential heating and decomposition of the oxide precursors in the presence of suitable capping agents.

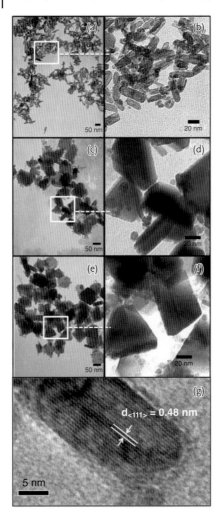

Figure 9.15 Influence of microwave cycles on the final iron oxide shape. TEM (a, c, e) and high magnification images (b, d, f) of the selected zone after 4 (a), 6 (c) and 8 (e) MW cycles. (g) HRTEM micrograph of the nanorod after 4 MW reduction [43].

In 2005, Rao *et al.* [150] studied the decomposition of ferric acetate in ethanol under reflux for 30 min under a microwave power of 160 W. The obtained Fe_2O_3 product was amorphous and was subsequently heated to 200–600 °C to crystallize.

Using a microwave solvothermal method, Zhang *et al* [151] prepared uniform Fe_3O_4 nanospheres with a diameter of 100 nm. The reaction takes place in the presence of poloxamer P123, $FeCl_3 \cdot 6H_2O$ and sodium acetate in ethylene glycol at 210 °C for 30 min with a microwave power of 150 W.

Magnetite nanoparticles of 3–5 nm are synthesized on multiwall carbon nanotubes by decomposition of Fe(acac)$_3$ in polyol solution at 200 °C in 15 min under microwave irradiation [152]. In the same manner, Xiao et al. presented a simple and rapid microwave-assisted polyol process to synthesize superparamagnetic Fe$_3$O$_4$ monodisperse nanoparticles in the size range from 4 to 8 nm by varying reaction parameters. The reactants were heated to 220 °C by microwave for 10 min [153].

Highly crystalline metal oxide nanoparticles in good yields were prepared within minutes and even seconds using a combination of non-aqueous sol–gel chemistry and microwave heating [154]. Thus, the use of microwave heating offers an immense reduction in reaction times in compared to traditional heating in an oil bath or autoclave. This fast heating to 200 °C leads to homogeneous nucleation.

Iron oxide nanocubes [155] were prepared by thermal decomposition of iron oleate complex in the presence of oleic acid via a microwave-assisted solvothermal method with a power of 400 W for 6 min, followed by Ostwald ripening procedures by aging at 180 °C for 10 h. During the microwave irradiation, the rapid nucleation produced relatively uniform small spherical iron oxide nanoparticles with average diameter of about 6 nm.

An increasing number of syntheses have been implemented in the MW device. Regarding iron oxide, and especially magnetite/maghemite nanoparticles, the field remains open. Indeed, the current state-of-art MW-prepared nanostructures are still far from those of colloids obtained under conventional heating even if MW heating technology has been proven to drastically decrease the reaction time and also the temperature.

MW treatment has shown a great degree of flexibility and a high level of refinement in:

- The composition: many iron oxide phases can be achieved and we have also reported a flash transformation in which the elongated shape is maintained.
- The size: a range of sizes are obtained, from nanometers to micrometers
- The shape: even if the number of articles is limited compared to conventional heating, the few reports existing present an impressive variety of shapes.

Nevertheless, the underlying phenomena in MW-assisted synthesis have to be explored; it is a great challenge in the microwave field.

9.5.2
Microwave-Assisted Functionalization of Nanoparticles

The use of microwaves is fairly well documented for the synthesis of iron oxide nanoparticles as well as other types of inorganic nanocrystals. Nevertheless, the input of microwave irradiation for the surface modification of these materials is only emerging over the past five years. Microwave irradiation has already been shown to be effective for surface functionalization of carbon nanostructured materials [156], polystyrene micro- and nano-particles [157] and silica nanoparticles

[158, 159], but relatively few results on inorganic materials have been described. In all cases, microwave irradiation reduces the reaction times and gives rise to products with higher degrees of functionalization than those obtained by conventional thermal methods.

9.5.2.1 Gold Nanoparticle Microwave Functionalization

The first report of the use of microwaves for inorganic nanoparticle surface modification was given by Sommer et al. [160] for the facile functionalization of gold nanoparticle via microwave-assisted 1,3-dipolar cycloaddition. Gold nanoparticles bearing azide functions on their surfaces were successfully reacted with a library of alkynes substrates, Figure 9.16.

The reaction was carried out in a closed microwave reactor and was completed in 10 min. at 100 W with a shut-off temperature of 100 °C. Unfortunately, the authors did not compare the results with non-microwave-activated conditions. In the microwave conditions described, no change on nanoparticles was noticeable, except when treated for a period in excess of 15 min when decomposition through particle aggregation occurred.

More recently, Bahadur et al. [161] described a rapid, simple and one-step method for preparing silica-coated gold nanoparticles with fine tunable silica shell thickness and surface functionalization of the prepared particles with different groups. Silica coating was carried out by mixing monodisperse citrated Au nanoparticles with tetraethoxysilane (TEOS) and ammonia under microwave irradiation. This methodology enables uniform silica coating of the nanoparticles without the use of silane coupling agent or the need for a pre-coating step. Moreover, the silica shell thickness can be tuned from 5 to 105 nm by only changing the TEOS

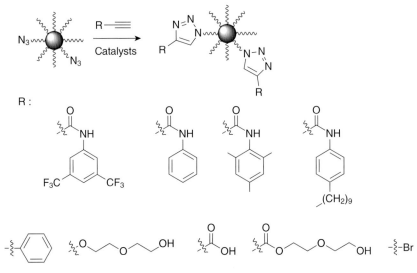

Figure 9.16 Library of alkynes used as substrates in the 1,3-dipolar cycloadditions and the functionalized nanoparticles before and after the transformation [160].

concentration. The use of microwaves permits simplification of the coating process and drastically decreases the reaction time (from a few hours to a day to 5 min). Moreover, compared to conventional methods size uniformity and monodispersity were found to be better.

9.5.2.2 Iron Oxide Nanoparticle Microwave Functionalization

In 2009 Nan *et al.* [162] described the first microwave surface functionalization of iron oxide nanoparticles. They studied the graft polymerization of ε-caprolactone onto pre-functional magnetite nanoparticles under microwave solvent-free conditions, Figure 9.17.

The microwave method presented in this study provides much shorter reaction times (maximum 7 min instead of 5 h at 130 °C as reported in the literature) and efficient formation of a polycaprolactone shell around the magnetite core. TEM revealed that the size of the magnetite nanoparticles was increased by the ring-opening polymerization from 13–16 nm to 22–27 nm. The obtained core–shell nanoparticles still present superparamagnetic behavior.

More recently, we proposed the use of the microwave technique to synthesize a multimodal nanoplatform with dual imaging properties [91]. Maghemite γ-Fe_2O_3 nanocrystals, with an average diameter of 10 nm were pre-functionalized with an antitumoral molecule having a bifunctional nature, alendronate or

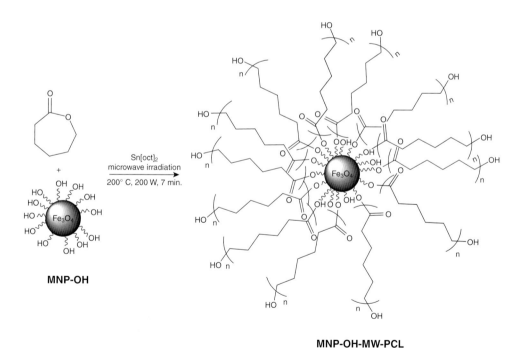

Figure 9.17 Surface-initiated ring opening polymerization of ε-caprolactone onto stabilized magnetite MNP-OH by microwave irradiation [163].

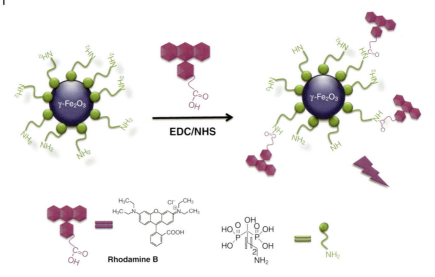

Figure 9.18 Schematic representation of the conjugation of γ-Fe$_2$O$_3$@ alendronate nanocrystals and Rhodamine B to obtain magneto-fluorescent nanoparticles.

4-amino-1-hydroxy-1-phosphonobutyl phosphonic acid (HMBP-(CH$_2$)$_3$–NH$_3^+$). This nanoplatform was used to study further surface functionalization with Rhodamine B, a fluorophore that serves as a model to study surface immobilization efficiencies, Figure 9.18.

Such coupling of fluorophores on nanoparticles was already shown to be hampered by their important aromatic structure, leading to poor functionalization yielding only around 10 molecules per nanoparticle [163, 164]. We studied the potential of the microwave technique to increase the conjugation of this molecule by carbodiimide coupling and compare it to grafting under the "classic" heating technique. Using microwave-assisted coupling the reaction time was decreased from 24 h to less than 20 min. Moreover, by increasing the Rhodamine excess per amine, the function grafting rate is gradually increased up to 600 molecules per nanoparticle (one function in two has reacted). Whereas, without microwave input, whatever the excess of Rhodamine, saturation corresponding to an average of 12 molecules per nanocrystal is quickly obtained. The obtained magneto-fluorescent nanocrystals have similar morphology, good colloidal stability, and retained superparamagnetic behavior. In vitro we also demonstrated the effectiveness of our multimodal nanoparticle in fluorescence nanoplatform imaging of cancer cells, Figure 9.19.

Using microwave technology we thus managed to have an approximately 50-fold increase in molecules per nanoparticles. Moreover, under microwave irradiation the number of Rhodamine molecules at the surface could be tuned by simply varying the ratio of reactants.

We further demonstrated recently the potency of microwave surface functionalization of iron oxide nanoparticles using a clickable superparamagnetic

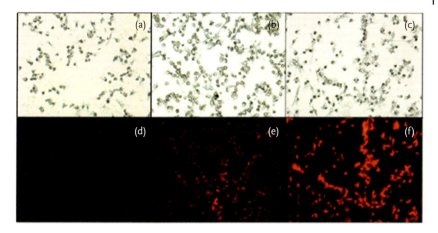

Figure 9.19 Images of MDA-MB-231 cells incubated for 2 h with nanocrystals labeling with Rhodamine. (a, b, c) Optical microscopy showing the uptake of nanocrystals (after Prussian blue reaction), (d, e, f) fluorescence microscopy. The hybrid nanocrystals were incubated at constant iron concentration and increasing the amount of Rhodamine grafted per nanoparticle (from d to f) [91].

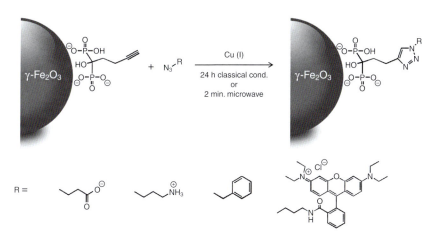

Figure 9.20 Click coupling on γ-Fe$_2$O$_3$@HMBPyne with a diverse array of functional species.

nanoplatform. This nanoplatform was obtained by coating iron oxide nanoparticles γ-Fe$_2$O$_3$, with a new bifunctional molecule (1-hydroxy-1-phosphonopentynyl) phosphonic acid (HMBPyne) bearing an alkyne group. Coupling of several azide functionalized molecules via 1,3-cycloaddition was evaluated with and without the input of microwave energy, Figure 9.20.

Still using Rhodamine B as a model molecule we once again proved that the coupling efficiency could be greatly enhanced using microwave-assisted reaction with an increase of 40-fold in functionalization with no harm to the nanoparticle

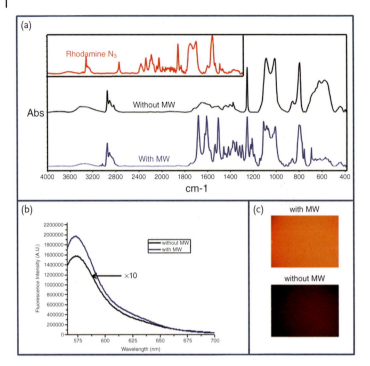

Figure 9.21 Comparison of the γ-Fe_2O_3@ HMBP-Rhodamine nanoparticles obtained by click chemistry with or without microwave irradiation (a) IR spectroscopy in KBr pellets. Insert: free azido-Rhodamine B. (b) Emission spectra (λ_{ex} = 573 nm) (c) photograph in fluorescence mode.

integrity, Figure 9.21. We also extended the reaction to functional molecules demonstrating the chemo-selectivity of the process.

9.5.2.3 Microwave-Assisted Silica Encapsulation of Iron Oxide Nanoparticles

The formation of a shell of silica on the surface of nanoparticles has been extensively used by several groups in order to increase their biocompatibility. Silca core–shell nanoparticles are described to be nontoxic and easy to functionalize. However, silica encapsulation through conventional methods requires long reaction times (hours to days) and it is still a challenge to achieve a uniform silica coating of monodisperse MNPs with a fast encapsulation rate, so microwave could provide an interesting alternative for silica encapsulation.

Hu et al. [165] reported the first one–two-step procedure for the synthesis of SiPEG grafted core–shell maghemite using microwaves. They compared a two-step procedure (maghemite synthesis followed by PEG grafting) to a one–two-step (synthesis and grafting at one time) procedure, assisted or not by microwaves. The procedure uses a modification of the Massart procedure including the addition of

Figure 9.22 Multi-steps protocol for silica encapsulation [166].

the grafting agent: silanated polyethylene glycol. Nanoparticles obtained by the one–two step-procedure with microwave or conventional heating were compared. The microwave heating procedure givess a decrease in the reaction time (82% shorter) for the same amount of PEG grafting. Moreover, Raman studies reveal that the microwave treatment tends to result in better crystallized particles.

Park et al. [166] proposed a silica encapsulation based on a multi-steps protocol, first the synthesis of nanoparticles in organic media followed by a phase transfer in water using tetramethylammonium hydroxide (TMAOH) as transfer agent and then a microwave-assisted silica encapsulation using tetraethyl orthosilicate (TEOS) as silanation agent in the presence of ammonia, Figure 9.22.

The mixture was irradiated at 200 W for 10 min at 70 °C varying the amount of TEOS; the results were compared to conventional heating methods. The resulting core–shell nanoparticles were highly spherical and monodisperse. The first advantage of the use of microwaves was the dramatically reduced time of the methodology. Moreover, it was demonstrated that the shells formed by microwave irradiation were more monodisperse than by conventional heating as only irregular and interconnected cluster networks were obtained through the conventional methodology.

9.5.2.4 Europium Oxide Nanoparticle Microwave Functionalization

Europium nanoparticles are very interesting for their uses as alternative labels for fluorometric assays. Unfortunately, natural or untreated Eu_2O_3 particles are insoluble in water and are easily dissolved by acid during conjugation, losing their desirable optical properties. Feng et al. [167] described a new microwave methodology for the coating of europium oxide labels that provides an amine group for conjugation and use in a model immunoassay. Microwave-assisted coating was done by treating commercially available Eu_2O_3 particles with (3-aminopropyl)-trimethoxysilanetetramethoxysilane, APTMS). The amine coverage was measured to be effective (0.1 nM mg^{-1}). The very attractive fluorescence properties of europium are preserved in this process, offering labels for biology that exhibit long fluorescence lifetime and sharp spectral features.

9.6 Conclusions

Magnetic nanoparticles are particularly attractive because their inherent superparamagnetic properties make them desirable for various applications, in particular in the medical field. With such applications on the horizon, synthetic routes for quick and reliable elaboration of nanoparticles with controlled size, shape and crystal structure are required. In addition to optimizing the physical properties of the magnetic nanoparticles, surface functionalization is a key factor. The need for easily functionalized and tunable materials, in terms of the number and variety of controlled and of attached species, is still a challenge. The microwave-assisted synthesis of inorganic nanoparticles as well as their surface functionalization could be a potential and powerful technology as significantly observed with enhanced publications over the past few years.

In this chapter, we have shown the expansion of microwave advantages as well as problems to be resolved in this research field for magnetic nanoparticle synthesis and surface functionalization processes.

In the first part, we focused on the medical potential applications of the most studied magnetic iron oxide nanoparticles. Then, after describing the many approaches that are used for their preparation (aqueous or non-aqueous conditions) and the strategies that are developed in order to chemically functionalize the nanoparticle surface (ligand exchange, carbodiimide chemistry, etc.), we focused on the use of microwave heating for these two objectives. Efforts by our group as well as others have centered on a promising microwave strategy. The use of microwave energy is relatively new in the field of nanotechnology. Compared with the expertise and attractiveness of this technology in organic chemistry synthesis, one can expect that in the next few years microwawe radiation could find increased application in the field of nanomedicine and, more widely, in nanotechnology.

References

1 Rodríguez, J.A., and Fernández-García, M. (2006) *Introduction to The World of Oxide Nanomaterials*, John Wiley & Sons, Inc., pp. 1–5.

2 Hyeon, T. (2003) Chemical synthesis of magnetic nanoparticles. *Chem. Commun.*, (8), 927–934.

3 Horng, H.E., Hong, C.-Y., Yang, S.Y., and Yang, H.C. (2001) Novel properties and applications in magnetic fluids. *J. Phys. Chem. Solids*, **62** (9–10), 1749–1764.

4 Tsang, S.C., Caps, V., Paraskevas, I., Chadwick, D., and Thompsett, D. (2004) Magnetically separable, carbon-supported nanocatalysts for the manufacture of fine chemicals. *Angew. Chem.*, **116** (42), 5763–5767.

5 Rana, S., White, P., and Bradley, M. (1999) Synthesis of magnetic beads for solid phase synthesis and reaction scavenging. *Tetrahedron Lett.*, **40** (46), 8137–8140.

6 Chou, L.Y.T., Ming, K., and Chan, W.C.W. (2011) Strategies for the intracellular delivery of nanoparticles. *Chem. Soc. Rev.*, **40** (1), 233–245.

7 Dave, S.R., and Gao, X. (2009) Monodisperse magnetic nanoparticles for biodetection, imaging, and drug delivery: a versatile and evolving technology. *Wiley Interdisc. Rev. Nanomed. Nanobiotechnol.*, **1** (6), 583–609.

8 Haun, J.B., Yoon, T.-J., Lee, H., and Weissleder, R. (2010) Magnetic nanoparticle biosensors. *Wiley Interdiscip. Rev. Nanomed. Nanobiotechnol.*, **2** (3), 291–304.

9 Liu, G., Swierczewska, M., Lee, S., and Chen, X. (2010) Functional nanoparticles for molecular imaging guided gene delivery. *Nano Today*, **5** (6), 524–539.

10 Motte, L., Benyettou, F., De Beaucorps, C., Lecouvey, M., Milesovic, I., and Lalatonne, Y. (2010) Multimodal superparamagnetic nanoplatform for clinical applications: immunoassays, imaging and therapy. *Faraday Discuss.*, **149** (0), 211–225.

11 Tran, N., and Webster, T.J. (2010) Magnetic nanoparticles: biomedical applications and challenges. *J. Mater. Chem.*, **20** (40), 8760–8767.

12 Kappe, C.O. (2004) Controlled microwave heating in modern organic synthesis. *Angew. Chem. Int. Ed.*, **43** (46), 6250–6284.

13 Bose, A.K., Manhas, M.S., Banik, B.K., and Robb, E.W. (1994) Microwave-induced organic reaction enhancement (more) chemistry: techniques for rapid, safe and inexpensive synthesis. *Res. Chem. Intermed.*, **20** (1), 1–11.

14 Fernandez-Paniagua, U.M., Illescas, B., Martin, N., Seoane, C., De La Cruz, P., De La Hoz, A., and Langa, F. (1997) Thermal and microwave-assisted synthesis of Diels'Alder adducts of [60] Fullerene with 2,3-Pyrazinoquinodimethanes: characterization and electrochemical properties. *J. Org. Chem.*, **62** (11), 3705–3710.

15 Caddick, S. (1995) Microwave-assisted organic reactions. *Tetrahedron*, **51** (38), 10403–10432.

16 Loupy, A., Petit, A., Hamelin, J., Texier-Boullet, F.O., Jacquault, P., and Mathā, D. (1998) New solvent-free organic synthesis using focused microwaves. *Synthesis*, (9), 1213–1234.

17 Zhang, W., Chen, C.H.-T., Lu, Y., and Nagashima, T. (2004) A highly efficient microwave-assisted Suzuki coupling reaction of aryl perfluorooctylsulfonates with boronic acids. *Org. Lett.*, **6** (9), 1473–1476.

18 Moghimi, S.M., Hunter, A.C., and Murray, J.C. (2005) Nanomedicine: current status and future prospects. *FASEB J.*, **19** (3), 311–330.

19 Xia, Y., Xiong, Y., Lim, B., and Skrabalak, S.E. (2009) Shape-controlled synthesis of metal nanocrystals: simple chemistry meets complex physics?, *Angew. Chem. Int. Ed.*, **48** (1), 60–103.

20 Peer, D., Karp, J.M., Hong, S., Farokhzad, O.C., Margalit, R., and Langer, R. (2007) Nanocarriers as an emerging platform for cancer therapy. *Nat. Nano*, **2** (12), 751–760.

21 Kim, B.Y.S., Rutka, J.T., and Chan, W.C.W. (2010) Nanomedicine. *N. Engl. J. Med.*, **363** (25), 2434–2443.

22 Yalow, R.S., and Berson, S.A. (1960) Immunoassay of endogenous plasma insulin in man. *J. Clin. Invest.*, **39** (7), 1157–1175.

23 Aytur, T., Foley, J., Anwar, M., Boser, B., Harris, E., and Beatty, P.R. (2006) A novel magnetic bead bioassay platform using a microchip-based sensor for infectious disease diagnosis. *J. Immunol. Methods*, **314**, 21–29.

24 De Montferrand, C., Lalatonne, Y., Bonnin, D., Lièvre, N., Lecouvey, M., Monod, P., Russier, V., and Motte, L. (2012) Size-dependent nonlinear weak-field magnetic behavior of maghemite nanoparticles. *Small*, **8** (12), 1945–1956

25 De Montferrand, C., Lalatonne, Y., Bonnin, D., Motte, L., and Monod, P. (2012) Non-linear magnetic behavior around zero field of an assembly of superparamagnetic nanoparticles. *Analyst*, **137**, 2304–2308.

26 Geinguenaud, F.D.R., Souissi, I.S., Fagard, R., Motte, L., and Lalatonne, Y. (2012) Electrostatic assembly of a DNA superparamagnetic nano-tool for simultaneous intracellular delivery and in situ monitoring. *Nanomed.*

Nanotechnol. Biol. Med., **8** (7), 1106–1115.

27 Lawaczeck, R., Menzel, M., and Pietsch, H. (2004) Superparamagnetic iron oxide particles: contrast media for magnetic resonance imaging. *Appl. Organomet. Chem.*, **18** (10), 506–513.

28 Tanimoto, A., and Kuribayashi, S. (2006) Application of superparamagnetic iron oxide to imaging of hepatocellular carcinoma. *Eur. J. Radiol.*, **58**, 200–216.

29 Enochs, W.S., Harsh, G., Hochberg, F., and Weissleder, R. (1999) Improved delineation of human brain tumors on MR images using a long-circulating, superparamagnetic iron oxide agent. *J. Magn. Reson. Imaging*, **9** (2), 228–232.

30 Perrault, S.D., Walkey, C., Jennings, T., Fischer, H.C., and Chan, W.C.W. (2009) Mediating tumor targeting efficiency of nanoparticles through design. *Nano Lett.*, **9** (5), 1909–1915.

31 Benyettou, F., Chebbi, I., Motte, L., and Seksek, O. (2011) Magnetoliposome for alendronate delivery. *J. Mater. Chem.*, **21** (13), 4813–4820.

32 Benyettou, F. (2011) A multimodal magnetic resonance imaging nanoplatform for cancer theranostics. *Phys. Chem. Chem. Phys.*, **13**, 10020–10027.

33 Chang, E., Alexander, H.R., Libutti, S.K., Hurst, R., Zhai, S., Figg, W.D., and Bartlett, D.L. (2001) Laparoscopic continuous hyperthermic peritoneal perfusion 11 No competing interests declared. *J. Am. Coll. Surg.*, **193** (2), 225–229.

34 Feldman, A.L., Libutti, S.K., Pingpank, J.F., Bartlett, D.L., Beresnev, T.H., Mavroukakis, S.M., Steinberg, S.M., Liewehr, D.J., Kleiner, D.E., and Alexander, H.R. (2003) Analysis of factors associated with outcome in patients with malignant peritoneal mesothelioma undergoing surgical debulking and intraperitoneal chemotherapy. *J. Clin. Oncol.*, **21** (24), 4560–4567.

35 Van Der Zee, J. (2002) Heating the patient: a promising approach?, *Ann. Oncol.*, **13** (8), 1173–1184.

36 Wust, P., Hildebrandt, B., Sreenivasa, G., Rau, B., Gellermann, J., Riess, H., Felix, R., and Schlag, P.M. (2002) Hyperthermia in combined treatment of cancer. *Lancet Oncol.*, **3** (8), 487–497.

37 Massart, R. (1981) Preparation of aqueous magnetic liquids in alkaline and acidic media Magnetics. *IEEE Trans. Magn.*, **17** (2), 1247–1248.

38 Vayssieres, L. (2009) On the effect of nanoparticle size on water-oxide interfacial chemistry. *J. Phys. Chem. C*, **113** (12), 4733–4736.

39 Vayssières, L., Chanéac, C., Tronc, E., and Jolivet, J.P. (1998) Size tailoring of magnetite particles formed by aqueous precipitation: an example of thermodynamic stability of nanometric oxide particles. *J. Colloid Interface Sci.*, **205** (2), 205–212.

40 Ozaki, M., and Matijevic, E. (1985) Preparation and magnetic properties of monodispersed spindle-type γ-Fe_2O_3 particles. *J. Colloid Interface Sci.*, **107** (1), 199–203.

41 Matijevic, E., and Scheiner, P. (1978) Ferric hydrous oxide sols: III. Preparation of uniform particles by hydrolysis of Fe(III)-chloride, -nitrate, and -perchlorate solutions. *J. Colloid Interface Sci.*, **63** (3), 509–524.

42 Bailey, J.K., Brinker, C.J., and Mecartney, M.L. (1993) Growth mechanisms of iron oxide particles of differing morphologies from the forced hydrolysis of ferric chloride solutions. *J. Colloid Interface Sci.*, **157** (1), 1–13.

43 Milosevic, I., Jouni, H., David, C., Warmont, F., Bonnin, D., and Motte, L. (2011) Facile microwave process in water for the fabrication of magnetic nanorods. *J. Phys. Chem. C*, **115** (39), 18999–19004.

44 Ge, S., Shi, X., Sun, K., Li, C., Uher, C., Baker, J.R., Banaszak Holl, M.M., and Orr, B.G. (2009) Facile hydrothermal synthesis of iron oxide nanoparticles with tunable magnetic properties. *J. Phys. Chem. C*, **113** (31), 13593–13599.

45 Qin, W., Yang, C., Yi, R., and Gao, G. (2011) Hydrothermal synthesis and characterization of single-crystalline α-Fe_2O_3 nanocubes. *J. Nanomater.*, **2011**.

46 Cho, S.-B., Noh, J.-S., Park, S.-J., Lim, D.-Y., and Choi, S.-H. (2007)

Morphological control of Fe_2O_3; particles via glycothermal process. *J. Mater. Sci.*, **42** (13), 4877–4886.

47 Hayashi, H., and Hakuta, Y. (2010) Hydrothermal synthesis of metal oxide nanoparticles in supercritical water. *Materials*, **3** (7), 3794–3817.

48 Jia, C.-J., Sun, L.-D., Yan, Z.-G., You, L.-P., Luo, F., Han, X.-D., Pang, Y.-C., Zhang, Z., and Yan, C.-H. (2005) Single-crystalline iron oxide nanotubes. *Angew. Chem. Int. Ed.*, **44** (28), 4328–4333.

49 Liang, M.-T., Wang, S.-H., Chang, Y.-L., Hsiang, H.-I., Huang, H.-J., Tsai, M.-H., Juan, W.-C., and Lu, S.-F. (2010) Iron oxide synthesis using a continuous hydrothermal and solvothermal system. *Ceram. Int.*, **36** (3), 1131–1135.

50 Duan, L., Jia, S., Wang, Y., Chen, J., and Zhao, L. (2009) Synthesis of Fe_2O_3; polyhedra by hydrothermal method: using arginine as precipitator. *J. Mater. Sci.*, **44** (16), 4407–4412.

51 Niederberger, M., and Pinna, M.A.N. (2009) *Metal Oxide Nanoparticles in Organic Solvents: Synthesis, Formation, Assembly and Application*, 1st edn, Springer Edition, p. 217.

52 Tadic, M., Markovic, D., Spasojevic, V., Kusigerski, V., Remskar, M., Pirnat, J., and Jaglicic, Z. (2007) Synthesis and magnetic properties of concentrated α-Fe_2O_3 nanoparticles in a silica matrix. *J. Alloy. Compd.*, **441** (1–2), 291–296.

53 Lalatonne, Y., Benyettou, F., Bonnin, D., Lièvre, N., Monod, P., Lecouvey, M., Weinmann, P., and Motte, L. (2009) Characterization of magnetic labels for bioassays. *J. Magn. Magn. Mater.*, **321** (10), 1653–1657.

54 Hao, J.J., Chen, H.L., Ren, C.L., Yan, N., Geng, H.J., and Chen, X.G. (2010) Synthesis of superparamagnetic nanocrystals in reverse microemulsion at room temperature. *Mater. Res. Innov.*, **14** (4), 324–326.

55 Pinna, N., Grancharov, S., Beato, P., Bonville, P., Antonietti, M., and Niederberger, M. (2005) Magnetite nanocrystals: nonaqueous synthesis, characterization, and solubility. *Chem. Mater.*, **17** (11), 3044–3049.

56 Fievet, F., Lagier, J.P., Blin, B., Beaudoin, B., and Figlarz, M. (1989) Homogeneous and heterogeneous nucleations in the polyol process for the preparation of micron and submicron size metal particles. *Solid State Ionics*, **32–33** (Part 1), 198–205.

57 Soumare, Y., Dakhlaoui-Omrani, A., Schoenstein, F., Mercone, S., Viau, G., and Jouini, N. (2010) Nickel nanofibers and nanowires: elaboration by reduction in polyol medium assisted by external magnetic field. *Solid State Commun.*, **151** (4), 284–288.

58 Piquemal, J.-Y., Viau, G., Beaunier, P., Bozon-Verduraz, F., and Fiévet, F. (2003) One-step construction of silver nanowires in hexagonal mesoporous silica using the polyol process. *Mater. Res. Bull.*, **38** (3), 389–394.

59 Ung, D., Viau, G., Fiévet-Vincent, F., Herbst, F., Richard, V., and Fiévet, F. (2005) Magnetic nanoparticles with hybrid shape. *Prog. Solid State Chem.*, **33** (2–4), 137–145.

60 Ammar, S., Jouini, N., Fiévet, F., Beji, Z., Smiri, L., Moliné, P., Danot, M., and Grenèche, J.-M. (2006) Magnetic properties of zinc ferrite nanoparticles synthesized by hydrolysis in a polyol medium. *J. Phys. Condens. Matter*, **18** (39), 9055–9070.

61 Ammar, S., Helfen, A., Jouini, N., Fievet, F., Rosenman, I., Villain, F., Molinie, P., and Danot, M. (2001) Magnetic properties of ultrafine cobalt ferrite particles synthesized by hydrolysis in a polyol medium. *J. Mater. Chem.*, **11** (1), 186–192.

62 Hyeon, T., Lee, S.S., Park, J., Chung, Y., and Na, H.B. (2001) Synthesis of highly crystalline and monodisperse maghemite nanocrystallites without a size-selection process. *J. Am. Chem. Soc.*, **123** (51), 12798–12801.

63 Park, S.-J., Kim, S., Lee, S., Khim, Z.G., Char, K., and Hyeon, T. (2000) Synthesis and magnetic studies of uniform iron nanorods and nanospheres. *J. Am. Chem. Soc.*, **122** (35), 8581–8582.

64 Wang, Y., and Yang, H. (2009) Synthesis of iron oxide nanorods and nanocubes in an imidazolium ionic liquid. *Chem. Eng. J.*, **147** (1), 71–78.

65 Sun, S., Zeng, H., Robinson, D.B., Raoux, S., Rice, P.M., Wang, S.X., and Li, G. (2003) Monodisperse M/Fe$_2$O$_4$ (M = Fe, Co, Mn) nanoparticles. *J. Am. Chem. Soc.*, **126** (1), 273–279.

66 Bhattacharyya, S., Salvetat, J.-P., Fleurier, R., Husmann, A., Cacciaguerra, T., and Saboungi, M.-L. (2005) One step synthesis of highly crystalline and high coercive cobalt-ferrite nanocrystals. *Chem. Commun.*, (38), 4818–4820.

67 Ahniyaz, A., Sakamoto, Y., and Bergstrã, L. (2007) Magnetic field-induced assembly of oriented superlattices from maghemite nanocubes. *Proc. Natl. Acad. Sci.*, **104** (45), 17570–17574.

68 Shavel, A., and Liz-Marzan, L.M. (2009) Shape control of iron oxide nanoparticles. *Phys. Chem. Chem. Phys.*, **11** (19), 3762–3766.

69 Palchoudhury, S., An, W., Xu, Y., Qin, Y., Zhang, Z., Chopra, N., Holler, R.A., Turner, C.H., and Bao, Y. (2011) Synthesis and growth mechanism of iron oxide nanowhiskers. *Nano Lett.*, **11** (3), 1141–1146.

70 Hao, R., Xing, R., Xu, Z., Hou, Y., Gao, S., and Sun, S. (2010) Synthesis, functionalization, and biomedical applications of multifunctional magnetic nanoparticles. *Adv. Mater.*, **22** (25), 2729–2742.

71 Weissleder, R., Kelly, K., Sun, E.Y., Shtatland, T., and Josephson, L. (2005) Cell-specific targeting of nanoparticles by multivalent attachment of small molecules. *Nat. Biotechnol.*, **23** (11), 1418–1423.

72 Hu, L., De Montferrand, C., Lalatonne, Y., Motte, L., and Brioude, A. (2011) Effect of cobalt doping concentration on the crystalline structure and magnetic properties of monodisperse Co$_x$Fe$_{3-x}$O$_4$ nanoparticles within nonpolar and aqueous solvents. *J. Phys. Chem. C*, **116** (7), 4349–4355.

73 Euliss, L.E., Grancharov, S.G., O'brien, S., Deming, T.J., Stucky, G.D., Murray, C.B., and Held, G.A. (2003) Cooperative assembly of magnetic nanoparticles and block copolypeptides in aqueous media. *Nano Lett.*, **3** (11), 1489–1493.

74 Yang, H., Li, X., Zhou, H., Zhuang, Y., Hu, H., Wu, H., and Yang, S. (2010) Monodisperse water-soluble Fe-Ni nanoparticles for magnetic resonance imaging. *J. Alloy. Compd.*, **509** (4), 1217–1221.

75 Wang, L., Yang, Z., Gao, J., Xu, K., Gu, H., Zhang, B., Zhang, X., and Xu, B. (2006) A biocompatible method of decorporation: bisphosphonate-modified magnetite nanoparticles to remove uranyl ions from blood. *J. Am. Chem. Soc.*, **128** (41), 13358–13359.

76 Lee, S.-Y., and Harris, M.T. (2006) Surface modification of magnetic nanoparticles capped by oleic acids: characterization and colloidal stability in polar solvents. *J. Colloid Interface Sci.*, **293** (2), 401–408.

77 De Palma, R., Peeters, S., Van Bael, M.J., Van Den Rul, H., Bonroy, K., Laureyn, W., Mullens, J., Borghs, G., and Maes, G. (2007) Silane ligand exchange to make hydrophobic superparamagnetic nanoparticles water-dispersible. *Chem. Mater.*, **19** (7), 1821–1831.

78 Robinson, I., Alexander, C., Tung, L.D., Fernig, D.G., and Thanh, N.T.K. (2009) Fabrication of water-soluble magnetic nanoparticles by ligand-exchange with thermo-responsive polymers. *J. Magn. Magn. Mater.*, **321** (10), 1421–1423.

79 Zhang, T., Ge, J., Hu, Y., and Yin, Y. (2007) A general approach for transferring hydrophobic nanocrystals into water. *Nano Lett.*, **7** (10), 3203–3207.

80 Salgueiriã-Maceira, V.N., Liz-Marzãn, L.M., and Farle, M. (2004) Water-based ferrofluids from FexPt1-x nanoparticles synthesized in organic media. *Langmuir*, **20** (16), 6946–6950.

81 Herranz, F., Morales, M.P., Roca, A.G., Vilar, R., and Ruiz-Cabello, J. (2008) A new method for the aqueous functionalization of superparamagnetic Fe$_2$O$_3$ nanoparticles. *Contrast Media Mol. Imaging*, **3** (6), 215–222.

82 Maity, D., and Agrawal, D.C. (2007) Synthesis of iron oxide nanoparticles under oxidizing environment and their stabilization in aqueous and non-aqueous media. *J. Magn. Magn. Mater.*, **308** (1), 46–55.

83 Swami, A., Kumar, A., and Sastry, M. (2003) Formation of water-dispersible gold nanoparticles using a technique based on surface-bound interdigitated bilayers. *Langmuir*, **19** (4), 1168–1172.

84 Liong, M., Lu, J., Kovochich, M., Xia, T., Ruehm, S.G., Nel, A.E., Tamanoi, F., and Zink, J.I. (2008) Multifunctional inorganic nanoparticles for imaging, targeting, and drug delivery. *ACS Nano*, **2** (5), 889–896.

85 Pellegrino, T., Manna, L., Kudera, S., Liedl, T., Koktysh, D., Rogach, A.L., Keller, S., Rädler, J., Natile, G., and Parak, W.J. (2004) Hydrophobic nanocrystals coated with an amphiphilic polymer shell: a general route to water soluble nanocrystals. *Nano Lett.*, **4** (4), 703–707.

86 Lee, J., Lee, Y., Youn, J.K., Na, H.B., Yu, T., Kim, H., Lee, S.-M., Koo, Y.-M., Kwak, J.H., Park, H.G., Chang, H.N., Hwang, M., Park, J.-G., Kim, J., and Hyeon, T. (2008) Simple synthesis of functionalized superparamagnetic magnetite/silica core/shell nanoparticles and their application as magnetically separable high-performance biocatalysts. *Small*, **4** (1), 143–152.

87 Kim, J., Lee, J.E., Lee, J., Yu, J.H., Kim, B.C., An, K., Hwang, Y., Shin, C.-H., Park, J.-G., Kim, J., and Hyeon, T. (2005) Magnetic fluorescent delivery vehicle using uniform mesoporous silica spheres embedded with monodisperse magnetic and semiconductor nanocrystals. *J. Am. Chem. Soc.*, **128** (3), 688–689.

88 Neirinck, B., Soccol, D., Fransaer, J., Biest, O.V.D., and Vleugels, J. (2010) Influence of short chain organic acids and bases on the wetting properties and surface energy of submicrometer ceramic powders. *J. Colloid Interface Sci.*, **348** (2), 654–660.

89 Lalatonne, Y., Paris, C., Serfaty, J.M., Weinmann, P., Lecouvey, M., and Motte, L. (2008) Bisphosphonates–ultra small superparamagnetic iron oxide nanoparticles: a platform towards diagnosis and therapy. *Chem. Commun.*, 22, 2553–2555.

90 Mutin, P.H., Guerrero, G., and Vioux, A. (2005) Hybrid materials from organophosphorus coupling molecules. *J. Mater. Chem.*, **15** (35–36), 3761–3768.

91 Benyettou, F., Guénin, E., Lalatonne, Y., and Motte, M. (2011) Microwave-assisted nanoparticle surface functionalization. *Nanotechnology*, **22** (5), 055102.

92 Benyettou, F., Lalatonne, Y., Sainte-Catherine, O., Monteil, M., and Motte, L. (2009) Superparamagnetic nanovector with anti-cancer properties: γ-Fe_2O_3/Zoledronate. *Int. J. Pharm.*, **379** (2), 324–327.

93 Veiseh, O., Gunn, J.W., and Zhang, M. (2010) Design and fabrication of magnetic nanoparticles for targeted drug delivery and imaging. *Adv. Drug Deliv. Rev.*, **62** (3), 284–304.

94 Hein, C., and Wang, L.X. (2008) Click chemistry, a powerful tool for pharmaceutical sciences. *Pharm. Res.*, **25** (10), 2216–2230.

95 Lutz, J.-F. (2007) 1,3-Dipolar cycloadditions of azides and alkynes: a universal ligation tool in polymer and materials science. *Angew. Chem. Int. Ed.*, **46** (7), 1018–1025.

96 Nandivada, H., Jiang, X., and Lahann, J. (2007) Click chemistry: versatility and control in the hands of materials scientists. *Adv. Mater.*, **19** (17), 2197–2208.

97 Pieters, R.J., Rijkers, D.T.S., and Liskamp, R.M.J. (2007) Application of the 1,3-Dipolar cycloaddition reaction in chemical biology: approaches toward multivalent carbohydrates and peptides and peptide-based polymers. *QSAR Comb. Sci.*, **26** (11–12), 1181–1190.

98 Kolb, H.C., Finn, M.G., and Sharpless, K.B. (2001) Click chemistry: diverse chemical function from a few good reactions. *Angew. Chem. Int. Ed.*, **40** (11), 2004–2021.

99 Huisgen, R. (1984) 1,3-Dipolar cycloadditions–introduction, survey, mechanism, in *1,3 -Dipolar Cycloaddition Chemistry*, vol. 1 (ed. A. Padwa), John Wiley & Sons, Inc., New York, pp. 1–176.

100 Rostovtsev, V.V., Green, L.G., Fokin, V.V., and Sharpless, K.B. (2002) A stepwise huisgen cycloaddition process: copper(I)-catalyzed regioselective "ligation" of azides and terminal

alkynes. *Angew. Chem. Int. Ed.*, **41** (14), 2596–2599.
101 Hawker, C.J., and Wooley, K.L. (2005) The convergence of synthetic organic and polymer chemistries. *Science*, **309** (5738), 1200–1205.
102 Hayashi, K., Moriya, M., Sakamoto, W., and Yogo, T. (2009) Chemoselective synthesis of folic acidâˆ'functionalized magnetite nanoparticles via click chemistry for magnetic hyperthermia. *Chem. Mater.*, **21** (7), 1318–1325.
103 Hayashi, K., Ono, K., Suzuki, H., Sawada, M., Moriya, M., Sakamoto, W., and Yogo, T. (2010) High-frequency, magnetic-field-responsive drug release from magnetic nanoparticle/organic hybrid based on hyperthermic effect. *ACS Appl. Mater. Inter.*, **2** (7), 1903–1911.
104 Von Maltzahn, G., Ren, Y., Park, J.-H., Min, D.-H., Kotamraju, V.R., Jayakumar, J., Fogal, V., Sailor, M.J., Ruoslahti, E., and Bhatia, S.N. (2008) In vivo tumor cell targeting with nanoparticles. *Bioconjug. Chem.*, **19** (8), 1570–1578.
105 Devaraj, N.K., Keliher, E.J., Thurber, G.M., Nahrendorf, M., and Weissleder, R. (2009) 18F labeled nanoparticles for in vivo PET-CT imaging. *Bioconjug. Chem.*, **20** (2), 397–401.
106 Cutler, J.I., Zheng, D., Xu, X., Giljohann, D.A., and Mirkin, C.A. (2010) Polyvalent oligonucleotide iron oxide nanoparticle conjugates. *Nano Lett.*, **10** (4), 1477–1480.
107 Thorek, D., Elias, D., and Tsourkas, A. (2009) Comparative analysis of nanoparticle-antibody conjugations: carbodiimide versus click chemistry. *Mol. Imaging*, **8** (4), 221–229.
108 Rana, S., Yeh, Y.-C., and Rotello, V.M. (2010) Engineering the nanoparticle-protein interface: applications and possibilities. *Curr. Opin. Chem. Biol.*, **14** (6), 828–834.
109 Hild, W.A., Breunig, M., and Goepferich, A. (2008) Quantum dots – nano-sized probes for the exploration of cellular and intracellular targeting. *Eur. J. Pharm. Biopharm.*, **68** (2), 153–168.
110 Katz, E., and Willner, I. (2004) Integrated nanoparticle–biomolecule hybrid systems: synthesis, properties, and applications. *Angew. Chem. Int. Ed.*, **43** (45), 6042–6108.
111 Gedye, R., Smith, F., Westaway, K., Ali, H., Baldisera, L., Laberge, L., and Rousell, J. (1986) The use of microwave ovens for rapid organic synthesis. *Tetrahedron Lett.*, **27** (3), 279–282.
112 Caddick, S., and Fitzmaurice, R. (2009) Microwave enhanced synthesis. *Tetrahedron*, **65** (17), 3325–3355.
113 Kappe, C., and Dallinger, D. (2009) Controlled microwave heating in modern organic synthesis: highlights from the 2004–2008 literature. *Mol. Divers.*, **13** (2), 71–193.
114 Perreux, L., and Loupy, A. (2001) A tentative rationalization of microwave effects in organic synthesis according to the reaction medium, and mechanistic considerations. *Tetrahedron*, **57** (45), 9199–9223.
115 Lidstrom, P., Tierney, J., Wathey, B., and Westman, J. (2001) Microwave-assisted organic synthesis:a review. *Tetrahedron*, **57** (45), 9225–9283.
116 Rosana, M.R., Tao, Y., Stiegman, A.E., and Dudley, G.B. (2012) On the rational design of microwave-actuated organic reactions. *Chem. Sci.*, **3** (4), 1240–1244.
117 Wathey, B., Tierney, J., Lidstrom, P., and Westman, J. (2002) The impact of microwave-assisted organic chemistry on drug discovery. *Drug Discov. Today*, **7** (6), 373–380.
118 Larhed, M., and Hallberg, A. (2001) Microwave-assisted high-speed chemistry: a new technique in drug discovery. *Drug Discov. Today*, **6** (8), 406–416.
119 Colombo, M., and Peretto, I. (2008) Chemistry strategies in early drug discovery: an overview of recent trends. *Drug Discov. Today*, **13** (15–16), 677–684.
120 Koopmans, C., Iannelli, M., Kerep, P., Klink, M., Schmitz, S., Sinnwell, S., and Ritter, H. (2006) Microwave-assisted polymer chemistry: heck-reaction, trans esterification, Baeyer-Villiger oxidation, oxazoline polymerization, acrylamides, and porous materials. *Tetrahedron*, **62** (19), 4709–4714.
121 Bardts, M., Gonsior, N., and Ritter, H. (2008) Polymer synthesis and

modification by use of microwaves. *Macromol. Chem. Phys.*, **209** (1), 25–31.

122 Pedersen, S.L., Tofteng, A.P., Malik, L., and Jensen, K.J. (2012) Microwave heating in solid-phase peptide synthesis. *Chem. Soc. Rev.*, **41** (5), 1826–1844.

123 Economopoulos, S.P., Rotas, G., Miyata, Y., Shinohara, H., and Tagmatarchis, N. (2011) Exfoliation and chemical modification using microwave irradiation affording highly functionalized graphene. *ACS Nano*, **4** (12), 7499–7507.

124 Baghbanzadeh, M., Carbone, L., Cozzoli, P.D., and Kappe, C.O. (2011) Microwave-assisted synthesis of colloidal inorganic nanocrystals. *Angew. Chem. Int. Ed.*, **50** (48), 11312–11359.

125 Wang, G., Whittaker, G., Harrison, A., and Song, L. (1998) Preparation and mechanism of formation of acicular goethite-magnetite particles by decomposition of ferric and ferrous salts in aqueous solution using microwave radiation. *Mater. Res. Bull.*, **33** (11), 1571–1579.

126 Rigneau, P., Bellon, K., Zahreddine, I., and Stuerga, D. (1999) Microwave flash-synthesis of iron oxides nanoparticles. *EPJ Appl. Phys.*, **7** (1), 41–43.

127 Katsuki, H., and Komarneni, S. (2001) Microwave-hydrothermal synthesis of monodispersed nanophase α-Fe_2O_3. *J. Am. Ceram. Soc.*, **84** (10), 2313–2317.

128 Caillot, T., Aymes, D., Stuerga, D., Viart, N., and Pourroy, G. (2002) Microwave flash synthesis of iron and magnetite particles by disproportionation of ferrous alcoholic solutions. *J. Mater. Sci.*, **37** (23), 5153–5158.

129 Katsuki, H., Shiraishi, A., Komarneni, S., Moon, W.J., Toh, S., and Kaneko, K. (2004) Rapid synthesis of monodispersed α-Fe_2O_3 nanoparticles from $Fe(NO_3)_3$ solution by microwave irradiation. *Nippon Seramikkusu Kyokai*, **112** (7), 4.

130 Li, Q., and Wei, Y. (1998) Study on preparing monodispersed hematite nanoparticles by microwave-induced hydrolysis of ferric salts solution. *Mater. Res. Bull.*, **33** (5), 779–782.

131 Jia, Z.-B., Wei, Y., and Wang, H.-M. (2000) Preparation of ultrafine hematite particles by microwave heating of ferric salt solutions. *J. Inorg. Mater.*, **15** (5), 926–928.

132 Sreeja, V., and Joy, P.A. (2007) Microwave hydrothermal synthesis of γ-Fe_2O_3 nanoparticles and their magnetic properties. *Mater. Res. Bull.*, **42** (8), 1570–1576.

133 Komarneni, S., Hu, W., Noh, Y.D., Van Orden, A., Feng, S., Wei, C., Pang, H., Gao, F., Lu, Q., and Katsuki, H. (2012) Magnetite syntheses from room temperature to 150°C with and without microwaves. *Ceram. Int.*, **38** (3), 2563–2568.

134 Baruwati, B., Nadagouda, M.N., and Varma, R.S. (2008) Bulk synthesis of monodisperse ferrite nanoparticles at water-organic interfaces under conventional and microwave hydrothermal treatment and their surface functionalization. *J. Phys. Chem. C*, **112** (47), 18399–18404.

135 Hu, X., Yu, J.C., Gong, J., Li, Q., and Li, G. (2007) α-Fe_2O_3 nanorings prepared by a microwave-assisted hydrothermal process and their sensing properties. *Adv. Mater.*, **19** (17), 2324–2329.

136 Hu, X., and Yu, J.C. (2008) Continuous aspect-ratio tuning and fine shape control of monodisperse α-Fe_2O_3 nanocrystals by a programmed microwave–hydrothermal method. *Adv. Funct. Mater.*, **18** (6), 880–887.

137 Cao, S.W., and Zhu, Y.J. (2009) Iron oxide hollow spheres: microwave-hydrothermal ionic liquid preparation, formation mechanism, crystal phase and morphology control and properties. *Acta Mater.*, **57**, 2154–2165.

138 Wang, W.-W., Zhu, Y.-J., and Ruan, M.-L. (2007) Microwave-assisted synthesis and magnetic property of magnetite and hematite nanoparticles. *J. Nanopart. Res.*, **9** (3), 419–426.

139 Hu, X., Yu, J.C., and Gong, J. (2007) Fast production of self-assembled hierarchical α-Fe_2O_3 nanoarchitectures. *J. Phys. Chem. C*, **111**, 11180–11185.

140 Ai, Z., Deng, K., Wan, Q., Zhang, L., and Lee, S. (2010) Facile microwave-assisted synthesis and magnetic and gas

sensing properties of Fe_3O_4 nanoroses. *J. Phys. Chem. C*, **114** (14), 6237–6242.

141 Muraliganth, T., Vadivel Murugan, A., and Manthiram, A. (2009) Facile synthesis of carbon-decorated single-crystalline Fe_3O_4 nanowires and their application as high performance anode in lithium ion batteries. *Chem. Commun.*, (47), 7360–7362.

142 Zhou, H., Yi, R., Li, J., Su, Y., and Liu, X. (2010) Microwave-assisted synthesis and characterization of hexagonal Fe_3O_4 nanoplates. *Solid State Sci.*, **12** (1), 99–104.

143 Wan, J., Chen, X., Wang, Z., Yang, X., and Qian, Y. (2005) A soft-template-assisted hydrothermal approach to single-crystal Fe_3O_4 nanorods. *J. Cryst. Growth*, **276** (3–4), 571–576.

144 Wang, J., Chen, Q., Zeng, C., and Hou, B. (2004) Magnetic-field-induced growth of single-crystalline Fe_3O_4 nanowires. *Adv. Mater.*, **16** (2), 137–140.

145 Lian, S., Wang, E., Kang, Z., Bai, Y., Gao, L., Jiang, M., Hu, C., and Xu, L. (2004) Synthesis of magnetite nanorods and porous hematite nanorods. *Solid State Commun.*, **129** (8), 485–490.

146 Itoh, H., and Sugimoto, T. (2003) Systematic control of size, shape, structure, and magnetic properties of uniform magnetite and maghemite particles. *J. Colloid Interface Sci.*, **265** (2), 283–295.

147 Feng, L., Jiang, L., Mai, Z., and Zhu, D. (2004) Polymer-controlled synthesis of Fe_3O_4 single-crystal nanorods. *J. Colloid Interface Sci.*, **278** (2), 372–375.

148 Kumar, R.V., Koltypin, Y., Xu, X.N., Yeshurun, Y., Gedanken, A., and Felner, I. (2001) Fabrication of magnetite nanorods by ultrasound irradiation. *J. Appl. Phys.*, **89** (11), 6324–6328.

149 Blesa, M.A., Mijalchik, M., Villegas, M., and Rigotti, G. (1986) Transformation of akaganeite into magnetite in aqueous hydrazine suspensions. *React. Solids*, **2** (1–2), 85–94.

150 Rao, K., Mahesh, K., and Kumar, S. (2005) A strategic approach for preparation of oxide nanomaterials. *Bull. Mater. Sci.*, **28** (1), 19–24.

151 Zhang, H., Zhong, X., Xu, J.-J., and Chen, H.-Y. (2008) Fe_3O_4/Polypyrrole/Au nanocomposites with core/shell/shell structure: synthesis, characterization, and their electrochemical properties. *Langmuir*, **24** (23), 13748–13752.

152 Chen, Y., and Gu, H. (2009) Microwave-assisted fast fabrication of Fe_3O_4-MWCNTs nanocomposites and their application as MRI contrast agents. *Mater. Lett.*, **67** (1), 49–51.

153 Xiao, W., Gu, H., Li, D., Chen, D., Deng, X., Jiao, Z., and Lin, J. (2012) Microwave-assisted synthesis of magnetite nanoparticles for MR blood pool contrast agents. *J. Magn. Magn. Mater.*, **324** (4), 488–494.

154 Bilecka, I., Djerdj, I., and Niederberger, M. (2008) One-minute synthesis of crystalline binary and ternary metal oxide nanoparticles. *Chem. Commun.*, (7), 886–888.

155 Jiang, F.Y., Wang, C.M., Fu, Y., and Liu, R.C. (2010) Synthesis of iron oxide nanocubes via microwave-assisted solvothermal method. *J. Alloy. Compd.*, **503** (2), L31–L33.

156 Economopoulos, S., Karousis, N., Rotas, G., Pagona, G., and Tagmatarchis, N. (2011) Microwave-assisted functionalization of carbon nanostructured materials. *Curr. Org. Chem.*, **15** (8), 1121–1132.

157 Thielbeer, F., Donaldson, K., and Bradley, M. (2011) Zeta potential mediated reaction monitoring on nano and microparticles. *Bioconjug. Chem.*, **22** (2), 144–150.

158 Hu, Y., Zhang, Y., and Tang, Y. (2010) One-step hydrothermal synthesis of surface organosilanized nanozeolite under microwave irradiation. *Chem. Commun.*, **46** (22), 3875–3877.

159 Garcia, N., Benito, E., Guzmaì, N.J., De Francisco, R., and Tiemblo, P. (2010) Microwave versus conventional heating in the grafting of alkyltrimethoxysilanes onto silica particles. *Langmuir*, **26** (8), 5499–5506.

160 Sommer, W.J., and Weck, M. (2007) Facile functionalization of gold nanoparticles via microwave-assisted 1,3 dipolar cycloaddition. *Langmuir*, **23** (24), 11991–11995.

161 Bahadur, N.M., Watanabe, S., Furusawa, T., Sato, M., Kurayama, F., Siddiquey, I.A., Kobayashi, Y., and Suzuki, N. (2011) Rapid one-step synthesis, characterization and functionalization of silica coated gold nanoparticles. *Colloids Surf. A Physicochem. Eng. Aspects*, **392** (1), 137–144.

162 Nan, A., Turcu, R., Craciunescu, I., Pana, O., Scharf, H., and Liebscher, J. (2009) Microwave-assisted graft polymerization of ε-caprolactone onto magnetite. *J. Polym. Sci. A Polym. Chem.*, **47** (20), 5397–5404.

163 Reddy, G.R., Bhojani, M.S., Mcconville, P., Moody, J., Moffat, B.A., Hall, D.E., Kim, G., Koo, Y.-E.L., Woolliscroft, M.J., Sugai, J.V., Johnson, T.D., Philbert, M.A., Kopelman, R., Rehemtulla, A., and Ross, B.D. (2006) Vascular targeted nanoparticles for imaging and treatment of brain tumors. *Clin. Cancer Res.*, **12** (22), 6677–6686.

164 Bertorelle, F., Wilhelm, C., Roger, J., Gazeau, F., Ménager, C., and Cabuil, V.R. (2006) Fluorescence-modified superparamagnetic nanoparticles: intracellular uptake and use in cellular imaging. *Langmuir*, **22** (12), 5385–5391.

165 Hu, L., Hach, D., Chaumont, D., Brachais, C.-H., Couvercelle, J.-P., and Percheron, A. (2009) Comparison of various methods of grafting of modified-PEG onto maghemite nanoparticles in aqueous medium including synthesis by microwave refluxing. *J. Sol-Gel Sci. Technol.*, **49** (3), 277–284.

166 Park, J.C., Gilbert, D.A., Liu, K., and Louie, A.Y. (2012) Microwave enhanced silica encapsulation of magnetic nanoparticles. *J. Mater. Chem.*, **22** (17), 8449–8454.

167 Feng, J., Shan, G., Maquieira, A., Koivunen, M.E., Guo, B., Hammock, B.D., and Kennedy, I.M. (2003) Functionalized europium oxide nanoparticles used as a fluorescent label in an immunoassay for atrazine. *Anal. Chem.*, **75** (19), 5282–5286.

10
Microwave-Assisted Continuous Synthesis of Inorganic Nanomaterials

Naftali N. Opembe, Hui Huang, and Steven L. Suib

10.1
Introduction and Overview

The major focus of this chapter is to review with examples the techniques, advantages, and drawbacks of using microwave-assisted continuous-flow synthesis for the synthesis of inorganic nanomaterials. We will explore the findings of the work published and available in the open literature dealing with the different aspects of continuous microwave synthesis of inorganic nanomaterials. Aspects inherent to microwave synthesis, such as frequency, power, heating mode, and residence time, among other parameters, will be discussed alongside salient characteristics of the produced nanomaterials.

Microwave-assisted synthesis as a tool for chemical synthesis has been embraced in academia since it is a fast synthesis technique that offers unique capabilities. High product yields and purities are some of the unique features found in materials synthesized using microwave heating. Additionally, uniform structural, textural, and novel physical properties have been obtained. Microwaves as a synthesis technique have the potential of contributing immensely to all areas of synthetic chemistry, including inorganic chemistry. The synthesis of nanomaterials and nanostructures is bound to benefit more from the uniform and rapid heating provided by microwaves.

Microwave chemistry relies on the unique ability of the reaction mixture to efficiently absorb microwave energy. This ability is derived from the phenomenon of "microwave dielectic heating". What happens to a material when exposed to microwaves is a consequence of the material's intrinsic properties in relation to how it interacts with microwaves (MW). Based on this interaction, materials can be classified into three classes. Generally, materials will either be MW reflectors, transmitters, or absorbers. MW reflectors are useful as core components of MW waveguides since they have the ability to reflect and therefore guide MW to a specific region. MW transmitters on the other hand allow passage of microwaves without interference and as such are useful in making MW cookware or reaction

Microwaves in Nanoparticle Synthesis, First Edition. Edited by Satoshi Horikoshi and Nick Serpone.
© 2013 Wiley-VCH Verlag GmbH & Co. KGaA. Published 2013 by Wiley-VCH Verlag GmbH & Co. KGaA.

vessels for example, Teflon is a good MW transmitter. MW transmitters are also known as MW transparent materials. MW absorbers are useful in chemistry. These materials can absorb microwaves causing them to be heated.

Microwave chemistry also relies on the efficient dielectric heating of a material. This efficiency is dictated by the material's ability to absorb microwaves and generate heat [1, 2]. At 2.45 GHz, the energy of a photon of microwave radiation (1 J mol^{-1}) [3] is hardly enough to even cleave a chemical bond, meaning microwave radiation does not induce any chemical reactions but rather provides an efficient heating means and thus enhances said chemical reaction. This heating involves two mechanisms: dipolar polarization and ionic conduction [1]. When a polar medium is heated, this leads to the alignment of dipoles in the same direction as the applied electromagnetic field. However, the applied electromagnetic field is an oscillating field, which causes the dipoles to continuously seek to re-align themselves along the direction of the applied field. Depending on the frequency of this phenomenon, heating occurs. However, if the response time is much slower than the oscillation frequency then no heating occurs [4]. Such is the case for the dipolar polarization mechanism of microwave heating. On the other hand, in the ionic conduction mechanism, dissolved ionic species will oscillate back and forth bumping and knocking into one another, thereby generating heat in the process. If the irradiated material is an electrical conductor, charge carriers are caused to move through the material under the influence of the applied field, which causes polarization. Heating then occurs as a result of electrical resistance. Figure 10.1 shows an illustration of these two heating mechanisms.

The dielectric properties of a material are described based on: (i) the dielectric constant, ε' which describes the ability of a material to be polarized by the applied electromagnetic field, and (ii) the dielectric loss factor, ε'' which describes the extent of conversion of the electromagnetic field into heat [4]. These two properties are related to each other by Eq. (10.1). Also known as the energy dissipation or

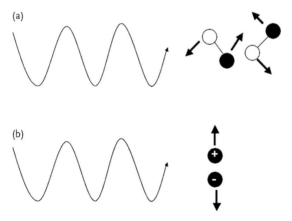

Figure 10.1 Microwave electromagnetic field interaction and heating mechanisms. (a) Dipolar–polar polarization and (b) ionic conduction mechanisms.

loss factor, Eq. (10.1) gives us a direct means of comparing any two materials in terms of their microwave absorption properties. Most reaction media (water and organic solvents) are classified for the sake of comparing their microwave suitability as reaction media based on this parameter. Table 10.1 shows $\tan \delta$ values for selected solvents. These are classified as strong absorbers ($\tan \delta \geq 0.5$), medium absorbers ($0.1 \leq \tan \delta \leq 0.5$), and poor absorbers ($\tan \delta \leq 0.1$). Solvents with high loss factors include ethylene glycol (1.350), ethanol (0.941) or dimethyl sulfoxide (DMSO) (0.825). The above compounds have a permanent dipole moment, whereas solvents without a permanent dipole moment (e.g., hexane) are, by definition, microwave transparent.

$$\tan \delta = \varepsilon'' / \varepsilon' \qquad (10.1)$$

10.2
Microwave-Assisted Continuous Synthesis of Inorganic Nanomaterials

Most reported MW-assisted techniques for synthesizing inorganic nanomaterials have been realized through the use of batch techniques [5–13]. However, batch operations have inherent disadvantages, such as limited scale-up possibilities, and time and effort in changing reagents in the batch processes, among others. Scale-up in batch synthesis is made even harder due to limited penetration depths of microwaves into most reaction media. As can be seen in Table 10.2, the penetration depth of microwaves is limited to about 1.4 cm at room temperature, increasing slightly to 5.7 cm when heated to 95 °C when water is used as a solvent. However, based on the penetration depths of some materials, for example, Teflon (9200 cm) or quartz (16 000 cm), a continuous process can be developed that utilizes these two as reaction vessels (microwave transparent) and if using water as

Table 10.1 Loss tangents ($\tan \delta$) of selected solvents (2.45 GHz, 20 °C).

Solvent	Dielectric constant	$\tan \delta$
Ethylene glycol	37.00	1.350
Ethanol	24.30	0.941
DMSO	46.68	0.825
2-Propanol	19.92	0.799
Formic acid	58.00	0.722
Methanol	32.70	0.659
NMP	32.20	0.275
DMF	36.71	0.161
Water	80.10	0.123
Toluene	2.38	0.040
Hexane	1.88	0.020

Note: DMSO refers to dimethyl sulfoxide, NMP to N-methylpyrrolidone and DMF to dimethyl formamide. (Adapted from ref. [2]).

Table 10.2 Microwave penetration depth (2.45 GHz) in selected materials.

Material	Penetration depth (mm)	Temperature (°C)
Water	14	25
Water	57	95
Glass (borosilicate)	350	25
Poly (vinyl chloride)	2 100	20
Teflon	92 000	25
Quartz glass	160 000	25

(Adapted from ref. [2]).

the reaction medium, restrict the penetration depth to a few centimeters (i.e., vessel diameters of a few centimeters). Additionally, reactions under microwave-assisted continuous synthesis can utilize MW-absorbing solvents to achieve the synthesis in a shorter time.

10.3
Types of Microwave Apparatus Used in Continuous Synthesis

MW-assisted synthesis of inorganic nanomaterials utilizes a wide variety of apparatuses. Most of these apparatuses are commercially available but others are custom-built. Pioneering work utilized domestic (kitchen-type) microwave ovens, which have the disadvantages of the lack of control of irradiation power, reaction temperature, and pressure inside the reaction vessels. Commercially available apparatuses on the other hand feature built-in temperature and pressure sensors as well as magnetic stirrers, among other features. Most of the commercially available apparatuses have provisions for adaptation to computer systems for control and monitoring of the reaction parameters. Most of these apparatuses operate at a standard frequency of 2.45 GHz which is a good compromise for a number of reasons. First, this frequency avoids interference with telecommunication, wireless or cellular phone networks, among other technologies relying on the use of electromagnetic frequencies. Secondly, the corresponding wavelength of 12.25 cm produced by this frequency is optimal for microwave penetration into foods. Lastly, production of magnetrons that produce this frequency is cheaper than for other frequencies.

To be useful in continuous synthesis, microwave apparatuses have to be adapted to receive reagents into the microwave cavity and discharge products from the cavity on a continuous basis. This can be achieved by customizing them to have entry and exit ports through which the charging and discharging can be affected. Nowadays, several manufacturers, in addition to making apparatuses that are dedicated to batch synthesis, also offer apparatuses that have continuous synthesis capabilities. For most, the presence of an entry/exit port is itself enough to enable

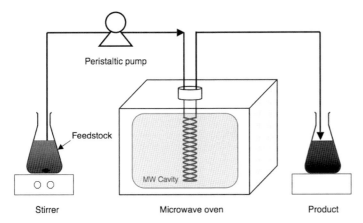

Figure 10.2 A schematic for a microwave apparatus showing the MW cavity and a charge/discharge port (Reproduced from: Opembe, N.N., King'ondu, C.K., Espinal, A.E., Chen, C.-H., Nyutu, E.K., Crisostomo, V.M., and Suib, S.L. (2010) *J. Phys. Chem. C*, **114**, 14417. Copyright (2012) American Chemical Society).

the continuous synthesis of nanomaterials. Such is the case with the CEM™ Mars 5 microwave apparatus that has an opening that enables tube reactors to be inserted into the cavity (see Figure 10.2 for a schematic). In this apparatus, a Teflon or quartz tubing can be used to charge and discharge materials from the apparatus.

Continuous microwave synthesis has been performed under hydrothermal conditions in a commercial microwave hydrothermal apparatus (Figure 10.3). This was achieved using a microwave-heated flow vertical tube reactor (MLS ETHOS CFR Continuous Flow Reactor). This system has a constant frequency of 2.45 GHz and is pressure controlled. The system can attain a maximum pressure of 30 bar. A Teflon reaction vessel is connected to a pressure transducer that monitors and controls the pressure during synthesis. Charging of reactants into the reaction vessel is pump controlled while the flow rate, pressure, and power are all computer controlled. The reactor is attached to a pump at the inlet end and passes into a wall-cooled heat exchanger at the effluent end. Reaction mixtures are rapidly cooled as they exit the irradiation zone.

A single-mode custom-built microwave apparatus that involves mixing reactants then spraying them into a microwave chamber has been reported [14–16]. As can be seen schematically in Figure 10.4, the apparatus is made up of three component parts all working to affect the continuous synthesis of inorganic nanomaterials. More importantly, this apparatus can be used to control purity levels and confine the size of the synthesized materials to the nanometer range. Two methods can be used. One involves just spraying reactants into a microwave-heated zone after passing through a nozzle and is termed nozzle microwave (NMW) and the other involves mixing of two reactant mixtures first and is termed *in-situ* nozzle microwave (INM). These are shown in Figure 10.4a and 10.4b, respectively. INM

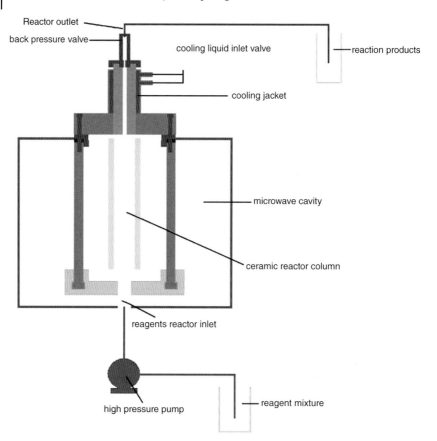

Figure 10.3 Schematic diagram of the continuous-flow microwave hydrothermal reactor. (Reprinted with permission from: Braun, I., Schulz-Ekloff, G., Wöhrle, D., and Lautenschläger, W. (1998) *Micropor. Mesopor. Mater.*, **23**, 79. Copyright (2012) Elsevier B.V.).

consists of three components. The first component is the *in-situ* mixing set-up, which relies on syringe pumps to continuously pump reactants. In this set-up, two solutions can be mixed essentially simultaneously at a T-connector before they arrive at the second component, which is an atomizer. The atomizer plays the role of nebulizing said reactants through the ultrasonic cavitation method into fine droplets. These misty reactants then pass down a quartz or Teflon tube under gravity and reach the third component, which is the reaction chamber. The chamber is under controlled microwave power, which controls the resultant temperature.

Continuous microwave synthesis using a plasma jet reactor has also been reported [17]. The plasma reactor can be operated starting from atmospheric pressure to moderate pressure (a few Torr). Power can be controlled from 300 W to 3 KW. A plasma zone of about 30–45 cm is produced in a quartz tube of 3–5 cm diameter. The plasma is ignited with a pointed metallic rod. This apparatus has

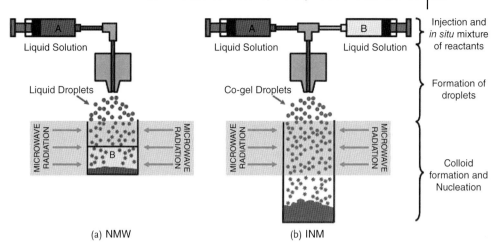

Figure 10.4 (a) Nozzle microwave and (b) *in situ* nozzle microwave apparatus (Reprinted with permission from: Espinal, L., Malinger, K.A., Espinal, A.E., Gaffney, A.M., and Suib, S.L. (2007) *Adv. Funct. Mater.* **17**, 2572. Copyright (2012) WILEY-VCH Verlag GmbH & Co. KGaA, Weinheim).

been used to accomplish syntheses where metallic powders were injected into the plasma cavity by means of mechanical dispensers or carried into it by means of a carrier gas. The powders interact with MW radiation as they pass through the plasma zone and the resultant products are collected at the bottom of the tube (see Figure 10.5).

10.4
Microwave Continuous Synthesis of Molecular Sieve Materials

The synthesis of molecular sieve nanomaterials has been performed using microwaves on a continuous basis [18–25]. The nature of the materials and synthesis conditions dictate the type of microwave apparatus used. Molecular sieves are important industrially. They have industrial uses such as in the separation of compounds based on their size or shape. The continuous synthesis of $AlPO_4^{-5}$ [18] has been realized in a commercial hydrothermal microwave apparatus (MLS ETHOS ContFLOW) shown schematically in Figure 10.3. The optimal synthesis gel was composed of $1.0Al_2O_3:1.0P_2O_5:1.5Pr_3N:150H_2O$ made by mixing Al_2O_3 as the aluminum source, P_2O_5 as the phosphorus source, and tri-n-propylamine as the template. The optimal process conditions were found to be: a flow rate of $900 cm^3 h^{-1}$ that translated to a residence time of 8 min and a temperature range of 180 to 190 °C. A 2 g synthesis gel produced 0.1 g of product that had been completely crystallized in 160 min, confirmed using X-ray diffraction (XRD) and scanning electron microscopy (SEM).

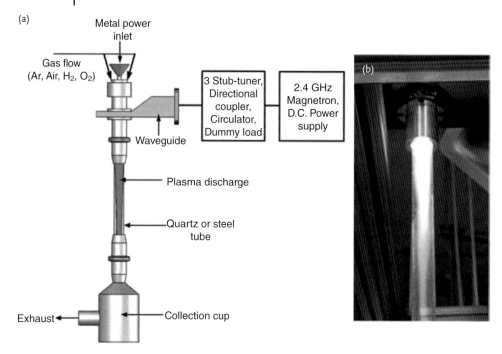

Figure 10.5 (a) Schematic diagram for the microwave plasma jet reactor showing the components, and (b) the actual plasma generated. (Reprinted with permission from: Kumar, V., Kim, J.H., Pendyala, C., Chernomordik, B., and Sunkara M.K. (2008) *J. Phys. Chem. C*, **112**, 17750. Copyright (2012) America Chemical Society).

Microporous zeolites have been synthesized by the continuous microwave synthesis technique. Molecular sieve ZSM-5, one of the most important industrial zeolite catalysts, was recently synthesized by Park et al. [19, 20]. A precursor mixture was prepared by mixing aluminosilicate gel with a nanoseed solution obtained under microwave irradiation in the first step. This precursor mixture was further pumped continuously into the microwave apparatus. Under microwave irradiation, 5 min was sufficient to crystallize ZSM-5 in a continuous process. This was proved by high-quality XRD patterns of ZSM-5. The morphology of ZSM-5 particles from continuous microwave synthesis exhibited regular and agglomerated spherical morphology (<0.3 μm). Batch-type syntheses under the same conditions were also conducted to compare the structural properties of these two as-synthesized particles, which were proven to be very similar to those of ZSM-5. The overall advantages of using microwave continuous flow to synthesize ZSM-5 materials were: shortened synthesis time, minimized use of a templating agent, improved throughput, balanced heat distribution in the reactor, and suppressed production of undesired phases.

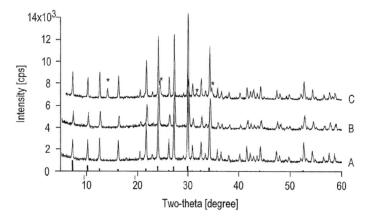

Figure 10.6 XRD patterns of zeolite A synthesized at: (a) 50 W, 48 kHz nozzle frequency, and 0.4 ml min^{-1} flow rate; (b) 100 W, 48 kHz nozzle frequency, and 0.4 ml min^{-1} flow rate, and (c) 200 W, 48 kHz nozzle frequency, and 0.4 ml min^{-1} flow rate. The hydroxysodalite phase is denoted by asterix. (Reprinted with permission from: Suib, S.L., Espinal, L., and Nyutu, E. K., US Patent 2006-0291827-A1).

Zeolites Y and A have been synthesized on a continuous basis using microwave heating. This was achieved in a custom-built microwave apparatus similar to the one in Figure 10.4. This involved spraying reactant solutions into either a receiving solution or into air at atmospheric pressure [16]. For the synthesis of zeolite Y, one method involved feeding reactant solutions into an ultrasonic nozzle, which generated a mist due to the ultrasonic frequency used. The mist passed through the microwave-heated zone under gravity, thus the process was referred to as nozzle microwave. The frequency of the ultrasonic nozzle could be varied from 40 to 120 kHz with higher frequencies generating finer droplets than the lower frequencies. Microwave power of 500 W was the best experimentally determined condition due to prevention of agglomeration of the particles. The synthesis of zeolite A involved mixing specific amounts of sodium aluminate (NaAl$_2$O$_3$) and sodium hydroxide (NaOH) in one clear solution (sol) and tetraethylorthosilicate (TEOS) in another. The two mixtures were then injected simultaneously via tubing to meet at a T-connection for onward transmission into an ultrasonicator [21]. The effects of altering the nozzle frequency, microwave power, and flow rate were studied. The ultimate goal of doing this was to control the particle size and limit aggregation of the product. The *in situ* nozzle microwave method led to smaller, uniformly distributed particles than was achieved by doing the same reaction under microwave hydrothermal treatment. By varying microwave power from 50 to 200 W, flow rates from 0.4 to 1 ml min^{-1}, and nozzle frequency from 48 to 120 kHz, the particle size could be controlled from 180 to 43 nm. As can be seen in the XRD patterns in Figure 10.6, a higher frequency of 120 kHz led to the presence of the impurity phase of hydroxysodalite but this phase was not detected at a lower frequency. Also, as can be seen in Figure 10.7, smaller particles were obtained by using a flow rate of 0.4 ml min^{-1} and frequency of 48 kHz.

Zeolite A has also been synthesized continuously in a fixed frequency multi-mode microwave apparatus similar to the one in Figure 10.2 [22]. In this apparatus, the reactor configuration involved using either a coiled reactor or non-coiled

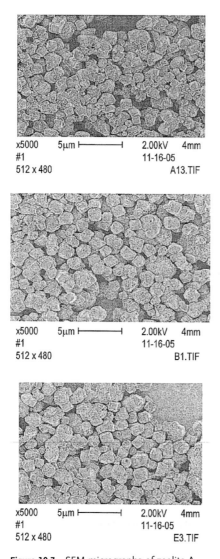

Figure 10.7 SEM micrographs of zeolite A synthesized at; (a) 100 W, 48 kHz nozzle frequency, 0.4 ml min^{-1} and unseeded, (b) 100 W, 48 kHz nozzle frequency, 0.4 ml min^{-1} and seeded, and (c) 200 W, 48 kHz nozzle frequency, 0.4 ml min^{-1} and seeded. (Reprinted with permission from: Suib, S.L., Espinal, L., and Nyutu, E.K., US Patent 2006-0291827-A1).

spherical Teflon tubing as reactors, while the introduction of reactant mixtures was achieved using a peristaltic pump. Microwave power levels of 180, 360, 900, 1260, and 1800 W were studied. The synthesis gel was diluted to correspond to a synthesis formula of 3.26 Na_2O: 1 Al_2O_3: 1.92 SiO_2: 146.88 H_2O due to problems of clogging in the tubes, encountered when using thicker gels. The synthesis also involved the use of 0.1 or 0.2 wt% of a seeding gel, which was compared to non-seeded gel. Prior to the synthesis the reaction mixture was aged for at least 30 min. The peristaltic pump was used to control the feeding rate at 23 and 10 ml min^{-1}, corresponding to residence times of 13 and 29 min, respectively. Different combinations of flow rate and power levels were tested. The unseeded gel did not crystallize zeolite A at either lower or higher microwave power, with the high microwave power yielding the hydroxysodalite phase. Microwave power, reagent flow rate and, hence, microwave residence time, seeding, and reactor configuration were all crucial parameters in the crystallization of zeolite A. These findings are summarized in Table 10.3. As can be seen, pure zeolite A was only crystallized when: (i) the residence time was 13 min, power 900 W, and 0.2 wt% seed was used, or (ii) when the residence time was 29 min, power 360 W and both 0.1 and 0.2 wt% seed were used. The produced zeolite A ranged in size from micron to sub-micron with a narrow particle size distribution, as seen in the published SEM micrographs.

Cryptomelane-type manganese oxide (K-OMS-2) nanomaterials have also been synthesized using a microwave continuous process [23]. K-OMS-2 materials have potential for molecular sieving based on their tunnel structure and porous nature. They are made of corner or edge shared MnO_6 octahedral units with molecule-sized tunnels balanced by potassium ions. This involved microwave synthesis in aqueous and non-aqueous systems. The chosen method utilizes the

Table 10.3 Experimental parameters and results for microwave continuous synthesis of zeolite A (LTA) using two reactor configurations.

Residence time/min	Output power, wt % seed, (phase obtained)			
	180 W	360 W	900 W	1260 W
Coiled reactor				
13	Low heat	Low heat	0% (SOD) 0.2% (LTA)	0% (SOD) 0.2% (SOD)
29	Low heat	0% (NC) 0.1% (SOD) 0.2% (SOD)	0% (SOD) 0.1% (–) 0.2% (LTA and SOD)	High heat
Spherical reactor				
13	Low heat	Low heat	0.2% (SOD)	None-optimal
29	None-optimal	0.2% (SOD)	None-optimal	None-optimal

Where: LTA refers to zeolite A (Linde type), SOD refers to hydroxysodalite phase. (Adapted from ref. [22]).

conproportionation by Mn^{2+} and Mn^{7+} ions present in manganese sulfate and potassium permanganate, respectively. In a typical reaction, 42 mmol (6.65 g) $KMnO_4$ was added to 100 ml distilled-deionized water (DDW) to make mixture A. In another flask, 59 mmol (9.9 g) of $MnSO_4 \cdot H_2O$ was added to 33 ml DDW to make mixture B. These mixtures were individually stirred until complete dissolution of the reagents was affected. To B, 3.4 ml of concentrated HNO_3 was added with further stirring. Solution A was transferred to a dropping funnel and added dropwise to solution B with vigorous stirring. The resultant mixture (C) was further stirred at room temperature for about 15 min before addition of different volumes of DMSO to make up various percentages (v/v). For this synthesis, a 15% DMSO level was used but comparison was made to a synthesis mixture that did not contain any DMSO. Reactants were continuously pumped through the coiled reactor in the microwave cavity at different flow rates, where they interacted with microwave irradiation for a given duration of time. Products were thereafter collected at the discharge end of the reactor into a collecting flask, washed with distilled-deionized water (DDW), centrifuged and dried overnight at 120 °C in an oven. Crystallization of K-OMS-2 was achieved under optimized conditions of reactant flow rate, volume of DMSO, microwave power, and using a coiled Teflon reactor.

DMSO was found to be a necessary constituent of the reactant medium for rapid crystallization of K-OMS-2. Reactions carried out in its absence did not yield any crystalline product. The importance of DMSO as a medium of reaction can be deduced from Table 10.1. The high $\tan\delta$ value exhibited by DMSO means that it is a strong absorber of microwaves, coupling well with microwaves and leading to rapid heating. At faster flow rates for example, 40 ml min^{-1}, no crystalline product was identified (Figure 10.8). This is because at this flow rate the residence time in the microwave cavity (50 s at this flow rate), was hardly enough for adequate exposure of reactant mixture to microwaves. The resultant product had a poorly ordered phase, as seen in the XRD patterns in Figure 10.8. Slower flow rates of 10 and 5 ml min^{-1}, with corresponding residence times of 4 and 8 min, respectively, yielded the desired phase. Longer residence times led to adequate exposure of microwaves to the reactant mixture and hence better coupling. A slightly slower flow rate of 20 ml min^{-1} required more DSMO (25%v/v) as compared to 10 and 5 ml min^{-1} rate that only required 10%v/v to obtain a pure phase (Figure 10.9).

These materials had small crystallite sizes, evidenced by calculations based on the Scherrer equation. These calculations gave sizes ranging from 1.6 to 1.8 nm. On average the fibers had diameters of less than 5 nm and lengths as small as 20 nm were identified (Figure 10.10). The synthesis was achieved at average temperatures of 90 °C at a power setting of 300 W. Interestingly, an FTIR study of these materials revealed the presence of DMSO on fibers synthesized with higher amounts of DMSO (Figure 10.11). This could be explained either by an adsorption phenomenon or by fiber growth processes. In the former case, DMSO could well have just adsorbed strongly onto the synthesized materials, this was manifested in the FTIR patterns, or DMSO could be involved in the fiber growth process by restricting lateral aggregation, as may be the case in the latter premise.

10.5 Microwave Continuous Synthesis of Metal Oxides and Mixed Metal Oxide Materials | 259

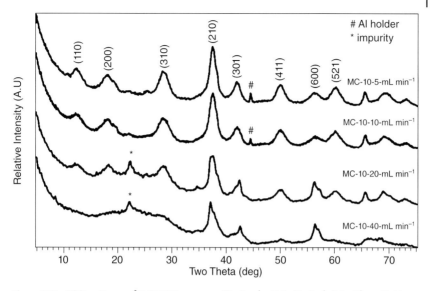

Figure 10.8 XRD patterns of K-OMS-2 showing the effect of reactant flow rates at 300 W and 10% DMSO. Reactant flow rate was varied from 5 to 40 ml min^{-1}. (Reproduced from: Opembe, N.N., King'ondu, C.K., Espinal, A.E., Chen, C.-H., Nyutu, E.K., Crisostomo, V.M., and Suib, S.L. (2010) *J. Phys. Chem. C*, **114**, 14417. Copyright (2012) America Chemical Society).

10.5
Microwave Continuous Synthesis of Metal Oxides and Mixed Metal Oxide Materials

Microwave-assisted continuous synthesis techniques have been applied in the synthesis of selected non-sieving metal oxides and mixed metal oxides. Spherical monodispersed zirconia (ZrO$_2$) nanoparticles were synthesized by Bondioli *et al.* [24] using a microwave-assisted continuous process. This synthesis process was performed in an MLS ETHOS ContFLOW system, shown schematically in Figure 10.3, with maximum pressure restricted to 15 bar and the flow rate varied from 50 to 100 ml min^{-1}. The nanoparticles were formed from the hydrolysis and condensation of tetra-*n*-propylzirconate under microwave irradiation. The flow rate was optimized to obtain non-agglomerated ZrO$_2$ nanoparticles. When the flow rate was 50 ml min^{-1}, the particles were spherical with a mean particle size of about 100 nm. The zirconia obtained was amorphous, based on XRD analysis. The amorphous zirconia had narrowly distributed particles, as seen in the published TEM micrographs.

Monodispersed silica (SiO$_2$) colloidal spherical nanoparticles have been synthesized from the hydrolysis and condensation of tetraethylorthosilicate (TEOS) by using a continuous microwave synthesis process [25]. This was achieved through a hydrothermal synthesis process using a commercial microwave-heated flow vertical tube reactor (MLS ETHOS CFR Continuous Flow Reactor). After carrying out preliminary tests, reactions were conducted at a pressure of 20 bar. The flow

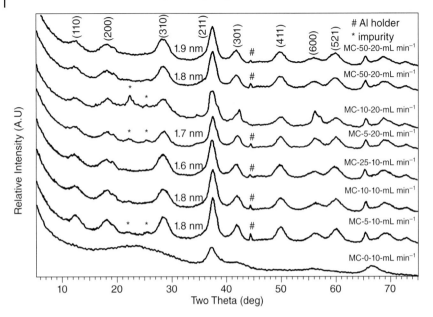

Figure 10.9 XRD patterns of K-OMS-2 prepared using different percentages of DMSO (0–50%) at either 10 or 20 ml min^{-1} with pure phases were matched to the Q-phase of Cryptomelane (KMn$_8$O$_{16}$; JCPDS 29-1020). (Reproduced from: Opembe, N.N., King'ondu, C.K., Espinal, A.E.; Chen, C.-H., Nyutu, E.K., Crisostomo, V.M., Suib, S.L. (2010) *J. Phys. Chem. C*, **114**, 14417. Copyright (2012) American Chemical Society).

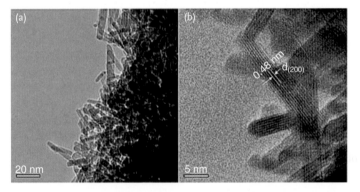

Figure 10.10 TEM micrographs at (a) low resolution and (b) high resolution of K-OMS-2 nanomaterials prepared by a continuous flow microwave technique at 20 ml min^{-1} in 25% DMSO (Reproduced from: Opembe, N.N., King'ondu, C.K., Espinal, A.E.; Chen, C.-H., Nyutu, E.K., Crisostomo, V.M., and Suib, S.L. (2010) *J. Phys. Chem. C*, **114**, 14417. Copyright (2012) American Chemical Society).

10.5 Microwave Continuous Synthesis of Metal Oxides and Mixed Metal Oxide Materials

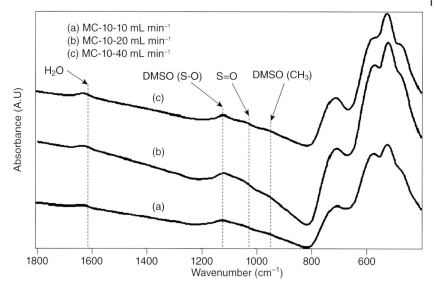

Figure 10.11 FTIR spectra of OMS-2 nanomaterials synthesized at 300 W with flow rates of: (a) MC-10-40 ml min^{-1}, (b) MC-10–20 ml min^{-1}, and (c) MC-10–10 ml min^{-1}. (Reproduced from: Opembe, N.N., King'ondu, C.K., Espinal, A.E.; Chen, C.-H., Nyutu, E.K., Crisostomo, V.M., and Suib, S.L. (2010) *J. Phys. Chem. C*, **114**, 14417. Copyright (2012) America Chemical Society).

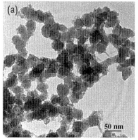 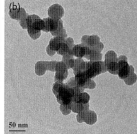 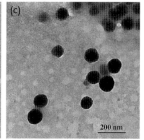

Figure 10.12 Transmission electron micrographs of ZrO$_2$ particles obtained by microwave continuous synthesis at (a) 101 ml min^{-1}, (b) 56 ml min^{-1}, (c) 43 ml min^{-1}. (Reprinted with permission from: Corradi, A.B., Bondioli, F., and Maria, A. (2006) *Powder Technol.* **167**, 45. Copyright (2012) Elsevier B.V.).

rate of the solution, and thus the residence time of the solution in the microwave oven cavity, was varied from 43 to 101 ml min^{-1}. TEM results (Figure 10.12) revealed that the flow rate affected both the dispersion and particle sizes of the resultant nanomaterials. As the flow rates decreased from 56 to 43 ml min^{-1}, that corresponded to a mean residence time of 3 and 5 min, respectively, the particle size of the spherical SiO$_2$ nanoparticles increased to nearly 100 nm in diameter. Moreover, the particles obtained by reducing the flow rate to 43 ml min^{-1}, and consequently by increasing the irradiation time of the suspensions, became monodispersed. Similarly, Ag–SiO$_2$ core–shell nanoparticles have also been synthesized on a continuous basis using microwaves [26]. This involved formation of Ag nanoparticles first, followed by their coating with a silica layer in a subsequent step. Both steps involved use of microwaves. This was realized using a custom-built microwave that operated at a fixed frequency of 2.5 GHz and operated at 100 W. The thickness of the shell material was increased with increased irradiation time. A variation of this process has also been reported in the microwave continuous synthesis of silica-coated zinc oxide (ZnO) nanomaterials [27].

A similar set-up used in the synthesis of molecular sieve nanomaterials K-OMS-2 was also used to synthesize γ-MnO$_2$ [28]. In the synthesis of γ-MnO$_2$ the recipe chosen proved to be very important. When the synthesis was done using a reported recipe [29], no crystalline product was obtained. Even addition of microwave absorbing solvents (DMSO) did not result in the crystallization of any phase. The clear liquid that was fed into the reactor emerged as a clear liquid product without any crystallization occurring. In the reported synthesis [29], persulfate ions were photolyzed in the presence of UV radiation under acidic conditions, creating powerful oxidizing radical species. Sulfate radicals are responsible for the oxidization of Mn^{2+} to Mn^{4+} and thus crystallizing MnO$_2$ in the process. The use of microwaves was futile, pointing to the limited role played by microwaves in the creation of the radical species. This can be explained from an energetic point of view. Microwaves are less energetic than UV radiation and thus are not able to cleave the chemical bonds involved to create free radicals. UV radiation being more energetic is able to achieve this easily.

Switching the recipe to a similar one used in [30] led to precipitation of γ-MnO$_2$. However, the approach used was a modified one. This approach involved varying the Mn^{2+}/Mn^{7+} ratios. When the ratio was exactly 1.4, a K-OMS-2 (α-MnO$_2$) phase was obtained. Adjusting the recipe upwards to 2.8, achieved by cutting the amount of Mn^{4+} in the first synthesis by half, resulted in the XRD patterns shown in Figure 10.13. The phase was still largely α-MnO$_2$.

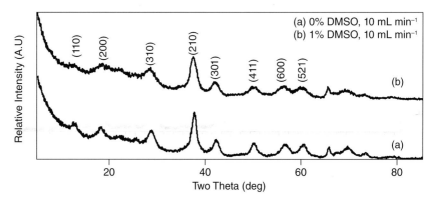

Figure 10.13 XRD patterns of materials continuously synthesized at Mn^{2+}/Mn^{7+} = 2.8, with 10 ml nitric acid, (a) without and (b) with DMSO.

When this ratio was re-adjusted downwards to exactly 2.3 by slightly increasing the amount of Mn^{7+}, the XRD patterns in Figure 10.14 were obtained. The different flow rates studied could crystallize the right phase. Based on the (110) reflection, the crystallites reduced in size since the width of this reflection is increasing Figure 10.14 a–c as the flow rate is increased. When the same recipe is left at 85 °C for 24 h under conventional heating, the materials with the pattern in Figure 10.14d are obtained. This pattern shows emergence of some impurity peaks not present in the MW-prepared patterns and with a high possibility of being α-MnO_2. Picking a flow rate of 10 ml min^{-1}, the effect of DMSO on the synthesis was studied (Figure 10.15). However, DMSO might not be playing as dominant a role in the crystallization of γ-MnO_2 as in the crystallization of K-OMS-2, since γ-MnO_2 could crystallized even in the absence of DMSO (Figure 10.15a). However, the contribution of DMSO becomes apparent when we look at the SEM micrograph images (Figure 10.16) of materials prepared with various amounts of DMSO. Here, the materials prepared with higher amounts of DMSO all lead to nanoparticles Figure 10.16 a,b while reducing the DMSO amount leads to spheres with flower-like morphology (Figure 10.16c). Whether this recipe can lead to the crystallization of γ-MnO_2 phase in the absence of microwaves was of interest. For this, after mixing all the reagents, synthesis mixture (without DMSO) was left at room temperature and samples collected at 2 and 24 h. Thereafter, XRD and SEM were performed on these samples after wash and dry procedures. The results are shown in Figures 10.17 and 10.18, respectively. At both times crystallization of the correct phase is observed without any impurities (Figure 10.17). Also, the phase does not change significantly, as can be seen in the SEM images. Compared to the materials synthesized under MW, differences in the SEM images are noted, which points to

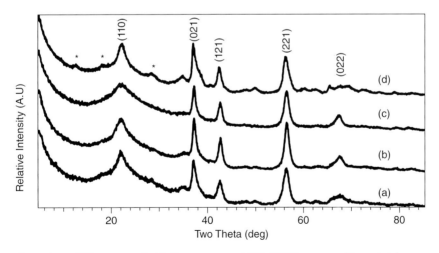

Figure 10.14 XRD patterns of γ-MnO_2 continuously synthesized at different flows: (a) 5 ml min^{-1}, (b) 10 ml min^{-1}, and (c) 20 ml min^{-1}, constant power of 300 W and 25% DMSO, compared to materials made by (d) oil bath treatment of the same recipe for 24 h at 85 °C with impurity phase shown as *.

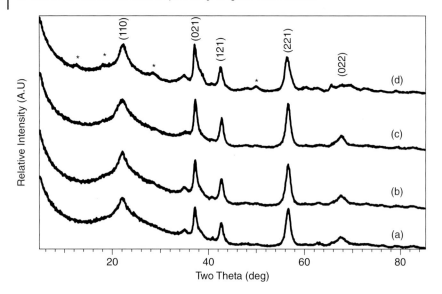

Figure 10.15 XRD patterns of γ-MnO$_2$ continuously synthesized with different volumes of DMSO: (a) 0%, (b) 1%, and (c) 25%, at constant power of 300 W and flow rate of 10 ml min^{-1} compared to materials made by oil bath treatment of the same recipe for 24 h at 85 °C (d) with impurity peaks shown by *.

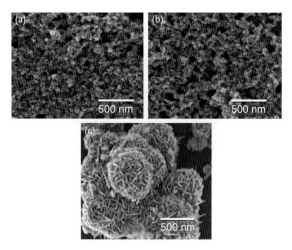

Figure 10.16 SEM micrographs of γ-MnO$_2$ continuously synthesized at (a) 20 ml min^{-1}, 25% DMSO and (b) 10 ml min^{-1}, 25% DMSO, and (c) 10 ml min^{-1}, 1% DMSO.

the role played by both microwaves and DMSO in the synthesis. From this study, other than providing a microwave-absorbing environment, DMSO also plays a role in controlling the morphology of the materials synthesized. DMSO provides an environment for nucleating species to individually crystallize, thus forming

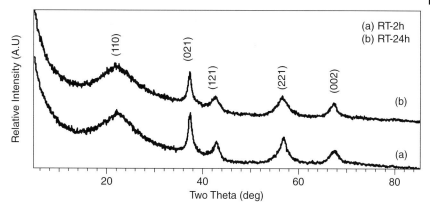

Figure 10.17 XRD patterns of γ-MnO$_2$ synthesized with Mn^{2+}/Mn^{7+} = 2.3 left at room temperature for (a) 2 and (b) 24 h.

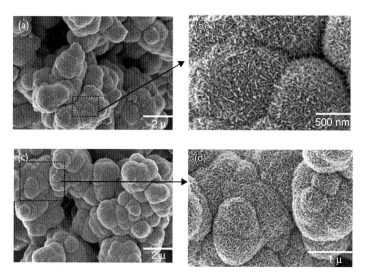

Figure 10.18 SEM micrographs of γ-MnO$_2$ room temperature synthesized for: (a) 2 h and (c) 24 h and their higher magnifications (b) and (d). respectively.

individual nanoparticles rather than aggregating to form spheres. This could be due to nucleating species being bound by DMSO and the resultant environment created repulsing other nucleating species from aggregating, purely a surface phenomenon.

Continuous gas-phase productions of ZnO, SnO$_2$, TiO$_2$, and Al$_2$O$_3$ nanoparticles and nanowires have been realized in a plasma jet microwave reactor [17]. These involved the direct oxidation of metal powders fed directly into the plasma cavity at atmospheric pressure, resulting in the production of nanoparticles or nanowires whose production depended directly on the particle size of the starting material, the microwave power, and the gas-phase composition. These results are summarized in Table 10.4.

Mixed metal oxides have also been continuously prepared using a continuous microwave process [14] via spraying a reactant solution into a receiving solution or into air under microwave radiation at atmospheric pressure. This was achieved in a microwave apparatus shown diagrammatically in Figure 10.4. A series of single/multiple metal/metal oxide alloys were prepared; including, Mo–V–O system, Mo–Nb–O system, Mo–V–Te–O system, Mo–V–Te–Nb–Pd–O system, Mo–V–Te–O system, and Mo–V–Te–Nb–O system. These nanomaterials are important catalysts for the highly catalytic partial oxidation of propane to acrylic acid. Both the nozzle microwave (NMW) and the *in situ* mixing nozzle microwave (INM) methods were employed. The INM method produced amorphous materials but these could be further crystallized into small (ca. 200 nm) particles with interesting morphologies after calcination. The materials prepared by this combined microwave-assisted continuous technique and calcination method showed smaller particle sizes, larger surface areas, and unique morphologies compared to those prepared under conventional hydrothermal methods. The INM method (Figure 10.4) was also explored by Nyutu *et al.* [15] to prepare spinel-type metal oxide materials. Small sized (6–20 nm), nanocrystalline spinel-type nickel ferrite and zinc aluminate particles with average particle sizes about 9 nm were prepared. The syntheses were carried out at ambient pressure (1 atm), microwave power (0–600 W), and an ultrasonic nozzle with resonant frequency of 48 or 120 kHz. The surface area of nickel ferrite ranged from 57 to 72 $m^2 g^{-1}$. The use of a lower-frequency ultrasonic nozzle *in situ* mixing (48 kHz) resulted in marginally higher surface area of nickel ferrite nanoparticles than with a 120-kHz nozzle.

Table 10.4 Experimental parameters and results for microwave continuous synthesis using plasma.

Metal precursor	MW power and gas flow rates				Characteristics of resulting metal oxide nanowires
	Power (W)	H$_2$ (sccm)	O$_2$ (sccm)	Air (slpm)	
Sn	1200–1500	100	500	10–15	10 µm length 20–100 nm diameter
Zn	1400–1700	100	500	10–15	5 µm length 50–100 nm diameter
Ti	700	100	500	8–10	1 µm length 100–250 nm diameter
Al	800	100	500	8–10	3–5 µm length 50–150 nm diameter

Adapted from ref [17].

10.6
Microwave Continuous Synthesis of Metallic Nanomaterials

The microwave continuous flow concept has been extended to the synthesis of metallic nanoparticles. Lin *et al.* [31] designed a continuous flow tubular microreactor for synthesizing silver (Ag) nanoparticles. The silver pentafluoropropionate was used as a single-phase reactant precursor, which was thermally reduced in isoamyl ether to form silver nanoparticles in the presence of trioctylamine. The produced Ag nanoparticles were narrowly distributed in size and had an average diameter of 8.7 ± 0.9 nm by size distribution. The moderate temperature for the formation of Ag nanoparticles is the key to this continuous flow synthesis process. A polyol continuous microwave process has also been reported for the synthesis of Ag nanoparticles [6].

In this process, silver nitrate ($AgNO_3$) and poly(*N*-vinylpyrrolidone) (PVP) dissolved in ethylene glycol were used, respectively, as a silver metal precursor and as a capping agent of the nanoparticles. Ethylene glycol on the other hand worked as the solvent and simultaneously as the reductant. Silver nanoparticles of narrow size distributions (9.8 ± 0.9 nm) were synthesized steadily for 5 h with average residence time of 2.8 s, maintaining almost constant yield and high quality. Ag nanoparticles were not formed under this flow rate by conventional heating. A narrower particle size distribution was achieved by increasing the flow rate of the reaction mixture.

Another polyol continuous microwave process for synthesizing Ag nanoparticles has also been reported by Dzido *et al.* [32]. In this process silver acetate was used as the silver precursor while ethylene glycol or glycerol acted as both the solvent and capping agent. PVP was used as a stabilizing as well as a capping agent. A domestic microwave apparatus operating at 2.45 GHz and adapted for continuous synthesis was used. Successful synthesis of Ag nanoparticles was realized at a temperature of 70 °C, residence time of 14–18 min, and flow rates between 0.25 and 0.5 l h^{-1}. The synthesized nanoparticles were narrowly distributed in size, as revealed by dynamic light scattering (DLS) analysis. Higher concentrations of the starting silver precursor led to development of two size distributions, both of them narrow.

Gold nanoparticles have been synthesized by Wagner *et al.* [33] with the use of a microfluidic set-up. The Au nanoparticles were directly synthesized via a redox reaction of a gold salt ($HAuCl_4$) and a reducing agent (ascorbic acid). The sizes of the as-synthesized nanoparticles ranged from 5 to 50 nm. The size distribution was an order of magnitude two times narrower than that obtained in a conventional synthesis when experimental parameters, such as pH, flow rate, the use of reducing agent and the use of surfactant were optimized.

Platinum nanoparticles have been synthesized by introduction of a polyol solution containing $H_2[PtCl_6] \cdot 6H_2O$ and PVP into a microwave reactor tube [34]. In these syntheses, ethylene glycol, glycerol, and 1,3-propanediol were used as reducing agents and as reaction solvents. These syntheses were achieved using a single-mode microwave flow reactor controlled with a temperature feedback module. The

MW reactor system is composed of a variable frequency generator, field intensity monitor, temperature sensor, and a cylindrical single-mode aluminum cavity. By tuning the flow rate and reaction temperature, this study realized silver nanoparticles with particle sizes ranging from 3.4 to 5.9 nm, depending on the chosen conditions.

10.7
Conclusions and Outlook

From the foregoing information, the continuous microwave synthesis of inorganic nanomaterials still lags behind its counterpart in organic material syntheses as well as microwave batch synthesis of inorganic nanomaterials. This is a research area that is yet to be explored to full potential. However, the unique physical properties, such as smaller particle sizes and high surface areas, exhibited by materials synthesized under microwave continuous techniques, that could also be a pointer to the potential uniqueness in chemical properties, provides a much needed impetus to the furtherance of this technology. Microwave continuous synthesis offers several advantages that microwave batch synthesis lack, such as opportunity for scaled synthesis where the only impediment is the size of both the feeding and receiving vessels, and more rapid heating due to the smaller sizes of reaction vessels (tubes) as opposed to lack of microwave penetration in larger batch vessels.

With examples, we have clearly demonstrated that microwave continuous synthesis is still an emerging technology that is slowly gaining use in the synthesis of inorganic nanomaterials. This technology has successfully been employed in the synthesis of microporous and mesoporous aluminosilicate nanomaterials, metal oxides and mixed metal oxide systems, as well as metallic systems. These syntheses have been accomplished using commercially available as well as custom-made apparatus. Both types of apparatus require a means of introducing the reactant materials into the microwave cavity. This is achieved using pumping devices for example, peristaltic pumps.

The only drawback to this technology may be presented in the form of nanomaterials that require extreme reaction conditions to crystallize. For instance, nanomaterials that require long aging periods for nucleation and crystallization and will take longer than say about 6 h to crystallize in a microwave environment may not benefit from continuous synthesis. Also, those nanomaterials that require extreme pressures may offer a challenge, not forgetting nanomaterials that tend to solidify (form thick gels) in the course of the reaction. If the above challenges are overcome, then there is great potential in using microwave continuous techniques in nanomaterials synthesis.

References

1 Kappe, C.O. (2004) *Angew. Chem. Int. Ed.*, **43**, 6250.
2 Kappe, C.O., Dallinger, D., and Murphree, S.S. (2009) *Practical Microwave Synthesis for Organic Chemists*, Wiley-VCH Verlag GmbH, Weinheim, Germany.
3 Nuchter, M., Ondruschka, B., Bonrath, W., and Gum, A. (2004) *Green Chem.*, **6**, 128.
4 Mingos, D.M.P., and Baghurst, D.R. (1991) *Chem. Soc. Rev.*, **20**, 1.
5 Nyutu, E.K., Chen, C.-H., Sithambaram, S., Crisostomo, V.M.B., and Suib, S.L. (2008) *J. Phys. Chem. C*, **112**, 6786.
6 Cides da Silva, L.C., Abate, G., Oliveira, N.A., Fantini, M.C.A., Masini, J.C., Mercuri, L.P., Olkhovyk, O., Jaroniec, M., and Matos, J.R. (2005) *Stud. Surf. Sci. Catal.*, **156**, 941.
7 Subramanian, V., Burke, W.W., Zhu, H., and Wei, B. (2008) *J. Phys. Chem. C*, **112**, 4550.
8 Zhang, Y., Feng, H., Wu, X., Wang, L., Zhang, A., Xia, T., Dong, H., and Liu, M. (2009) *Electrochim. Acta*, **54**, 3206.
9 Lee, K.S., Myung, S.T., and Sun, Y.K. (2007) *Chem. Mater.*, **19**, 2727.
10 Luo, W., Wang, D., Wang, F., Liu, T., Cai, J., Zhang, L., and Liu, Y. (2009) *Appl. Phys. Lett.*, **94**, 202507/1.
11 Abbaspour, A., and Ghaffarinejad, A. (2009) *Anal. Chem.*, **81**, 3660.
12 Gharibeh, M., Tompsett, G.A., Yngvesson, K.S., and Conner, W.C. (2009) *J. Phys. Chem. B*, **113**, 8930.
13 Tompsett, G.A., Conner, W.C., and Yngvesson, K.S. (2006) *Chemphyschem*, **7**, 296.
14 Espinal, L., Malinger, K.A., Espinal, A.E., Gaffney, A.M., and Suib, S.L. (2007) *Adv. Funct. Mater.*, **17**, 2572.
15 Nyutu, E.K., Conner, W.C., Auerbach, S.M., Chen, C.-H., and Suib, S.L. (2008) *J. Phys. Chem. C*, **112**, 1407.
16 Suib, S.L., Espinal, L., and Nyutu, E.K. (2006) US Patent 2006-0291827-A1.
17 Kumar, V., Kim, J.H., Pendyala, C., Chernomordik, B., and Sunkara, M.K. (2008) *J. Phys. Chem. C*, **112**, 17750.
18 Braun, I., Schulz-Ekloff, G., Wöhrle, D., and Lautenschläger, W. (1998) *Micropor. Mesopor. Mater.*, **23**, 79.
19 Park, S.-E., Chang, J.-S., Hwang, Y.K., Kim, D.S., Jhung, S.H., and Hwang, J.S. (2004) *Catal. Surv. Asia*, **8**, 91.
20 Park, S.-E., Kim, D.S., Chang, J.-S., and Kim, J.M. (2001) US Patent 2001-0054549-A1.
21 Corbin, D.R., Sacco, A.J., and Suib, S.L. (2006) U.S. Patent 7,014,838.
22 Bonaccorsi, L., and Proverbio, E. (2008) *Micropor. Mesopor. Mater.*, **112**, 481.
23 Opembe, N.N., King'ondu, C.K., Espinal, A.E., Chen, C.-H., Nyutu, E.K., Crisostomo, V.M., and Suib, S.L. (2010) *J. Phys. Chem. C*, **114**, 14417.
24 Bondioli, F., Corradi, A.B., Ferrari, A.M., and Leonelli, C. (2008) *J. Am. Ceram. Soc.*, **91**, 3746.
25 Corradi, A.B., Bondioli, F., and Maria, A. (2006) *Powder Technol.*, **167**, 45.
26 Nishioka, M., Miyakawa, M., Kataoka, M., Koda, H., Sato, K., and Suzuki, T.M. (2011) *Nanoscale*, **3**, 2621.
27 Takeshi, F., Yu-ichiro, H., Yusaku, K., Masahide, S., Noboru, S., and Fumio, K. (2010) *Mater. Technol.*, **28**, 244.
28 Opembe, N.N., King'ondu, C.K., and Suib, S.L. (2012) Efficient oxidation of 2,3,6-trimethyl phenol non-exchanged and H+ exchanged manganese oxide octahedral molecular sieves (K-OMS-2 and H-K-OMS-2) as catalysts. *Catal. Lett.*, **142**, 427–432.
29 King'ondu, C.K., Iyer, A., Njagi, E., Opembe, N., Genuino, H., Huang, H., Ristau, R., and Suib, S. (2011) *J. Am. Chem. Soc.*, **133**, 4186.
30 Jin, L., Chen, C.-H., Crisostomo, V.M.B., Xu, L., Son, Y.-C., and Suib, S.L. (2009) *Appl. Catal. A Gen.*, **355**, 169.
31 Lin, X.Z., Terepka, A.D., and Yang, H. (2004) *Nano Lett.*, **4**, 2227.
32 Dzido, G., and Jarzebski, A.B. (2005) *J. Nanopart. Res.*, **13**, 2533.
33 Wagner, J., and Kohler, J.M. (2005) *Nano Lett.*, **5**, 685.
34 Nishioka, M., Miyakawa, M., Daino, Y., Kataoka, H., Koda, H., Sato, K., and Suzuki, T.M. (2011) *Chem. Lett.*, **40**, 1327.

11
Microwave Plasma Synthesis of Nanoparticles: From Theoretical Background and Experimental Realization to Nanoparticles with Special Properties

Dorothée Vinga Szabó

11.1
Introduction

The synthesis of nanoparticles using various microwave-assisted methods is an established technology nowadays [1–11]. However, due to the complexity of microwave interactions with materials, simply placing a solution into a microwave oven, and expecting to heat it efficiently will rarely lead to high quality products. Microwave methods offer some unique benefits, especially with respect to producing particles with nanosize and controlled compositions, and a greater process flexibility by taking advantage of several combinations of volumetric, rapid, and selective heating conditions, which are not possible by conventional means. All of these microwave heating advantages can be used to process and tailor extremely fine powders by controlled reactions in sol–gel processing, solution evaporation/ decomposition, hydrothermal reactions, or in gas-phase synthesis.

A very special field of microwave-assisted technologies is the synthesis of high quality nanoparticles using a plasma which is generated by microwaves. Here, gas-phase synthesis methods are combined with microwave and plasma technology. This topic is an interdisciplinary field of research, development, and technology, interplaying with microwave and plasma technology, chemistry, physics, and nanotechnology. It also refers to topics of electrical engineering, process technology, thermodynamics, aerosol technology, chemical engineering, and materials science.

Since the early 1990s microwave plasma methods for the synthesis of various nanoscaled materials have become increasingly popular. The first conference papers were published by Metha *et al.* [12] reporting on TiN and TiO_2, by Sickafus *et al.* [13] reporting on ZrO_2 and Al_2O_3, and by Vollath and coworkers [14–16] reporting on nanoscaled oxides and on technical equipment. The first scientific papers [17–21] were presented, reporting on different metals, and oxides, on coated ceramic nanoparticles and on special magnetic properties of nanoparticles

Microwaves in Nanoparticle Synthesis, First Edition. Edited by Satoshi Horikoshi and Nick Serpone.
© 2013 Wiley-VCH Verlag GmbH & Co. KGaA. Published 2013 by Wiley-VCH Verlag GmbH & Co. KGaA.

synthesized in a microwave plasma. Whereas in the beginning mainly the feasibility of the various materials was the focus of research, later, theoretical considerations on microwave plasmas, combined with a deeper understanding of the processes occurring in the plasma and the interaction between different species in the plasma, were of interest in several groups [22–27]. Finally, the properties and application potential of the resulting nanomaterials have seen very much increased interest [28–37].

This chapter will focus on the application of microwave plasmas for the synthesis of high quality nanoparticles from the viewpoint of a materials scientist, using a physical method to synthesize materials with special properties. In this context "high quality nanoparticles" means nanoparticles with size-dependent physical properties, or core–shell nanoparticles, both mainly characterized by particle sizes below around 20 nm and with narrow particle size distribution. The chapter will mainly rely on application relevant items: why and how microwave plasmas are applied for the synthesis of bare nanoparticles with particle sizes mainly below around 10 nm and extremely narrow particle size distribution, as well as for the synthesis of core–shell nanoparticles of the same order of magnitude. First, a general overview on plasma synthesis methods will be given. Then, the fundamentals of microwave plasma processes, the relevant process parameters with their interdependence, and their influence on the resulting nanoparticles will be explained. Finally, several examples of interesting physical properties of nanoparticles, resulting from the small particle size and narrow particle size distribution, as well as application potentials of synthesized nanomaterials for technical use will be presented.

11.2
Using Microwave Plasmas for Nanoparticle Synthesis

Syntheses of nanoparticles using microwave-assisted methods can be divided into two general groups. The first is the field of wet-chemical methods with different sub-groups, applying microwaves for heating of a liquid with subsequent decomposition. These methods are topics of several other chapters of this book. The second group is the area of gas-phase reactions, where microwaves are used to generate plasma in which chemical reactions take place. Therefore, in this subsection a closer look into the background of plasmas is presented.

11.2.1
General Comments on Plasmas

In physics and chemistry an ionized gas, comprising molecules, atoms, ions, electrons and photons, typically, is denoted as "plasma". Because of their unique properties, plasmas are often considered to be a separate state of matter. The term "ionized" refers to the presence of one or more free electrons. These electrons are not bound to an atom or a molecule. A plasma is electrically conductive due to free electric charges (electrons and ions), so that it responds strongly to electromagnetic fields. Plasmas can be accelerated and steered by electric and/or magnetic fields, which allows control of the plasma.

The classification of plasmas in the literature is sometimes imprecise. To give a rough overview, plasma types have been categorized according to several criteria: the type of power source, the operating pressure, the type of coupling, the temperature relationship, and finally their application. An overview of the different plasma categories, including some relevant literature, is presented in Table 11.1.

Thermal plasmas are characterized by a thermodynamic equilibrium, meaning that all species (electrons, ions, neutral species) have the same temperature (= energy). In non-thermal plasmas the temperature of the electrons ranges between several electron volts (eV), whereas the temperature of the positively charged ions and neutral species is significantly colder (around room temperature) [32, 82], leading to a quite low overall temperature. Therefore, non-thermal plasmas, also called non-equilibrium plasmas, are favorable for the synthesis of

Table 11.1 Overview on different plasma types, according to the operating pressure, the temperature relationship, the type of power source, discharge, coupling, and application, including relevant publications.

	Operating pressure		Temperature relationship	
	Low-pressure plasma	Atmospheric plasma	Thermal plasma (jet, torch)	Non-thermal plasma
Type of power source				
Microwave (MW)	[24, 38–41]	[29, 34, 42–48]	[29, 34, 42, 43, 47–49]	[19, 20, 38–40, 46, 50–52]
Radio frequency (RF)	[53, 54]	[55, 56]	[56–59]	[60, 61]
Direct current (DC) or Alternating current (AC)	[62]	[63]	[55, 63–65]	[66]
Type of discharge				
Arc discharge		[55]	[55, 59, 63]	[66]
Corona discharge		[46]		[46]
Dielectric barrier discharge (DBD)		[46, 59, 67]		[46, 67]
Type of coupling				
Inductive coupled (ICP)	[53, 54, 68]	[55]	[55, 58, 59]	
Capacitive coupled (CCP)	[69, 70]			
Type of application				
CVD	[71–74]	[75, 76]	[77]	[78–80]
Particle synthesis	[24, 38–41, 53, 54, 81]	[29, 34, 42–47, 55, 56, 63]	[29, 34, 42–44, 47–49, 57]	[19, 20, 38–40, 50–52, 67]

nanoparticles at low temperatures. For industrial processes atmospheric plasmas are favorable, as they need smaller experimental effort and thus have lower costs. Low-pressure plasmas in contrast require more expensive vacuum-components and more sophisticated set-ups.

Excellent overviews on the varieties of plasmas and their applications are given by several authors and research groups: Kortshagen [32] on non-thermal plasmas, Bárdos and Báranková [46] on cold atmospheric plasmas, Belmonte *et al.* [59] on non-equilibrium plasmas at atmospheric pressure, and Tendero *et al.* [83] on atmospheric pressure plasmas. A detailed description of thermal plasma methods has been published by Ishigaki [84]. A general review paper on plasma synthesis of nanopowders was presented by Vollath [85].

The following considerations of this chapter will focus on non-thermal plasmas, using microwaves as the generating source for chemical synthesis of nanoparticles, and are mainly based on papers from Vollath and Szabó [39] and Vollath [85, 86].

11.2.2
Considerations in a Microwave Plasma

The energy E, transferred to a charged species of a mass m in an oscillating electrical field with frequency f is proportional to the charge Q, and inversely proportional to its mass m and the squared frequency f

$$E \propto \frac{Q}{m \cdot f^2} \qquad (11.1)$$

and is a measure for the temperature. The charge Q may either be the charge of an elementary species, or the electric charge of a particle, which then is directly proportional to the particle diameter. This equation is fundamental, as it shows the dependence on the frequency used in the power source, and the dependence on the mass of the charged species: as the mass of the electrons is small compared to that of ions or neutral species, at a given frequency a substantially larger amount of energy is transferred to the electrons, as compared to the energy transferred to the ions. With increasing frequency the transferred energy decreases, meaning a decreasing temperature. Due to their low mass, the energy of the free electrons, meaning their "temperature", is significantly higher than the "temperature" of the ions. Additionally, the "temperature" of the neutral gas molecules, obtaining their energy from collisions with the charged ones, is even lower. This leads to a significantly lower "overall temperature" of a gas passing a microwave plasma compared to a DC (direct current) or RF (radio frequency) plasma, where temperatures in the range from 5000 to 15 000 K are obtained.

Besides ions and free electrons, a microwave plasma consists of neutral gas species, as well as dissociated gas and precursor molecules. Therefore, collisions between charged (electrons, ions) and uncharged species (molecules, atoms, or particles) influence the energy transfer to the particles. In this case, the collision frequency z has also to be considered:

$$E \propto \frac{Q}{m} \frac{z}{f^2 + z^2} \qquad (11.2)$$

This additional variable does not alter the principal charge and mass relationship of the transferred energy; however, it introduces an additional dependence of the collision frequency in the plasma (the collision frequency z is proportional to the gas pressure p). According to Eq. (11.2) the collision frequency has a significant influence on the synthesis as the collisions between energy-rich charged particles and uncharged ones control the energy transfer. As the collision frequency increases with increasing gas pressure, the transferred energy is also a function of the gas pressure. With increasing gas pressure the transferred energy increases for $z < f$, exhibits a maximum at $f = z$, and decreases for $z > f$. Figure 11.1a shows

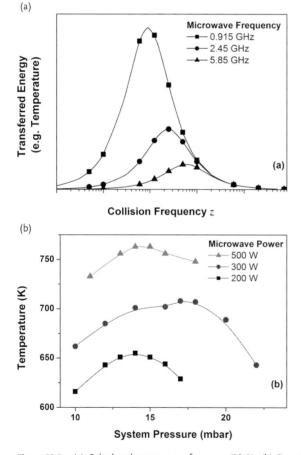

Figure 11.1 (a) Calculated energy transfer on elementary species as a function of collision frequency at different commercially used microwave frequencies, according to Eq. (11.2). (b) Experimentally determined temperature as a function of system pressure using 2.45 GHz microwaves at different microwave power.

schematically the calculated relationship between collision frequency z and transferred energy E on an elementary species, applying the three common industrial microwave frequencies, 915 MHz, 2.45 GHz and 5.85 GHz. The measured temperatures as a function of system pressure (using a reaction gas of 80% Ar with 20% O_2, but without thermal evaporation of a precursor and without chemical reaction) show experimental evidence for the theoretical curve progression for one selected frequency (Figure 11.1b). It can be seen that the real temperatures (measured directly behind the plasma) can be set below 1000 K in 2.45 GHz systems. Even lower temperatures in the region of 500 K can be realized under certain conditions, being favorable for *in situ* coating of nanoparticles with organic compounds.

From these figures several implications can be drawn:

- With increasing frequency the transferred energy decreases: in consequence for nanoparticle synthesis in a microwave plasma, a lower microwave frequency leads to higher reaction temperatures. Lower synthesis temperatures can be realized with higher frequencies.

- The case of $z \ll f$ is the range of low gas pressure (several mbar). In this regime the energy of the electrons may be some keV, due to the large free path length. Electrons of such high energies are able to ionize particles, which then carry positive charges.

- The case of $z \gg f$ is the range of higher gas pressure (several 10 to 100 mbar). In this regime the energy of the electrons is lower (only some eV), because the free path length is shorter, and the number of collisions is higher. Electrons of such low energy may attach at the surface of particles, which then carry negative charges.

- The case $z \approx f$ is an intermediate range. In this regime the particles may carry charges of both signs.

In contrast, applying electromagnetic waves with radio frequency further increases the transferred energy (temperature) by several orders of magnitude. For RF-plasmas it is known that the charging of particles depends on the ion temperature, the electron temperature, and the particle diameter [87], which obviously are influenced by the pressure. Matsoukas and Russel [87] suggested that particles synthesized in a plasma must exhibit a smaller particle size distribution compared to conventional gas-phase synthesis methods, because collisions between larger particles of the same (charge) distribution are unlikely due to strong repulsion. Most of the theoretical considerations of charged particles have been developed for non-thermal low-pressure RF-plasmas. Analogous considerations for microwave plasmas are rare. Nevertheless, it is clear that charging of particles in a microwave plasma obviously depends on the pressure [86]. As experimental evidence for narrow size distribution was shown by several groups applying microwave plasma synthesis methods [24, 38, 39, 88–90], it is most likely that findings for RF-plasmas are transferable to microwave plasmas.

11.2 Using Microwave Plasmas for Nanoparticle Synthesis

In consequence, the selected microwave frequency of the equipment, the used gas pressure (mainly regulated by the gas flow, or by using a throttle valve), the precursor concentration, and the energy input (microwave power) are the main tools for proper setting of experimental conditions to achieve optimum synthesis conditions for each material. To produce small nanoparticles with narrow size distribution, pressure conditions should be selected in the range of equal charges, so that particles can repel each other to thwart particle growth.

11.2.3
Particle Formation

As the microwave plasma process is a gas phase process, a vaporized precursor is the starting material. Unfortunately, besides SiH_4 (silane), few available precursors are gases, so that the precursors (either liquid or solid) usually have to be evaporated outside and prior to the reaction zone. Similarly to thermal gas-phase reactions, the nanoparticle formation mechanism is assumed to occur in the following steps:

1) In the reaction zone (plasma) the gaseous components (vaporized precursor, carrier gas, and reaction gas) are ionized and dissociated.
2) The dissociated components react, forming molecules (e.g., oxide molecules). As the components are ionized or dissociated, thermal activation barriers do not have to be overcome for the reaction. The components carry electrical charges and are accelerated in the electric field of the microwaves.
3) Homogeneous nucleation of clusters by random collision of molecules happens.
4) Growth of clusters by further collision with molecules or clusters takes place.
5) The growth by coagulation of particles, occurring in thermal gas-reactions, is reduced in a microwave plasma, due to electric charging of the particles.

These steps last a few milliseconds only. Then, particles leave the reaction zone, and lose their electric charges. The growth process in plasmas is more complex than in conventional gas-phase synthesis, due to the charging of particles in the plasma. The charging of particles in the plasma is the most important mechanism, as it governs the interaction between the particles and the plasma [27]. Figure 11.2 shows schematically the relevant particle formation steps.

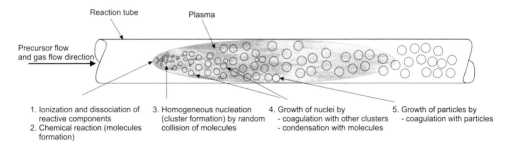

Figure 11.2 Mechanism of particle formation in the plasma.

The critical step for particle formation is the nucleation. The variation of particle size is correlated with changes in the number concentration of nuclei in the plasma. For a given amount of precursor a high number concentration of nuclei leads to small particle sizes, whereas a small number concentration of nuclei results in larger particles.

11.2.4
Characterization of Nanoparticles

Characterization of nanoparticles is a challenging task. Many comprehensive methods can be used for the characterization of nanoparticles regarding their size, structure, morphology, chemistry, surface properties, or physical properties. In the context of this chapter only those methods relevant to explain or illustrate statements regarding particle size, crystal structure, shape, or morphology are briefly described. They are frequently used methods to characterize the structure and morphology of "as-produced" nanoparticles and nanocomposites, and therefore are very important tools for the characterization of nanoparticles.

X-ray diffraction (XRD) is a common technique used to characterize the crystallographic structure, crystallite size (grain size), and preferred orientation in polycrystalline or powdered solid samples [91, 92]. The interaction of X-rays (incoming and reflected X-rays) with a geometric structure (the crystal lattice of nanoparticles), is evaluated using the Bragg equation [93, 94]. Powder diffraction is commonly used to identify unknown substances by comparing diffraction data against a database maintained by the International Center for Diffraction Data.

Transmission electron microscopy (TEM) is a microscopy technique, roughly similar to light microscopy. Instead of light, accelerated electrons are used to generate the images. Lenses are, instead of glass, electromagnetic lenses. The accelerated electron beam passes through an ultra-thin, electron transparent sample and interacts with the sample. The transmitted electrons are used to determine the size, shape and morphology of particles, or in the case of high-resolution TEM, to generate lattice images. Details about this method can be found in textbooks, for example, [95]. Due to the short wavelengths compared to light (200 keV electrons: λ = 0.00251 nm, visible light: λ = 380–750 nm), a resolution around 0.3 nm can be obtained in a conventional transmission electron microscope. This resolution is in the size regime of atomic distances. Due to technical improvements in highly sophisticated transmission electron microscopes a resolution even better than 0.1 nm can be reached [96, 97]. Elastically scattered electrons are used for electron diffraction, and employed to determine the crystal structure, and the crystallite size of particles [98, 99]. TEM and XRD are both *ex situ* methods.

In contrast, a very common method for the *in situ* determination of particle sizes, originated in aerosol science, is the so-called particle mass spectrometry (PMS) [100, 101]. The nanoparticles are sampled directly after leaving the synthesis zone (*in situ*) and without prior dilution in a molecular beam. The charged nanoparticles are then deflected by an electric field to determine their mass according to the time-of-flight principle.

Finally, scanning electron microscopy (SEM) is used to characterize the morphology and topography of nanogranular layers. In a scanning electron microscope images are generated by scanning a sample with a beam of accelerated electrons in a raster scan pattern.

11.3
Experimental Realization of the Microwave Plasma Synthesis

Experimental set-ups for the synthesis of nanoparticles can be realized with standard industrial frequencies of 0.915 (wavelength $\lambda \sim 32\,cm$), and 2.45 GHz ($\lambda \sim 12\,cm$, household kitchen frequency). The frequency of 5.85 GHz ($\lambda \sim 5\,cm$) is not very common in plasma synthesis. With increasing microwave frequency the dimensions of the components decrease, so that the equipment becomes smaller. Microwave generators, magnetrons, appropriate waveguides, and passive components such as isolators, directional couplers, tri-stub tuners, or sliding shorts are commercially available worldwide. Isolators are a combination of a circulator, allowing transmission in only one direction, and a water load, absorbing reflected power. This device protects the magnetron from reflected power. Directional couplers measure the transmitted and reflected power. The tri-stub tuners maximize the field-strength at the desired position, therefore, they are usually located as close as possible to the applicator. Sliding shorts are tuning devices, allowing tuning for resonance in an applicator. They are positioned after the applicator. Resonant cavities (applicators) are also commercially available to some extent. Examples of commercially available cavities are downstream sources, plasma torches, surfatrons, or surfaguides. In many cases, cavities are a special and individual custom design, depending on the needs of the customer. They are mainly developed in collaboration with microwave companies.

11.3.1
Custom-Made Applicators

One example of a custom-made cavity is the rectangular applicator in TE_{01} mode (transverse electrical mode), producing a standing wave at the intersection of a quartz glass tube and the waveguide. This standing wave has to be adjusted to maximum resonance by a sliding short, and tuned for maximum field strength for each pressure as the burning plasma changes the impedance of the system significantly. Therefore, operation in optimum pressure conditions is problematic, as in the case of plasma extinction a re-ignition of the plasma is difficult. From this point of view untuned cavities are more advantageous, as plasma re-ignition is easier. On the other hand, untuned cavities may lead to plasma quenching if the operating conditions are changed.

Another special cavity design [102] uses the rotating TE_{11} mode. Here also plasma is ignited at the intersection of the quartz glass tube and the cavity. The cut-off tubes attenuate the microwaves and prevent microwave leakage. This type

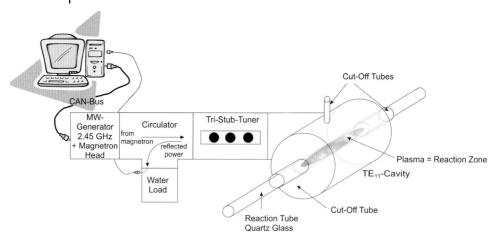

Figure 11.3 Schematic microwave component arrangement of a microwave-plasma synthesis equipment used in the Karlsruhe microwave plasma process.

of cavity has the advantage of easy plasma ignition with high microwave efficiency over a broad range of gas pressure [25]. Furthermore, the length of the reaction zone can be adjusted to a multiple of the half wavelength of the plasma. This feature is interesting for significant increase in residence time, as will be shown later. Figure 11.3 shows the schematic microwave component arrangement at 2.45 GHz, as it is used for the synthesis of nanoparticles in the Karlsruhe microwave plasma process (KMPP) [25, 39, 103], operating at low pressures. Using two or more consecutive branches of this set-up enables the synthesis of core–shell or core–shell-1–shell-2 nanoparticles.

11.3.2
Coated Nanoparticles and Particle Collection

Many applications require core–shell or nanocomposite materials. This is, for example, the case for specific SnO_2-SiO_2 nanoparticle-based gas sensors [104], or in the case of Li-ion batteries, where Sn or SnO_2 nanoparticles combined with a carbon phase are used as anode material [105–113]. Applications in biology, medicine or diagnostics often require bi-functional or hybrid nanoparticles [114–119]. Such particles can be synthesized in a microwave plasma when they are "core–shell" nanoparticles.

One necessary prerequisite for the formation of core–shell nanoparticles with a ceramic core and a ceramic shell is, that there is no (or only very low) solubility, or phase formation, between the phases forming the core and the shell. Technically, two microwave cavities are aligned consecutively. A second precursor is evaporated and introduced into the system behind the area where nanoparticles leave the first plasma zone. The reaction of the second compound is performed in the second plasma. As the gas kinetic cross section of particles is at least two orders of magnitude larger than the gas kinetic cross section of atoms or molecules, nanoparticles formed in the first step via homogeneous nucleation act as condensation seeds for the second phase forming the shell. The core particles from the first step are not yet agglomerated, and coating of individual nanoparticles is therefore possible. The precursor feeding rate of the second precursor has a straight influence on the shell thickness [37, 104]. This set-up is shown schematically in Figure 11.4a.

A modified set-up is used for the synthesis of hybrid core–shell nanoparticles. These are nanoparticles consisting of an inorganic core and an organic shell. The evaporation of an organic compound is performed after the first plasma, and the condensation happens outside the plasma. This is because most of the organic compounds (e.g., monomers or dyes) are decomposed in the plasma. For the coating of nanoparticles with organic compounds low temperatures are required. Otherwise the organic coating material may degrade during the coating process. The arrangement of such a set-up is shown in Figure 11.4b.

One of the major challenges of microwave plasma synthesis is the "collection" of particles with diameters below 10 nm. In aerosol technology and gas phase processes, a thermophoretic particle deposition is common [120–122]. Thermophoresis is described as a physical phenomenon in which aerosol particles, subjected to a temperature gradient, move from high- to low-temperature zones of the gas. With increasing temperature gradient and with increasing particle diameter the thermophoretic forces on a particle increase linearly [122, 123]. Additional effects on thermophoretic forces are found when the particles carry electrical charges [123]. This method is principally suited for the collection of nanoparticles synthesized in a microwave plasma. A cooling finger, which is separated from the heated surface by a gap of a few millimeters, is used as the thermophoretic particle collector. The hot gas stream containing the nanoparticles is directed to the cooled surface where the nanoparticles are deposited. The collection efficiency decreases rapidly for decreasing particle sizes in the range around 2–3 nm. For such particles the collection efficiency may be around 20% of the conversion, for particles with sizes around 8 nm about 60% collection efficiency may be reached. Commercial particle collectors do not exist, so that most particle collectors are experimental. After the collection, the powder is scraped of the cooling finger.

A second method of particle collection is the thermophoretic deposition on substrates [31, 113, 124]. This deposition method results in porous nanoparticle films. The substrate size depends on the lateral dimension of the system components, and the thickness of the films depends on the deposition time. The particles exhibit cauliflower and club-like morphology, which can be varied by preheating the substrates [31]. Finally, a deposition of nearly agglomerate free particles is possible in liquids such as glycol, resins, monomers, or reactive diluters [125]. This collection method is interesting when the nanoparticles are used as fillers to modify or tune polymer properties, and is realized by passing the gas flow through the desired liquid.

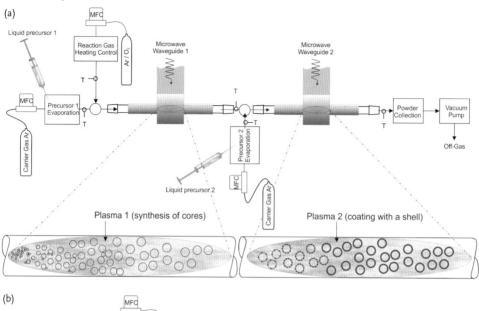
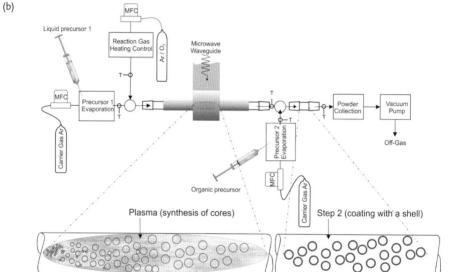

Figure 11.4 Typical set-ups for microwave plasma synthesis of (a) ceramic core–shell nanoparticles and (b) hybrid core–shell nanoparticles.

11.4
Influence of Experimental Parameters

The discussion of experimental parameters in microwave plasma synthesis of nanoparticles is quite complex: it comprises as well as microwave frequency and cavity design (both related to the equipment design), the system temperature, the microwave power, the system pressure, the residence time, the precursor type, and the precursor concentration in the system. Many parameters are interdependent. This means for example, increasing the gas flow of the reaction gas increases not only the gas pressure, but also the temperature (see Figure 11.1b), in turn influencing the collision conditions as well as residence time in the reaction zone. Furthermore, the temperature of the system can be influenced directly by the applied microwave power (see Figure 11.1b). These interdependences are overlapped by the influence of the pressure on particle charging.

A straightforward influence of microwave frequency is on the temperature (see Eqs. [11.1] and [11.2], and Figure 11.1a). As the frequency is a constant parameter depending on the particular experimental set-up in use, it will not be discussed further in detail. Nevertheless, some remarks on this parameter are presented. By changing the microwave frequency from 2.45 GHz to 915 MHz, the geometrical dimensions increase, because the wavelength increases. Therefore, larger reaction tube diameters and longer reaction zones are used; higher gasflows, higher pressures and higher temperatures are obtained using a 915 GHz equipment. Additionally, ionization conditions are changed, adding complexity to the system, and changing experimental boundary conditions. In consequence, a direct comparison of the influence of microwave frequency on particle size cannot be made, as the relevant synthesis parameters (e.g., microwave power, pressure, temperature, gas velocity, and precursor concentration) are not equal. The two TEM images in Figure 11.5 visualize exemplarily the difference in size and crystallinity of the obtained nanoscaled powders synthesized using the two different frequencies. The

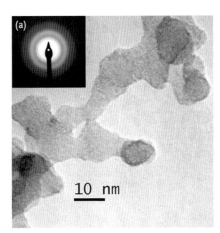
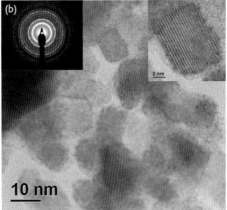

Figure 11.5 Comparison of two different WO$_3$-powders synthesized using a microwave plasma generated with 2.45 GHz (a) and 915 MHz (b), respectively. Using 2.45 GHz equipment results in interconnected amorphous particles, whereas the lower frequency equipment yields single crystalline particles, exhibiting clear lattice fringes (inset right). The difference in crystallinity and crystallite size can be deduced from the electron diffraction images (insets left, respectively).

difference in particle size and structure cannot be attributed only to the different frequencies. It is a combined result of different temperatures, pressure, residence time in the plasma, precursor concentration and different frequencies.

To reduce the complexity, considerations and the following examples from our own research will focus on a 2.45 GHz system, using a brazen TE_{11} cavity [102] of 0.09 m length, and a quartz-glass reaction tube with an inner diameter of 26 mm. This is the equipment which is used for the synthesis of nanoparticles in the KMPP. For the parameter evaluation, the measurement of temperature and pressure is performed using the reaction gas mixture of 80% Ar with 20% O_2, but without chemical reaction of a precursor. Additionally, data from the literature, using 2.45 GHz plasmas will be analyzed. The influence of the different synthesis parameters on, for example, particle size, modification of the resulting phase, or incomplete/complete precursor reaction will be discussed exemplarily in this section.

11.4.1
Precursor Selection

The considerations for precursor selection are general and may apply for each desired nanoparticle synthesis. They will be explained exemplarily for the most relevant and common liquid precursor materials: water-free chlorides, metal-carbonyls, and different organometallic precursors. Liquid precursors are favorable, as they can be fed accurately via syringe pumps. The most relevant liquid precursors for the synthesis of SnO_2, Fe_2O_3, TiO_2, and ZrO_2 nanoparticles – the scientific and technically most interesting nanoparticles – are summarized in Table 11.2. Similar considerations are valid for any solid and volatile precursors.

The remarks show, that all precursors have to be handled under safe conditions, as all are either toxic or very toxic, or at least irritant or harmful to human health, and also partly sensitive to the atmosphere. Thus, toxicity is certainly one aspect one has to consider for precursor selection. Another point of interest is, besides the boiling point, the theoretical conversion of product per unity of precursor. Therefore, one has to look at the chemical reactions which may occur in the plasma during a complete or incomplete reaction:

Precursor 1: $SnCl_4 + O_2 \rightarrow SnO_2 + 2\,Cl_2$

Precursor 2: $TiCl_4 + O_2 \rightarrow TiO_2 + 2\,Cl_2$

Precursor 2: $2\,Fe(CO)_5 + 6.5\,O_2 \rightarrow Fe_2O_3 + 10\,CO_2$

Precursor 4: $Sn(C_4H_9)_4 + 26\,O_2 \rightarrow SnO_2 + 16\,CO_2 + 18\,H_2O$ (complete reaction)
$Sn(C_4H_9)_4 + O_2 \rightarrow SnO_2 + [C_{16}H_{36}]$ (incomplete reaction)

Precursor 5: $Ti(OC_4H_9)_4 + 24\,O_2 \rightarrow TiO_2 + 16\,CO_2 + 18\,H_2O$ (complete reaction)
$Ti(OC_4H_9)_4 + 0.5\,O_2 \rightarrow TiO_2 + [C_{16}H_{36}]$ (incomplete reaction)

Precursor 6: $Zr(OC_4H_9)_4 + 24\,O_2 \rightarrow ZrO_2 + 16\,CO_2 + 18\,H_2O$ (complete reaction)
$Zr(OC_4H_9)_4 + 0.5\,O_2 \rightarrow ZrO_2 + [C_{16}H_{36}]$ (incomplete reaction)

Table 11.2 Selection of different commercially available, liquid and volatile precursors, typically used for oxide nanoparticle synthesis.

Nr.	Chemical formula [CAS-Number]	Name	Melting point/ Boiling point (°C) (from data sheet)	Remarks
1	$SnCl_4$ [7646-78-8]	Tin(IV)chloride	−33/114	Moisture and air sensitive; corrosive to metals and tissues; harmful; irritant; releases Cl
2	$TiCl_4$ [7550-45-0]	Titanium(IV)chloride	−25/136	Moisture sensitive; corrosive to metals and tissues; irritant; releases Cl
3	$Fe(CO)_5$ [13463-40-6]	Ironpentacarbonyl	−20/103	Air sensitive; highly flammable; very toxic
4	$Sn(C_4H_9)_4$ [1461-25-2]	Tetra-n-butyltin	−97/145/10 mm Hg pressure	Harmful; toxic
5	$Ti(OC_4H_9)_4$ [5593-70-4]	Titanium(IV)-n-butoxide	−55/206/10 mm Hg pressure	Moisture sensitive; flammable; irritant
6	$Zr(OC_4H_9)_4$ [2081-12-1]	Zirconium(IV)-t-butoxide	3/90/5 mm Hg pressure	Moisture sensitive; irritant

It can be seen, that all organometallic precursors release more or less large amounts of CO_2 and H_2O, or alternatively hydrocarbons during the reaction. For a complete oxidation of a precursor these reactions need much more oxygen than reactions using water-free chlorides. In the case of incomplete oxidation residual hydrocarbons (denoted as $[C_{16}H_{36}]$) may be a by-product. On the other hand, the chloride precursors release Cl_2, which might be a problem for the vacuum components, and also a problem in some applications if Cl is adsorbed on the surface of the nanoparticles. The theoretical conversion of oxide per ml of inserted precursor, the oxygen needed for a complete reaction, and the released by-products for the selected six precursors are shown in Figure 11.6a. Comparing, for example, $SnCl_4$ with $Sn(C_4H_9)_4$ means that, even when using identical precursor concentrations, the chloride precursor contains 2.4 times more reacting Sn-cations than the corresponding organometallic precursor. Therefore, an influence on particle size is also expected depending on the precursor. Figure 11.6b shows the theoretical average excess oxygen in the system, assuming a complete oxidation, for four different synthesis conditions (reaction-gas feeding rate 5 or $10 \, l \, min^{-1}$, precursor feeding rate 1 or $5 \, ml \, h^{-1}$). These diagrams show that careful precursor selection is necessary for each system, in combination with all reaction possibilities. An unfavorable combination of precursor, precursor feeding rate, reaction-gas flow and microwave power may result in an incomplete chemical reaction. The real suitability of a precursor for particle synthesis, however, is affected by

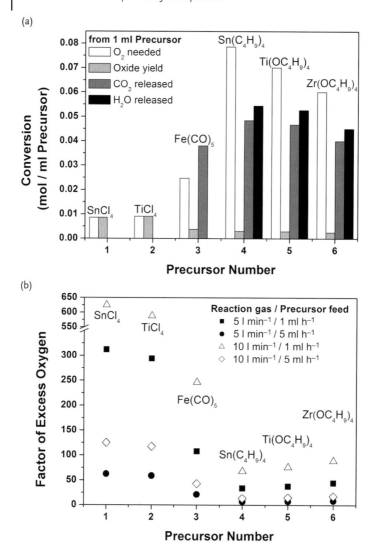

Figure 11.6 Conversion (a) and excess oxygen (b) for different precursors used for oxide nanoparticle synthesis. The numbers correspond to those from Table 11.2.

the capability of nuclei formation under the selected experimental conditions. Therefore, experiments to test if a precursor is really suited or not are always necessary.

From these two figures it is clear that chlorides are favorable as precursors when high oxide conversion per ml of precursor are wanted, and if Cl_2 as a by-product is not a problem. Organometallic precursors produce high amounts of by-products (mainly CO_2, and H_2O, or hydrocarbons), and therefore need low feeding rates combined with higher reaction-gas flow to operate the system in significant oxygen excess. On the other hand, the *in situ* formation of hydrocarbons due to incomplete chemical reactions can also be exploited for the one-step coating of SnO_2 nanoparticles with a carbonaceous phase, as is required in Li-ion battery materials. The carbon coating improves the specific capacities of SnO_2-based nanomaterials [37] and will be shown in Section 11.5.3 in more detail.

11.4.2
Influence of Precursor Concentration

Changing the precursor concentration mainly means changing the collision frequency between atoms, reactive molecules, ions, and electrons in the plasma. Therefore a strong impact of this parameter on the number concentration, nuclei formation, and particle size is expected. Here several research groups, using different 2.45 GHz equipment find comparable behavior for the dependence of nanoparticle sizes on the precursor concentration for different synthesized materials. Most investigated are the particle sizes of Fe_2O_3 [38, 126], Si [26, 41], SiO_2 [126], GaN [80] and SnO_2 [127] nanoparticles as a function of precursor concentration. In all cases the observed particle crystallite sizes do not exceed 12 nm. The tendency for increasing particle size with increasing precursor concentration is also confirmed by theoretical reflections for the formation of SiC in an atmospheric pressure plasma [47]: a strong positive impact of precursor concentration on particle size is calculated. The summarized literature data are shown in Figure 11.7.

For the dependence of particle size on precursor concentration general rules can be stated, independent of the material synthesized, the research groups, and even the pressure in the microwave plasma: with increasing precursor concentration an increase in particle size is observed. At very low concentrations particle sizes around 2 nm can be realized, whereas an increase of several orders of magnitude in concentration is necessary to produce particles with sizes around 10 nm.

Theoretically the conversion of a precursor can be considered somehow as a concentration effect (see Figure 11.6a). Using identical synthesis parameters, but different Sn-precursors ($SnCl_4$ and $Sn(C_4H_9)_4$) results in different particle sizes: the particle size of SnO_2 made from $SnCl_4$ is in the region of 4 nm, whereas the particle size of SnO_2 made from $Sn(C_4H_9)_4$ is around 3 nm (both determined from TEM images). The organometallic precursor contains 2.4 times less Sn-cations than the chloride precursor, being responsible for this "concentration effect".

The influence of precursor concentration on the particle size of a material can be observed during the synthesis of carbides [128] in a low-pressure microwave plasma. This is shown using the example of SiC, synthesized from methyltrichlorosilane (MTS), CH_3SiCl_3 in Ar/4% H_2 atmosphere at a pressure of 10 mbar and a temperature of around 860 K. As these experiments are also combined with a significant variation in the residence time by changing the length of the cavity, they will be discussed in the context of Section 11.4.4.

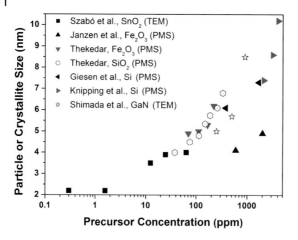

Figure 11.7 Experimentally determined particle crystallite sizes as a function of precursor concentration. The data are taken from different research groups for different materials [26, 38, 41, 80, 126, 127].

11.4.3
Interdependence of Microwave Power, Pressure, Temperature, and Gas Velocity

First, the dependences of the parameters have to be defined. The residence time t of a particle in the plasma is only a few milliseconds, and depends on the system pressure p_2, the system temperature T_2, the gas-flow V_1 (or volume) used in the experiment, the diameter d of the reaction tube, and the length L of the reaction zone (usually the cavity). It can be calculated, using the equation of states:

$$t = \frac{1}{4}\pi d^2 L \frac{p_2}{p_1}\frac{T_1}{T_2}\frac{1}{V_1} \tag{11.3}$$

where T_1 is the temperature, and p_1 is the pressure at standard conditions. Equation (11.3) shows that the residence time is influenced by the temperature, the gas flow (or volume), and the pressure used for the experiment. These parameters are directly connected to the microwave power and to the collision frequency z (Eq. [11.2]), and are overlapped by particle charging effects. Furthermore, a linear relationship is given with the length L of the cavity.

The temperature in the system can be influenced either by varying the microwave power, or by changing the system pressure. This was already shown for a system without particles formation in Figure 11.1b. The residence time t is inversely proportional to the system temperature T_2 (Eq. [11.3]). Figure 11.8 depicts experimentally determined system temperatures T_2 as a function of microwave power, but without chemical reaction of a precursor. In this case the increase in pressure was realized by increasing the gas flow. A more or less linear relationship is found between microwave power and temperature. Therefore, a principal decrease in particle sizes due to shorter residence time with increasing microwave

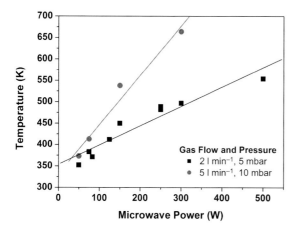

Figure 11.8 Measured system temperatures T_2 as a function of microwave power for different system pressures p_2.

Table 11.3 Calculated residence times for different experimental conditions.

Microwave power (W)	5 l min^{-1} 10 mbar	7 l min^{-1} 14 mbar	9 l min^{-1} 17 mbar	Maximum difference (%)
200	2.55 ms	2.40 ms	2.36 ms	−7.5
300	2.37 ms	2.24 ms	2.10 ms	−11
500	2.14 ms	2.06 ms	1.98 ms	−7.5
Maximum difference (%)	−16	−14	−16	

power should be expected. On the other hand, the residence time t is proportional to the system pressure p_2. Therefore an increasing particle size should be expected with increasing system pressure. This effect is complemented by a parallel temperature increase (Figure 11.8), leading again to residence times in the opposite direction. Furthermore, three different methods can be applied to increase the system pressure. The first is just increasing the gas-flow. This reduces the concentration ratio between precursor and plasma gas and, therefore, may suffer from additional effects. The second possibility is, to increase the gas flow in parallel with the feeding rate (concentration) of the precursor. In this case, concentration effects do not play an additional role. The third possibility is to use a throttle valve. In this case, concentration relationships are also kept constant and the collision probability is increased.

Table 11.3 shows calculated residence times in the plasma zone from the above measured gas temperatures and pressures, and the above-mentioned cavity. All residence times are in the region of a few milliseconds, showing the general suitability of this process for the formation of small nanoparticles.

It can clearly be seen that with increasing microwave power at constant pressure the maximum difference in residence time is around 16%. Also, with increasing pressure at constant microwave power a slight decrease in residence time is observed, with a maximum difference of about 11%. These differences seem marginal. Therefore, a strong influence of microwave power and system pressure on particle size should not be expected. A more significant influence is expected by increasing the cavity length, as will be shown in Section 11.4.4.

Astonishingly, data in the literature show a strong influence of microwave power as well as a pronounced influence of system pressure on particle size (or volume) for several nanoparticle systems. Janzen et al. [38] showed, that by doubling the microwave power from 80 to 160 W, the mean particle size of Fe_2O_3 nanoparticles decreases from 5.3 to 4.1 nm. Although the size difference seems marginal, this is a reduction in the particle volume by a factor of approximately two. A similar relationship is found for the same material from another research group [126]: an increase in microwave power from 255 to 425 W (factor 1.6) decreases the particle size from 5.3 to 4.9 nm. This corresponds to a reduction of the volume by a factor of approximately 1.3. Synthesizing ZnO [88] using different microwave powers leads to a similar dependence. Increasing the microwave power from 60 to 100 W reduces the particle size from approximately 5.1 to 4.35 nm. This is a factor of approximately 1.6 for both microwave power and particle volume. Shimada et al. [80] found a similar relationship for GaN nanoparticles, and Knipping et al. [41] for Si nanoparticles. The results for all these materials are summarized in Figure 11.9. Taking into account the influence of microwave power on the residence time (Table 11.3), the residence time in the plasma zone is not the dominating factor for particle formation and growth. The reactivity of the plasma at low power results in a smaller number of nuclei. The molecules which are formed by the chemical reaction condense on the few nuclei, leading to larger

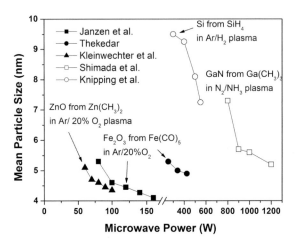

Figure 11.9 Particle size as a function of microwave power. Data are taken from the literature [38, 41, 80, 88, 126].

particles. In the case of higher microwave power more nuclei are formed, resulting in smaller particles.

Increasing the gas pressure by increasing the gas flow should lead to slightly smaller nanoparticles when the parameters summarized in Table 11.3 are dominating. Experimental evidence is the opposite to what is expected when just considering the residence time in the plasma: with increasing pressure a linear increase in the diameter of the nanoparticles is observed for Si [41], ZnO [88], SiO_2 [126], and Al_2O_3 [129]. Literature data from different groups, investigating the influence of pressure on particle size are shown in Figure 11.10. This behavior can be explained by the increased collision probability, leading to enhanced nanoparticle growth. The quite large Al_2O_3 nanoparticles were synthesized in a pure oxygen plasma, where Al-powder was combusted to Al_2O_3.

As the results stem from different research groups, using at least different cavities, and synthesizing different materials, it seems to be a general rule that increasing microwave power leads to smaller particles. This statement, in addition, is supported by theoretical calculations in an atmospheric microwave plasma for SiC formation [47]. The second general rule is that with increasing system pressure in a low pressure plasma (realized by keeping concentration ratio constant) increased particle sizes are observed. Obviously the residence time in the plasma zone is not the dominating effect for particle growth.

Besides its influence on particle size, microwave power may also affect the degree of chemical conversion, especially when using organometallic precursors with a high organic fraction. This was observed for the synthesis of SnO_2 made from $Sn(C_4H_9)_4$ under several experimental conditions [37]. A higher microwave power yielded a whiter powder, containing less unreacted organic residuals at the surface of the particles, whereas lower microwave power resulted in a gray powder. Analysis by different analytical methods revealed residual hydrocarbons on the

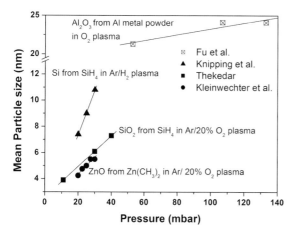

Figure 11.10 Particle size as a function of system pressure for different materials [41, 88, 126, 129].

surface of the nanoparticles [37]. This effect must not be neglected or underestimated, as residual contamination stemming from incompletely reacted organometallic precursor influences the surface properties of the nanoparticles. Depending on the desired application, this contamination may be utilized as a positive effect (e.g., in the case of the Li-ion battery, where C or carbonaceous materials are needed for the conductivity), or it may be a problem (e.g., in the case of gas sensor, or optical applications).

11.4.4
Influence of Residence Time in the Plasma on Particle Size

As was shown in Section 11.4.3 and in Table 11.3, the variation of residence time due to experimental parameters, such as microwave power, pressure, or temperature, is only marginal. Therefore a strong influence of residence time on particle size cannot be expected. A significant increase in residence time can only be realized by either increasing the length of the reaction zone, or by using two or more plasma zones consecutively, as shown by Schlabach et al. [128]. Figure 11.11 shows XRD data from SiC powders, synthesized under different conditions (variation of precursor feeding rate and length of plasma zone, with microwave power and gas flow constant). The influence of the length of the plasma zone and the precursor concentration on particle modification can be seen. The synthesis of amorphous SiC is straightforward. Amorphous SiC shows no features in the diffraction image. With increasing residence time, or with increasing precursor feeding rate (concentration), a peak starts to evolve, indicating the beginning of the presence of ß-SiC crystallites. The significant increase in residence time by a factor of 5, or by using two consecutive plasma stages, also shows a noticeable change in modification: a pronounced ß-SiC peak evolves with increasing residence time. Evaluating the diffraction data with the Scherrer formula, a crystallite size of around 4 nm can be estimated for the feeding rate of $20\,\text{ml}\,\text{h}^{-1}$ and the $15/4\ \lambda$ plasma length, or the two consecutive plasma zones.

11.4.5
Summary of Experimental Parameters

The investigation of different process parameters shows that microwave power, gas flow, pressure, temperature, residence time in the plasma and gas velocity are interdependent parameters, overlapped by particle charging. Therefore, straightforward statements concerning their influence on particle size or modification are difficult. As there is no "adjustable screw" for setting the particle size in synthesis equipment, an understanding of the interrelation between the different parameters is absolutely necessary.

Nevertheless, some general remarks can be made. Due to the low reaction temperatures (usually below 800 K) and short residence times of a few milliseconds, particle sizes of nanoparticles synthesized in microwave plasma are usually below 10 nm. This is shown by several, independent groups for different materials

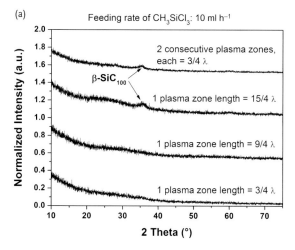

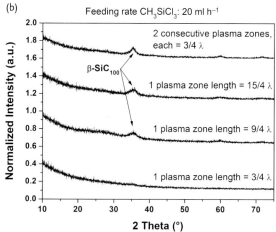

Figure 11.11 Variation of the residence time of nanoparticles in the plasma by increasing the length of reaction zone from 3/4 λ to 15/4 λ at a feeding rate of (a) 10 and (b) 20 ml h^{-1}. A strong influence of residence time on crystallization of SiC nanoparticles can be seen. In parallel, an influence of the precursor concentration is also visible.

(oxides, nitrides, and metals). A variation in particle size is possible to some extent by varying experimental conditions. This was explained using examples from the literature (Fe_2O_3, ZnO, SiO_2, Al_2O_3, GaN, Si) and our own experiments (SnO_2, SiC). In some cases, depending on the particle size, the modification of the nanoparticles may change, as was shown for SiC.

As many parameters are interdependent, each system needs its own optimization. The low-pressure non-thermal microwave plasma synthesis of nanoparticles with particle sizes larger than 10 nm is also possible, but needs a lot of effort and experimental tricks.

11.5
Nanoparticle Properties and Application

To clarify the advantages of nanoscaled powders synthesized in a low-pressure non-thermal microwave plasma compared to commercially available nanopowders it is necessary to have a closer look at the powders. Figure 11.12 shows two TEM images of commercial (a) and microwave plasma synthesized (b) SnO_2 nanoscaled powders. The most striking differences are the particle size and the size distribution. The commercial powder contains nanoparticles of sizes ranging from around 5 to 200 nm, whereas the powder synthesized in microwave plasma is characterized by a very homogeneous particle size distribution of around 4 nm. Similar statements can be done for other commercially available nanoscaled powders, like Fe_2O_3, ZrO_2, or TiO_2.

In the following section several examples will explain, why and how proper selection of experimental parameters is necessary to influence nanoparticle properties with respect to application relevant properties. As many properties are size dependent, particle size and particle size distribution are very important factors. Going into detail on the elementary consequences of small particle sizes would extend the scope of this book, and therefore will only be mentioned where it is necessary. Detailed information on nanomaterials and their size-dependent properties can be found elsewhere, for example, Vollath [130]. Also examples will be presented, where the possibility of *in situ* coating of the particles plays an important role for application.

11.5.1
Ferrimagnetic Nanoparticles

Magnetic properties of ferritic nanoparticles are known to be size dependent [131–134]. With decreasing particle size and particle volume, the saturation

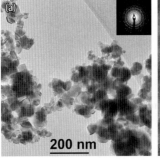 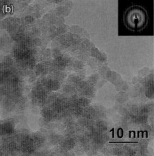

Figure 11.12 Comparison of commercially available (a) and microwave plasma synthesized (b) SnO_2 nanoparticles. The most striking differences between the two particle batches are the particle sizes and the particle size distribution. This can be seen in the TEM images and the electron diffraction (insets).

magnetization decreases drastically. This is because the ferritic nanoparticles are covered by a nonmagnetic surface layer of around 0.5 to 1 nm thickness [135, 136]. With decreasing particle size the amount of specific surface area increases, leading to reduction in saturation magnetization because of the increasing nonmagnetic surface layer. Figure 11.13a shows magnetization data from the literature as a function of particle size for several ferritic nanoparticles. The saturation magnetization in all cases is below the saturation magnetization of the bulk material.

The magnetic properties of Fe_2O_3-based nanoparticles can be influenced by selecting appropriate synthesis conditions. This is shown for Fe_2O_3–PMMA

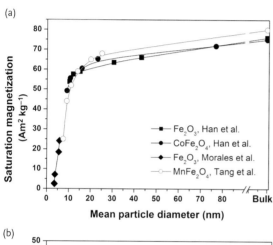

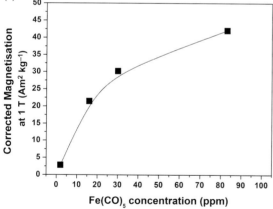

Figure 11.13 (a) Saturation magnetization (literature data from [131, 132, 134]) as a function of particle size for different nanoparticles. (b) Measured saturation magnetization of Fe_2O_3–PMMA core–shell nanoparticles as a function of $Fe(CO)_5$ precursor concentration. The ratio between $Fe(CO)_5$ and monomer feeding rates is constant, respectively. The saturation magnetization is corrected for the net Fe_2O_3 content.

core–shell nanoparticles synthesized in a TE_{11} cavity and a 2.45 GHz system in a set-up as shown in Figure 11.4b, using different precursor concentrations. The ratio between the $Fe(CO)_5$ precursor and the monomer methyl methacrylic acid is kept constant, and the concentration variation is realized by changing the precursor feeding rates. The polymer coating prevents or reduces the magnetic interaction between the magnetic particles. As was explained in Section 11.4.2, the precursor concentration has a significant influence on the resulting particle size, which itself influences the magnetization of the particles. This is shown in Figure 11.13b. With increasing precursor concentration increasing magnetization (corrected for the polymer content) is observed.

Comparing Figure 11.13a with 11.13b shows that the saturation magnetization measured for low precursor concentration corresponds to particle sizes around 5 nm and smaller, whereas the saturation magnetization measured for higher concentrations corresponds to particle sizes around 10 nm.

The residence time in the plasma can also be used to tailor the particle size, with respect to the magnetic properties of a material: using the second cavity just for "heat treatment" and a polymer coating step thereafter, a significant increase in susceptibility (this is the slope of the linear part around the zero field of the magnetization curve) and saturation magnetization is observed (Figure 11.14). This is due to the increasing particle size provoked by the "heat treatment" using a second cavity before coating with the polymer. Both materials are free of hysteresis.

Coating of ferritic nanoparticles with an organic dye (e.g., anthracene, pyrene) plus a polymer coating finally results in bifunctional nanoparticles, exhibiting magnetic properties plus fluorescence [137, 138]. Such particles may be interesting for applications in biology, medicine or diagnostics.

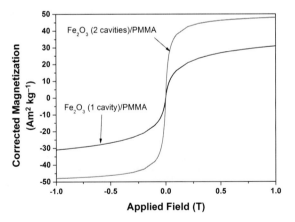

Figure 11.14 Saturation magnetization of Fe_2O_3–PMMA nanoparticles synthesized using one cavity compared to Fe_2O_3–PMMA using two cavities. Besides the number of cavities used, the other relevant experimental parameters (microwave power, reaction-gas flow, and precursor feeding rate) were identical for both experiments. The saturation magnetization is corrected for the net Fe_2O_3 content.

11.5.2
Gas-Sensing Nanoparticles

Tin dioxide, SnO_2, is a wide band gap n-type semiconducting metal oxide, commonly used as a gas sensitive material. Usually, sensor layers are produced by sputtering techniques [139–141], leading to a thin film with grain sizes around 30 to 300 nm. Gas analytical microsystems, such as the Karlsruhe micro-nose (KAMINA) [141] can be improved toward higher sensitivity, when using SnO_2 nanoparticle thin films instead of well-established sputtered layers. The SnO_2 particles produced with the microwave plasma process are crystalline; they exhibit tetragonal cassiterite structure. The size of the nanoparticles is usually 3–4 nm. Particles of that size have approximately 50 to 70% of their atoms at the surface. Thus they seem ideal for application in sensor technology. However, a problem of particle growth during sensor application at typical operating temperatures of 570–620 K exists. Annealing experiments show that uncoated SnO_2 nanoparticles feature a slight particle growth to approximately 5 to 6 nm after 3 months. This can be overcome by depositing core–shell SnO_2–SiO_2 nanoparticles instead of bare particles, using the SiO_2 shell as a growth barrier [142]. Open structures with low in-depth diffusion would be expected to have short response times. Usually nanoparticles are applied via spin-coating [143], drop-coating [144, 145] or screen-printing [145, 146] to form the sensing layers. However, deposition processes using colloidal solutions of nanoparticles pose the problem of particle growth, as one has to eliminate the binding phase by a temperature treatment. A further problem may be due to residuals of the binding phase.

These problems can be avoided by the direct deposition of microwave plasma synthesized SnO_2 nanoparticles on a substrate [30, 31, 142]. The signal strength of metal oxide gas-sensors is successfully improved further by decreasing the crystallite size of the metal oxide, as shown by Kennedy *et al.* [147], and also by Schumacher *et al.* [31]. Figure 11.15 shows the general design of a SnO_2 or SnO_2–SiO_2 nanoparticle porous layer, deposited via the mask technique on a Si-substrate

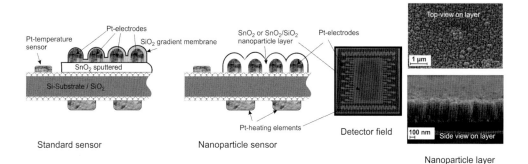

Figure 11.15 Schematic set-up of a sputtered gas-sensor microarray, compared with a nanoparticle-coated gas-sensor microarray, showing a real detector field and the nanoparticle layer morphology.

already equipped with an electrode array consisting of 38 single detector fields, in comparison to a conventional KAMINA gas sensor element. The nanoparticle layer is deposited on top of the Pt-electrodes. The porous morphology is shown on the right-hand side of the image. For long term stability, of course particle growth/agglomeration has to be avoided, or at least reduced, by keeping the electrical conductivity and gas permeability through grain–grain contacts. This can be realized by coating the nanoparticles with a second phase [20, 148] to produce "silica-pinned" nanoscaled SnO_2 [149]. Optimization of the open-pored SiO_2-shell is therefore necessary. Figure 11.16 compares the sensitivity of a sputtered SnO_2 gas sensor thin film (KAMINA) with the sensitivity of two different nanoparticle films. The data for this graph are taken from Schumacher [150] and Fuchs [151]. The sensitivity of both nanoparticle films, the bare SnO_2 and the core–shell SnO_2–SiO_2, are several orders of magnitude better than the sensitivity of the conventionally fabricated SnO_2 layer. This can be explained by the small particle size (see also Figure 11.12), and hence large specific surface area of the layer, and the fact, that a porous SiO_2-shell does not separate electrically the SnO_2 particles from each other. The response time of core–shell SnO_2–SiO_2 nanoparticles is of the same order of magnitude as for sputtered SnO_2 layers, and the time of regeneration is 40 to 50% longer for medium concentrations, and slightly better for higher or lower concentrations compared to a conventional system [151]. Thus, open-pored structures and high specific surface area do not exhibit disadvantages compared to a sputtered detector field. Concerning long term stability, the bare nanoparticles show a decrease in performance after 3 weeks [150], whereas the core–shell nanoparticles showed an improved stability even after 12 weeks [151].

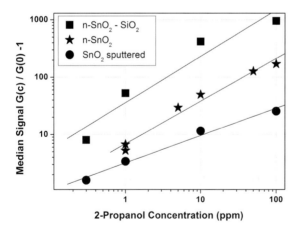

Figure 11.16 Sensitivity versus 2-propanol concentration for three different sensor types. The sensitivity of the bare SnO_2 nanoparticle sensor is one order of magnitude higher than for the sputtered reference sensor. The sensitivity is excellent and reproducible even at concentrations around 1 ppm. The sensitivity of the core–shell nanoparticle sensor is even better.

11.5.3
Nanoparticles for Anodes in Li-Ion Batteries

Sn and SnO$_2$-based materials are widely discussed as interesting anode materials for Li-ion batteries [105, 111, 113, 152–156]. The maximum theoretical reversible specific capacity of SnO$_2$, indicated as around 780 mAh g^{-1}, is more than twice as high as the capacity of the graphite anodes used nowadays, featuring a specific reversible capacity of 372 mAh g^{-1} [111, 156, 157]. SnO$_2$ is a conversion material, forming metallic Sn in a first, irreversible step, followed by a reversible alloying and de-alloying process of Li$^+$ with metallic Sn. However, the conversion is accompanied by the side effect of volume expansion larger than 200%, leading to strong capacity fading due to crack formation and loss of electrical contact between particles during charging/discharging. Using nanomaterials may exhibit the advantage of enhanced Li- insertion/removal, due to the short distances in small particles. Furthermore, nanomaterials are expected to overcome the problem of electrode deterioration [158, 159]. Improved capacity of nanosized SnO$_2$ compared to micron-sized SnO$_2$ has been demonstrated, for example, by Chen and coworkers [160]. Encapsulation of SnO$_2$-nanoparticles with a carbon-containing scaffold [113], or application of either SnO$_2$–graphene nanocomposites [110, 112] or SnO$_2$–C core–shell nanoparticles [109] seems a promising way toward improved capacity.

For the utilization of microwave-plasma generated SnO$_2$-based nanoparticles as negative electrode material in Li-ion batteries the nanoparticles are deposited *in situ* as a porous nanoparticle film on preheated Ni-substrates according to the method described in previous work [31, 113]. This concept does not use carbon black as a binder, as carbonaceous material can be used either *in situ* from incompletely reacted organometallic precursor, or by coating the SnO$_2$ nanoparticles in a second step (see Figure 11.4a) with a carbon-containing organic material [37]. The *in situ* deposited nanoparticle films (SnO$_2$ and core–shell SnO$_2$–"C") are then directly assembled in Swagelock half-cells, without any additional carbon black or binder, in an Ar-filled glove box using a Li metal counter electrode, a porous separator, and a standard electrolyte of 1 M LiPF$_6$ solution in a 1 : 1 mixture of ethylene carbonate (EC) and dimethyl carbonate (DMC).

Although SnCl$_4$ has a significantly higher SnO$_2$ conversion per ml of precursor, in this case Sn(C$_4$H$_9$)$_4$ is selected as the appropriate precursor. This is because residual Cl adheres to the surface of the nanoparticles, and may deteriorate the properties of the half-cells. As already pointed out in Section 11.4.1 a pronounced influence of synthesis parameters is found on the specific capacities of anode materials. Beyond the influence of microwave power on the particle, with respect to crystallite size, an influence on residual organometallic precursor is observed. Thermogravimetric analyses combined with X-Ray photoelectron spectroscopy show that decreasing microwave power results in an increase in carbonaceous residual material (C$_x$H$_y$). Usually, precursor residuals are unwanted; in the case of anode material this effect has a positive impact on the electrochemical target property. Table 11.4 shows the relevant parameters and results of two samples,

Table 11.4 Applied microwave power and important resulting characteristics of SnO_2 nanoparticles synthesized from $Sn(C_4H_9)_4$ as anode material for Li-ion batteries.

#	MW Power (W)	Organic residual (wt%)	Mean crystallite size (nm)	Calculated thickness of organic (nm)	2nd Cycle discharge capacity (Ah kg^{-1})
1	600	16.57	4.1	0.42	819
2	1200	12.43	3.0	0.23	605

synthesized at two different microwave powers, with all other parameters constant.

Table 11.4 shows, that with increasing microwave power the particle size (volume) decreases. Similar to the materials presented in Section 11.4.3, also in the case of SnO_2 a doubling in microwave power reduces the particle volume by a factor of around two. Furthermore, with increasing microwave power a decrease in organic residual is measured. This can even be seen by eye, as the powder synthesized using 600 W is beige to light brown, whereas the powder synthesized at 1200 W is white. For the target property, the specific capacity, the less pure material is even better than the "cleaner" material. This is, because carbonaceous material is necessary for conductivity.

Of course charge and discharge capacities are only one property to be optimized when developing anode materials. Much more relevant is cycling stability. Therefore, optimizing particles and nanoparticle layers in different directions (particle size, coating thickness, composition, phases, morphology, and nanoparticle layer thickness) as well as reproducible manufacturing steps for battery assembly are necessary. The presented results are exemplary, and incomplete, just demonstrating the influence of selected synthesis parameters on one selected property.

11.6
Summary

This chapter focused on the application of microwave plasmas for the synthesis of nanoparticles exhibiting special properties. It was explained why the application of microwave plasma is so exciting for the synthesis of nanoparticles, and examples of experimental realization were presented. A detailed discussion of experimental parameters, based on literature data and our own experimental evidence was presented. From this discussion some general rules with regard to particle sizes can be stated. First, microwave plasmas at low pressure are very well suited to produce nanoparticles with small sizes, below around 10 nm, and narrow particle size distribution. By a proper choice of experimental parameters the resulting particle size and size distribution can be influenced to a certain extent. An increase in precursor concentration yields larger particle diameters, whereas an increase in microwave power produces smaller particles and narrower particle size distribution. An increase in system pressure also increases the particle size. These statements are supported by several experiments from different research groups and, furthermore, by theoretical calculations. Finally, nanoparticles with interesting physical or chemical properties and their application potential were shown.

References

1 Hu, X., and Yu, J.C. (2006) Microwave-assisted synthesis of a superparamagnetic surface-functionalized porous Fe_3O_4/C nanocomposite. *Chem. Asian J.*, **1** (4), 605–610.

2 Nyutu, E.K., Cheng, C.-H., Dutta, P.K., and Suib, S.L. (2008) Effect of microwave frequency on hydrothermal synthesis of nanocrystalline tetragonal barium titanate. *J. Phys. Chem. C*, **112** (26), 9659–9667.

3 Qiu, G., Huang, H., Genuino, H., Opembe, N., Stafford, L., Dharmarathna, S., and Suib, S.L. (2011) Microwave-assisted hydrothermal synthesis of nanosized α-Fe_2O_3 for catalysts and adsorbents. *J. Phys. Chem. C*, **115** (40), 19626–19631.

4 Opembe, N.N., King'ondu, C.K., Espinal, A.E., Chen, C.-H., Nyutu, E.K., Crisostomo, V.M., and Suib, S.L. (2010) Microwave-assisted synthesis of manganese oxide octahedral molecular sieve (OMS-2) nanomaterials under continuous flow conditions. *J. Phys. Chem. C*, **114** (34), 14417–14426.

5 Yamauchi, T., Tsukahara, Y., Yamada, K., Sakata, T., and Wada, Y. (2011) Nucleation and growth of magnetic Ni-Co (core-shell) nanoparticles in one-pot reaction under microwave irradiation. *Chem. Mater.*, **23** (1), 75–85.

6 Benyettou, F., Guenin, E., Lalatonne, Y., and Motte, L. (2011) Microwave assisted nanoparticle surface functionalization. *Nanotechnology*, **22** (5), 055102.

7 Barge, A., Tagliapietra, S., Binello, A., and Cravotto, G. (2011) Click chemistry under microwave or ultrasound irradiation. *Curr. Org. Chem.*, **15** (2), 189–203.

8 Visentin, S., Medana, C., Barge, A., Giancotti, V., and Cravotto, G. (2010) Microwave-assisted Maillard reactions for the preparation of advanced glycation end products (AGEs). *Org. Biomol. Chem.*, **8** (10), 2473–2477.

9 Horikoshi, S., Osawa, A., Abe, M., and Serpone, N. (2011) On the generation of hot-spots by microwave electric and magnetic fields and their impact on a microwave-assisted heterogeneous reaction in the presence of metallic Pd nanoparticles on an activated carbon support. *J. Phys. Chem. C*, **115** (46), 23030–23035.

10 Horikoshi, S., Matsubara, A., Takayama, S., Sato, M., Sakai, F., Kajitani, M., Abe, M., and Serpone, N. (2010) Characterization of microwave effects on metal-oxide materials: zinc oxide and titanium dioxide. *Appl. Catal. B Environ.*, **99** (3–4), 490–495.

11 Horikoshi, S., Abe, H., Torigoe, K., Abe, M., and Serpone, N. (2010) Access to small size distributions of nanoparticles by microwave-assisted synthesis. Formation of Ag nanoparticles in aqueous carboxymethylcellulose solutions in batch and continuous-flow reactors. *Nanoscale*, **2** (8), 1441–1447.

12 Mehta, P., Singh, A.K., and Kingon, A.I. (1991) Nonthermal microwave plasma synthesis of crystalline titanium oxide and titanium nitride nanoparticles, in: *MRS Fall Meeting 1991: Synthesis and Processing of Ceramics: Scientific Issues*, Boston, MA, USA, vol. 249, (eds Y. Chen, R.J. Gottschall, W.E. Rhine, and T.M. Shaw), Materials Research Society, Pittsburgh, PA, pp. 153–159.

13 Sickafus, K.E., Vollath, D., and Varma, R. (1992) Electron-microscopy study on zirconia and alumina ceramic powders

synthesized by microwave plasma pyrolysis, in *MRS Spring Meeting 1992: Microwave Processing of Materials III, San Francisco, CA, USA*, vol. 269, (eds R.L. Beatty, W.H. Sutton, and M.F. Iskander), Materials Research Society, Pittsburgh, PA, pp. 363–369.

14 Vollath, D., Varma, R., and Sickafus, K.E. (1992) Synthesis of nanocrystalline powders for oxide ceramics by microwave plasma pyrolysis, in *MRS Spring Meeting 1992: Microwave Processing of Materials III, San Francisco, CA, USA*, vol. 269, (eds R.L. Beatty, W.H. Sutton, and M.F. Iskander), Materials Research Society, Pittsburgh, PA, pp. 379–384.

15 Vollath, D. (1994) A cascaded microwave plasma source for synthesis of ceramic nanocomposite powders, in *MRS Spring Meeting 1994: Microwave Processing of Materials IV, San Francisco, CA, USA*, vol. 347, (eds M.F. Iskander, R.J. Lauf, and W.H. Sutton), Materials Research Society, Pittsburgh, PA, pp. 629–634.

16 Vollath, D., and Szabó, D.V. (1994) Synthesis of ceramic nanocomposites in a microwave plasma, in: Proceedings of 8. CIMTEC (28 June–4 July 1994, Firenze, Italy) Advances in Science and Technology vol. 3B, edited by P. Vincencini, Firenze 1995, pp. 1261–1268. Techna Group, Faenza, Italy.

17 Chou, C., and Phillips, J. (1992) Plasma production of metallic nanoparticles. *J. Mater. Res.*, **7** (08), 2107–2113.

18 Vollath, D., and Sickafus, K.E. (1993) Synthesis of ceramic oxide powders by microwave plasma pyrolysis. *J. Mater. Sci.*, **28** (21), 5943–5948.

19 Vollath, D., and Sickafus, K.E. (1993) Synthesis of ceramic oxide powders in a microwave plasma-device. *J. Mater. Res.*, **8** (11), 2978–2984.

20 Vollath, D., and Szabó, D.V. (1994) Nanocoated particles: a special type of ceramic powder. *Nanostruct. Mater.*, **4** (8), 927–938.

21 Vollath, D., Szabó, D.V., and Willis, J.O. (1996) Magnetic properties of nanocrystalline Cr_2O_3 synthesized in a microwave plasma. *Mater. Lett.*, **29** (4–6), 271–279.

22 Aliev, Y.M., Maximov, A.V., Kortshagen, U., Schlüter, H., and Shivarova, A. (1995) Modeling of microwave discharges in the presence of plasma resonances. *Phys. Rev. E*, **51** (6), 6091–6103.

23 Kortshagen, U., and Bhandarkar, U. (1999) Modeling of particulate coagulation in low pressure plasmas. *Phys. Rev. E*, **60** (1), 887.

24 Janzen, C., Kleinwechter, H., Knipping, J., Wiggers, H., and Roth, P. (2002) Size analysis in low-pressure nanoparticle reactors: comparison of particle mass spectrometry with in situ probing transmission electron microscopy. *J. Aerosol Sci.*, **33** (6), 833–841.

25 Vollath, D., and Szabó, D.V. (2002) Synthesis of nanopowders by the microwave plasma process – basic considerations and perspectives for scaling up, in *Innovative Processing of Films and Nanocrystalline Powders*, 1st edn (ed. K.-L. Choy), Imperial College Press, London, pp. 219–251.

26 Giesen, B., Wiggers, H., Kowalik, A., and Roth, P. (2005) Formation of Si-nanoparticles in a microwave reactor: comparison between experiments and modelling. *J. Nanopart. Res.*, **7** (1), 29–41.

27 Gatti, M., and Kortshagen, U. (2008) Analytical model of particle charging in plasmas over a wide range of collisionality. *Phys. Rev. E*, **78** (4), 046402.

28 Vollath, D., Lamparth, I., and Szabó, D.V. (2002) Fluorescence from coated oxide nanoparticles, in *MRS Fall Meeting 2001: Symposium on Nanophase and Nanocomposite Materials IV, Boston, MA, USA*, vol. 703 (eds S. Komarneni, J.C. Parker, R.A. Vaia, G.Q. Lu, and J.I. Matsushita), Materials Research Society, Pittsburgh, PA, pp. 303–308.

29 Li, S.-Z., Hong, Y.C., Uhm, H.S., and Li, Z.-K. (2004) Synthesis of nanocrystalline iron oxide particles by microwave plasma jet at atmospheric pressure. *Jpn J. Appl. Phys.*, **43** (11A), 7714–7717.

30 Schumacher, B., Ochs, R., Tröße, H., Schlabach, S., Bruns, M., Szabo, D.V., and Haußelt, J. (2006) Electronic micro

nose equipped with nano-structured gas sensitive SnO$_2$. *MST News*, **6** (4), 10–12.

31 Schumacher, B., Ochs, R., Tröße, H., Schlabach, S., Bruns, M., Szabó, D.V., and Haußelt, J. (2007) Nanogranular SnO$_2$ layers for gas sensing applications by in situ deposition of nanoparticles produced by the Karlsruhe microwave plasma process. *Plasma Process. Polym.*, **4** (S1), S865–S870.

32 Kortshagen, U. (2009) Nonthermal plasma synthesis of semiconductor nanocrystals. *J. Phys. D Appl. Phys.*, **42** (11), 113001.

33 Mahendra Kumar, S., Deshpande, P.A., Krishna, M., Krupashankara, M.S., and Madras, G. (2010) Photocatalytic activity of microwave plasma-synthesized TiO$_2$ nanopowder. *Plasma Chem. Plasma Process.*, **30** (4), 461–470.

34 Hong, Y.C., Lho, T., Lee, B.J., Uhm, H.S., Kwon, O.P., and Lee, S.H. (2011) Synthesis of titanium dioxide in O$_2$/Ar/SO$_2$/TiCl$_4$ microwave torch plasma and its band gap narrowing. *Curr. Appl. Phys.*, **11** (3), 517–520.

35 Kautsch, A., Brossmann, U., Krenn, H., Hofer, F., Szabó, D.V., and Würschum, R. (2011) Structural and optical properties of nanoparticulate Y$_2$O$_3$:Eu$_2$O$_3$ made by microwave plasma synthesis. *Appl. Phys. A Mater. Sci. Process.*, **105** (3), 709–712.

36 Nadeem, K., Krenn, H., Traussnig, T., Würschum, R., Szabó, D.V., and Letofsky-Papst, I. (2011) Effect of dipolar and exchange interactions on magnetic blocking of maghemite nanoparticles. *J. Magn. Magn. Mater.*, **323** (15), 1998–2004.

37 Szabó, D.V., Kilibarda, G., Schlabach, S., Trouillet, V., and Bruns, M. (2012) Structural and chemical characterization of SnO$_2$-based nanoparticles as electrode material in Li-ion batteries. *J. Mater. Sci.*, **47** (10), 4383–4391.

38 Janzen, C., Wiggers, H., Knipping, J., and Roth, P. (2001) Formation and in situ sizing of gamma-Fe$_2$O$_3$ nanoparticles in a microwave flow reactor. *J. Nanosci. Nanotechnol.*, **1** (2), 221–225.

39 Vollath, D., and Szabó, D.V. (2006) The Microwave plasma process – a versatile process to synthesise nanoparticulate materials. *J. Nanopart. Res.*, **8** (3–4), 417–428.

40 Lin, H., Gerbec, J.A., Sushchikh, M., and McFarland, E.W. (2008) Synthesis of amorphous silicon carbide nanoparticles in a low temperature low pressure plasma reactor. *Nanotechnology*, **19** (32), 325601.

41 Knipping, J., Wiggers, H., Rellinghaus, B., Roth, P., Konjhodzic, D., and Meier, C. (2004) Synthesis of high purity silicon nanoparticles in a low pressure microwave reactor. *J. Nanosci. Nanotechnol.*, **4** (8), 1039–1044.

42 Shin, D.H., Hong, Y.C., and Uhm, H.S. (2005) Production of nanocrystalline titanium nitride powder by atmospheric microwave plasma torch in hydrogen/nitrogen gas. *J. Am. Ceram. Soc.*, **88** (10), 2736–2739.

43 Hong, Y.C., and Uhm, H.S. (2006) Synthesis of MgO nanopowder in atmospheric microwave plasma torch. *Chem. Phys. Lett.*, **422** (1–3), 174–178.

44 Shin, D.H., Bang, C.U., Hong, Y.C., and Uhm, H.S. (2006) Preparation of vanadium pentoxide powders by microwave plasma-torch at atmospheric pressure. *Mater. Chem. Phys.*, **99** (2–3), 269–275.

45 García, M., Varo, M., and Martínez, P. (2010) Excitation of species in an expanded argon microwave plasma at atmospheric pressure. *Plasma Chem. Plasma Process.*, **30** (2), 241–255.

46 Bárdos, L., and Baránková, H. (2010) Cold atmospheric plasma: sources, processes, and applications. *Thin Solid Films*, **518** (23), 6705–6713.

47 Vennekamp, M., Bauer, I., Groh, M., Sperling, E., Ueberlein, S., Myndyk, M., Mäder, G., and Kaskel, S. (2011) Formation of SiC nanoparticles in an atmospheric microwave plasma. *Beilstein J. Nanotechnol.*, **2**, 665–673.

48 Cho, S.C., Hong, Y.C., and Uhm, H.S. (2006) TeO$_2$ nanoparticles synthesized by evaporation of tellurium in atmospheric microwave-plasma torch-flame. *Chem. Phys. Lett.*, **429** (1–3), 214–218.

49 Kim, J.H., Hong, Y.C., and Uhm, H.S. (2007) Synthesis of oxide nanoparticles

via microwave plasma decomposition of initial materials. *Surf. Coating. Tech.nol.*, 201 (9–11), 5114–5120.

50 Brenner, J.R., Harkness, J.B.L., Knickelbein, M.B., Krumdick, G.K., and Marshall, C.L. (1997) Microwave plasma synthesis of carbon-supported ultrafine metal particles. *Nanostruct. Mater.*, 8 (1), 1–17.

51 Chau, J.L.H., Hsu, M.-K., Hsieh, C.-C., and Kao, C.-C. (2005) Microwave plasma synthesis of silver nanopowders. *Mater. Lett.*, 59 (8–9), 905–908.

52 Brooks, D.J., Douthwaite, R.E., Brydson, R., Calvert, C., Measures, M.G., and Watson, A. (2006) Synthesis of inorganic fullerene (MS_2, M = Zr, Hf and W) phases using H_2S and N_2/H_2 microwave-induced plasmas. *Nanotechnology*, 17 (5), 1245.

53 Okada, K., Komatsu, S., and Matsumoto, S. (1999) Preparation of microcrystalline diamond in a low pressure inductively coupled plasma. *J. Mater. Res.*, 14 (02), 578–583.

54 Shen, Z., Kim, T., Kortshagen, U., McMurry, P.H., and Campbell, S.A. (2003) Formation of highly uniform silicon nanoparticles in high density silane plasmas. *J. Appl. Phys.*, 94 (4), 2277–2283.

55 Fauchais, P., Vardelle, A., and Denoirjean, A. (1997) Reactive thermal plasmas: ultrafine particle synthesis and coating deposition. *Surf. Coating. Technol.*, 97 (1–3), 66–78.

56 Hollabaugh, C.M., Hull, D.E., Newkirk, L.R., and Petrovic, J.J. (1983) RF-plasma system for the production of ultrafine, ultrapure silicon carbide powder. *J. Mater. Sci.*, 18 (11), 3190–3194.

57 Bystrzejewski, M., Huczko, A., Lange, H., Baranowski, P., Cota-Sanchez, G., Soucy, G., Szczytko, J., and Twardowski, A. (2007) Large scale continuous synthesis of carbon-encapsulated magnetic nanoparticles. *Nanotechnology*, 18 (14), 145608.

58 Jia, L., and Gitzhofer, F. (2009) Nano-particle sizing in a thermal plasma synthesis reactor. *Plasma Chem. Plasma Process.*, 29 (6), 497–513.

59 Belmonte, T., Arnoult, G., Henrion, G., and Gries, T. (2011) Nanoscience with non-equilibrium plasmas at atmospheric pressure. *J. Phys. D Appl. Phys.*, 44 (36), 363001.

60 Mangolini, L., Thimsen, E., and Kortshagen, U. (2005) High-yield plasma synthesis of luminescent silicon nanocrystals. *Nano Lett.*, 5 (4), 655–659.

61 Kortshagen, U., Mangolini, L., and Bapat, A. (2007) Plasma synthesis of semiconductor nanocrystals for nanoelectronics and luminescence applications. *J. Nanopart. Res.*, 9 (1), 39–52.

62 Park, H., and Choe, W. (2010) Parametric study on excitation temperature and electron temperature in low pressure plasmas. *Curr. Appl. Phys.*, 10 (6), 1456–1460.

63 Kulkarni, N.V., Karmakar, S., Banerjee, I., Sahasrabudhe, S.N., Das, A.K., and Bhoraskar, S.V. (2009) Growth of nano-particles of Al_2O_3, AlN and iron oxide with different crystalline phases in a thermal plasma reactor. *Mater. Res. Bull.*, 44 (3), 581–588.

64 Kim, D.-W., Kim, T.-H., Park, H.-W., and Park, D.-W. (2011) Synthesis of nanocrystalline magnesium nitride (Mg_3N_2) powder using thermal plasma. *Appl. Surf. Sci.*, 257 (12), 5375–5379.

65 Allaire, F., Parent, L., and Dallaire, S. (1991) Production of submicron SiC particles by D.C. thermal plasma: a systematic approach based on injection parameters. *J. Mater. Sci.*, 26 (15), 4160–4165.

66 Moreno-Couranjou, M., Monthioux, M., Gonzalez-Aguilar, J., and Fulcheri, L. (2009) A non-thermal plasma process for the gas phase synthesis of carbon nanoparticles. *Carbon*, 47 (10), 2310–2321.

67 Vons, V., Creyghton, Y., and Schmidt-Ott, A. (2006) Nanoparticle production using atmospheric pressure cold plasma. *J. Nanopart. Res.*, 8 (5), 721–728.

68 Harbec, D., Gitzhofer, F., and Tagnit-Hamou, A. (2011) Induction plasma synthesis of nanometric spheroidized glass powder for use in cementitious materials. *Powder Technol.*, 214 (3), 356–364.

69 Qin, C., and Coulombe, S. (2006) Synthesis of organic layer-coated copper

nanoparticles in a dual-plasma process. *Mater. Lett.*, **60** (16), 1973–1976.

70 Watanabe, Y. (2006) Formation and behaviour of nano/micro-particles in low pressure plasmas. *J. Phys. D Appl. Phys.*, **39** (19), R329.

71 Yu, J., Zhang, Q., Ahn, J., Yoon, S.F., Li, R.Y.B., Gan, B., Chew, K., and Tan, K.H. (2002) Synthesis of carbon nanostructures by microwave plasma chemical vapor deposition and their characterization. *Mater. Sci. Eng. B*, **90**, 16–19.

72 Malesevic, A., Vitchev, R., Schouteden, K., Volodin, A., Zhang, L., van Tendeloo, G., Vanhulsel, A., and van Haesendonck, C. (2008) Synthesis of few-layer graphene via microwave plasma-enhanced chemical vapour deposition. *Nanotechnology*, **19** (30), 305604.

73 Hisada, D., Fujiwara, Y., Sato, H., Jimbo, M., Kobayashi, T., and Hata, K. (2011) Structure and magnetic properties of FeCo nanoparticles encapsulated in carbon nanotubes grown by microwave plasma enhanced chemical vapor deposition. *J. Magn. Magn. Mater.*, **323** (24), 3184–3188.

74 Funer, M., Wild, C., and Koidl, P. (1998) Novel microwave plasma reactor for diamond synthesis. *Appl. Phys. Lett.*, **72** (10), 1149–1151.

75 Pajkic, Z., and Willert-Porada, M. (2009) Atmospheric pressure microwave plasma fluidized bed CVD of AlN coatings. *Surf. Coating. Technol.*, **203** (20–21), 3168–3172.

76 Pfuch, A., and Cihar, R. (2004) Deposition of SiO_x thin films by microwave induced plasma CVD at atmospheric pressure. *Surf. Coating. Technol.*, **183** (2–3), 134–140.

77 Zhang, L., Ma, Z., and Wu, L. (2011) Growth and characterization of diamond films deposited at high-pressure using a low-power microwave plasma reactor. *Inorg. Mater.*, **47** (3), 255–261.

78 Hemawan, K.W., Grotjohn, T.A., Reinhard, D.K., and Asmussen, J. (2010) Improved microwave plasma cavity reactor for diamond synthesis at high-pressure and high power density. *Diam. Relat. Mater.*, **19** (12), 1446–1452.

79 Marcinek, M., Song, X., and Kostecki, R. (2007) Microwave plasma chemical vapor deposition of nano-composite C/Pt thin-films. *Electrochem. Commun.*, **9** (7), 1739–1743.

80 Shimada, M., Wang, W.-N., and Okuyama, K. (2010) Synthesis of gallium nitride nanoparticles by microwave plasma-enhanced CVD. *Chem. Vapor Depos.*, **16** (4–6), 151–156.

81 Cernetti, P., Gresback, R., Campbell, S.A., and Kortshagen, U. (2007) Nonthermal plasma synthesis of faceted germanium nanocrystals. *Chem. Vapor Depos.*, **13** (6–7), 345–350.

82 MacDonald, A.D. (1966) *Microwave Breakdown in Gases*, Wiley Series in Plasma Physics. John Wiley & Sons, Inc., New York, USA.

83 Tendero, C., Tixier, C., Tristant, P., Desmaison, J., and Leprince, P. (2006) Atmospheric pressure plasmas: a review. *Spectrochim. Acta B At. Spectrosc.*, **61** (1), 2–30.

84 Ishigaki, T. (2008) Synthesis of ceramic nanoparticles with non-equilibrium crystal structures and chemical compositions by controlled thermal plasma processing. *J. Ceram. Soc. Jpn.*, **116** (1351), 462–470.

85 Vollath, D. (2008) Plasma synthesis of nanopowders. *J. Nanopart. Res.*, **10**, 39–57.

86 Vollath, D. (2011) Estimation of particle size distributions obtained by gas phase processes. *J. Nanopart. Res.*, **13** (9), 3899–3909.

87 Matsoukas, T., and Russell, M. (1995) Particle charging in low-pressure plasmas. *J. Appl. Phys.*, **77** (9), 4285–4292.

88 Kleinwechter, H., Janzen, C., Knipping, J., Wiggers, H., and Roth, P. (2002) Formation and properties of ZnO nanoparticles from gas-phase synthesis processes. *J. Mater. Sci.*, **37**, 4349–4360.

89 Baumann, W., Thekedar, B., Paur, H.R., and Seifert, H. (2006) Characterization of nanoparticles synthesized in the microwave plasma discharge process by particle mass spectrometry and transmission electron microscopy. Paper

presented at the AIChE Fall and Annual Meeting, San Francisco, CA, USA, November 12–17, 2006.

90 Baumann, W., Thekedar, B., Paur, H.R., and Seifert, H. (2007) Comparison of size distribution of iron oxide nanoparticles measured with particle mass spectrometer and transmission electron microscopy. Paper presented at the International Congress on Particle Technology (PARTEC 2007) Nürnberg, Germany, March 27–29, 2007.

91 Klug, H.P., and Alexander, L.E. (1974) *X-Ray Diffraction Procedures: For Polycrystalline and Amorphous Materials*, 2nd edn, John Wiley & Sons, New York.

92 Warren, B.E. (1990) *X-Ray Diffraction*, Dover Publications, New York.

93 Jauncey, G.E.M. (1924) The scattering of X-rays and Bragg's law. *Proc. Natl. Acad. Sci.*, **10** (2), 57–60.

94 Cowley, J.M. (1975) *Diffraction Physics*, North-Holland, Amsterdam; The Netherlands.

95 Williams, D.B., and Carter, C.B. (2009) *Transmission Electron Microscopy: A Textbook for Materials Science*, Springer Science + Business Media, New York, NY, USA.

96 Freitag, B., Kujawa, S., Mul, P.M., Ringnalda, J., and Tiemeijer, P.C. (2005) Breaking the spherical and chromatic aberration barrier in transmission electron microscopy. *Ultramicroscopy*, **102** (3), 209–214.

97 Rose, H. (2005) Prospects for aberration-free electron microscopy. *Ultramicroscopy*, **103** (1), 1–6.

98 Bendersky, L.A., and Gayle, F.W. (2001) Electron diffraction using transmission electron microscopy. *J. Res. Natl Inst. Stand. Technol.*, **106**, 997–1012.

99 Schamp, C.T., and Jesser, W.A. (2005) On the measurement of lattice parameters in a collection of nanoparticles by transmission electron diffraction. *Ultramicroscopy*, **103** (2), 165–172.

100 Roth, P., and Hospital, A. (1994) Design and test of a particle mass spectrometer (PMS). *J. Aerosol Sci.*, **25** (1), 61–73.

101 Baumann, W., Paur, H.R., Mätzing, H., and Seifert, H. (2005) Mass spectrometry of particles in the nanometer size regime. Paper presented at the ACCENT Workshop on European On-Line Particle Mass Spectrometry, Ispra, Italy, November 29–30, 2005.

102 Mühleisen, M., and Möbius, A. (1997) Vorrichtung zur reflektionsarmen Absorption von Mikrowellen. Germany, Patent DE 195 28 343 C2.

103 Szabó, D.V., Schlabach, S., Ochs, R., Schumacher, B., and Bruns, M. (2007) The Karlsruhe Microwave Plasma Process: a non-thermal plasma method for the synthesis of nanoparticles. *Cfi-Ceram. Forum Int.*, **13**, 7–11.

104 Fuchs, M., Breitenstein, D., Fartmann, M., Grehl, T., Kayser, S., Koester, R., Ochs, R., Schlabach, S., Szabó, D.V., and Bruns, M. (2010) Characterization of core/shell nanoparticle thin films for gas analytical applications. *Surf. Interface Anal.*, **42** (6–7), 1131–1134.

105 Wang, Y., and Lee, J.Y. (2005) Microwave-assisted synthesis of SnO_2-graphite nanocomposites for Li-ion battery applications. *J. Power Sources*, **144** (1), 220–225.

106 Marcinek, M., Hardwick, L.J., Richardson, T.J., Song, X., and Kostecki, R. (2007) Microwave plasma chemical vapor deposition of nano-structured Sn/C composite thin-film anodes for Li-ion batteries. *J. Power Sources*, **173** (2), 965–971.

107 Yuan, L., Wang, J., Chew, S.Y., Chen, J., Guo, Z.P., Zhao, L., Konstantinov, K., and Liu, H.K. (2007) Synthesis and characterization of SnO_2-polypyrrole composite for lithium-ion battery. *J. Power Sources*, **174** (2), 1183–1187.

108 Doh, C.-H., Park, C.-W., Shin, H.-M., Kim, D.-H., Chung, Y.-D., Moon, S.-I., Jin, B.-S., Kim, H.-S., and Veluchamy, A. (2008) A new SiO/C anode composition for lithium-ion battery. *J. Power Sources*, **179** (1), 367–370.

109 Qiao, H., Zheng, Z., Zhang, L., and Xiao, L. (2008) SnO_2@C core-shell spheres: synthesis, characterization, and performance in reversible Li-ion storage. *J. Mater. Sci.*, **43** (8), 2778–2784.

110 Yao, J., Shen, X., Wang, B., Liu, H., and Wang, G. (2009) In situ chemical

synthesis of SnO$_2$-graphene nanocomposite as anode materials for lithium-ion batteries. *Electrochem. Commun.*, **11** (10), 1849–1852.

111 Aifantis, K.E., Brutti, S., Hackney, S.A., Sarakonsri, T., and Scrosati, B. (2010) SnO$_2$/C nanocomposites as anodes in secondary Li-ion batteries. *Electrochim. Acta*, **55** (18), 5071–5076.

112 Du, Z., Yin, X., Zhang, M., Hao, Q., Wang, Y., and Wang, T. (2010) In situ synthesis of SnO$_2$/graphene nanocomposite and their application as anode material for lithium ion battery. *Mater. Lett.*, **64** (19), 2076–2079.

113 Ochs, R., Szabó, D.V., Schlabach, S., Becker, S., and Indris, S. (2011) Development of nanocomposites for anode materials in Li-ion batteries. *Phys. Status Solidi A*, **208** (2), 471–473.

114 Cho, H.-S., Dong, Z., Pauletti, G.M., Zhang, J., Xu, H., Gu, H., Wang, L., Ewing, R.C., Huth, C., Wang, F., and Shi, D. (2010) Fluorescent, superparamagnetic nanospheres for drug storage, targeting, and imaging: a multifunctional nanocarrier system for cancer diagnosis and treatment. *ACS Nano*, **4** (9), 5398–5404.

115 Corr, S.A., O'Byrne, A., Gun'ko, Y.K., Ghosh, S., Brougham, D.F., Mitchell, S., Volkov, Y., and Prina-Mello, A. (2006) Magnetic-fluorescent nanocomposites for biomedical multitasking. *Chem. Commun.*, (43), 4474.

116 Gao, J., Gu, H., and Xu, B. (2009) Multifunctional magnetic nanoparticles: design, synthesis, and biomedical applications. *Acc. Chem. Res.*, **42** (8), 1097–1107.

117 Rahimi, M., Wadajkar, A., Subramanian, K., Yousef, M., Cui, W., Hsieh, J.-T., and Nguyen, K.T. (2010) *In vitro* evaluation of novel polymer-coated magnetic nanoparticles for controlled drug delivery. *Nanomed. Nanotechnol. Biol. Med.*, **6** (5), 672–680.

118 Ramakrishna, S., Mayer, J., Wintermantel, E., and Leong, K.W. (2001) Biomedical applications of polymer-composite materials: a review. *Compos. Sci. Technol.*, **61** (9), 1189–1224.

119 Xiong, H.-M., Xu, Y., Ren, Q.-G., and Xia, Y.-Y. (2008) Stable aqueous ZnO@Polymer core/shell nanoparticles with tunable photoluminescence and their application in cell imaging. *J. Am. Chem. Soc.*, **130** (24), 7522–7523.

120 Gonzalez, D., Nasibulin, A.G., Baklanov, A.M., Shandakov, S.D., Brown, D.P., Queipo, P., and Kauppinen, E.I. (2005) A new thermophoretic precipitator for collection of nanometer-sized aerosol particles. *Aerosol Sci. Technol.*, **39** (11), 1064–1071.

121 Walsh, J.K., Weimer, A.W., and Hrenya, C.M. (2006) Thermophoretic deposition of aerosol particles in laminar tube flow with mixed convection. *J. Aerosol Sci.*, **37** (6), 715–734.

122 Wen, J., and Wexler, A.S. (2007) Thermophoretic sampler and its application in ultrafine particle collection. *Aerosol Sci. Technol.*, **41** (6), 624–629.

123 Li, W., and Davis, E.J. (1995) The effects of gas and particle properties on thermophoresis. *J. Aerosol Sci.*, **26** (7), 1085–1099.

124 Mädler, L., Roessler, A., Pratsinis, S.E., Sahm, T., Gurlo, A., Barsan, N., and Weimar, U. (2006) Direct formation of highly porous gas-sensing films by in situ thermophoretic deposition of flame-made Pt/SnO$_2$ nanoparticles. *Sens. Actuators B Chem.*, **114** (1), 283–295.

125 Schlabach, S., Ochs, R., Hanemann, T., and Szabó, D.V. (2011) Nanoparticles in polymer-matrix composites. *Microsyst. Technol.*, **17** (2), 183–193.

126 Thekedar, B. (2006) Size characterization of nanoparticles synthesized in microwave plasma discharge process. Master Thesis, University of Magdeburg, Magdeburg.

127 Szabó, D.V., Schlabach, S., and Ochs, R. (2007) Analytical TEM investigations of size effects in SnO$_2$ nanoparticles produced by microwave plasma synthesis, in *Microscopy Conference 2007, Saarbrücken, Germany, Microscopy and Analysis*, vol. 13, Supplement 3 (eds T. Gemming, U. Hartmann, P. Mestres, and P. Walther), Cambridge University Press, pp. 430–431.

128 Schlabach, S., Szabó, D.V., Shi, Z., Wang, D., and Vollath, D. (2004) Synthesis of nanoparticulate SiC in a

129 Fu, L., Johnson, D.L., Zheng, J.G., and Dravid, V.P. (2003) Microwave plasma synthesis of nanostructured g-Al_2O_3 powders. *J. Am. Ceram. Soc.*, **86** (9), 1635–1637.

130 Vollath, D. (2008) *Nanomaterials: An Introduction to Synthesis, Properties and Applications*, 1st edn, Wiley-VCH Verlag GmbH, Weinheim, Germany.

131 Tang, Z.X., Sorensen, C.M., Klabunde, K.J., and Hadjipanayis, G.C. (1991) Size-dependent magnetic-properties of manganese ferrite fine particles. *J. Appl. Phys.*, **69** (8), 5279–5281.

132 Han, D.H., Wang, J.P., and Luo, H.L. (1994) Crystallite size effect on saturation magnetization of fine ferrimagnetic particles. *J. Magn. Magn. Mater.*, **136** (1–2), 176–182.

133 Chen, Q., and Zhang, Z.J. (1998) Size-dependent superparamagnetic properties of $MgFe_2O_4$ spinel ferrite nanocrystallites. *Appl. Phys. Lett.*, **73** (21), 3156–3158.

134 Morales, M.P., Veintemillas-Verdaguer, S., and Serna, C.J. (1999) Magnetic properties of uniform g-Fe_2O_3 nanoparticles smaller than 5 nm prepared by laser pyrolysis. *J. Mater. Res.*, **14** (7), 3066–3072.

135 Batlle, X., and Labarta, A. (2002) Finite-size effects in fine particles: magnetic and transport properties. *J. Phys. D Appl. Phys.*, **35** (6), R15.

136 Goya, G.F., and Leite, E.R. (2003) Ferrimagnetism and spin canting of $Zn^{57}Fe_2O_4$ nanoparticles embedded in ZnO matrix. *J. Phys. Condens. Matter*, **15**, 641–651.

137 Vollath, D., and Szabó, D.V. (2004) Synthesis and properties of nanocomposites. *Adv. Eng. Mater.*, **6** (3), 117–127.

138 Vollath, D. (2010) Bifunctional nanocomposites with magnetic and luminescence properties. *Adv. Mater.*, **22** (39), 4410–4415.

139 Chowdhuri, A., Gupta, V., and Sreenivas, K. (2003) Fast response H_2S gas sensing characteristics with ultra-thin CuO islands on sputtered SnO_2. *Sens. Actuators B Chem.*, **93** (1–3), 572–579.

140 Lee, D.S., Rue, G.H., Huh, J.S., Choi, S.D., and Lee, D.D. (2001) Sensing characteristics of epitaxially-grown tin oxide gas sensor on sapphire substrate. *Sens. Actuators B Chem.*, **77** (1–2), 90–94.

141 Althainz, P., Goschnick, J., Ehrmann, S., and Ache, H.J. (1996) Multisensor microsystem for contaminants in air. *Sens. Actuators B Chem.*, **33** (1–3), 72–76.

142 Schumacher, B., Szabó, D.V., Schlabach, S., Ochs, R., Müller, H., and Bruns, M. (2005) Nanoparticle SnO_2 films as gas sensitive membranes. *MRS Online Proc. Libr.*, **900**, O08-06.

143 Gong, J.W., Chen, Q.F., Fei, W.F., and Seal, S. (2004) Micromachined nanocrystalline SnO_2 chemical gas sensors for electronic nose. *Sens. Actuators B Chem.*, **102** (1), 117–125.

144 Sahm, T., Mädler, L., Gurlo, A., Barsan, N., Pratsinis, S.E., and Weimar, U. (2004) Flame spray synthesis of tin dioxide nanoparticles for gas sensing. *Sens. Actuators B Chem.*, **98** (2–3), 148–153.

145 Sahm, T., Mädler, L., Gurlo, A., Barsan, N., Pratsinis, S.E., and Weimar, U. (2004) Flame spray synthesis of tin oxide nanoparticles for gas sensing. *MRS Online Proc. Libr.*, **828**, A1.3.1.

146 Carotta, C.M., Giberti, A., Guidi, V., Malagù, C., Vendemiati, B., and Martinelli, G. (2004) Gas sensors based on semiconductor oxides: basic aspects onto materials and working principles. *MRS Online Proc. Libr.*, **828**, A4.6.

147 Kennedy, M.K., Kruis, F.E., Fissan, H., Mehta, B.R., Stappert, S., and Dumpich, G. (2003) Tailored nanoparticle films from monosized tin oxide nanocrystals: particle synthesis, film formation, and size-dependent gas-sensing properties. *J. Appl. Phys.*, **93** (1), 551–560.

148 Alan, V.C., Shelley, L.P.S., Luke, A.O.D., and Mark, E.S. (2006) Keeping it small – restricting the growth of nanocrystals. *J. Phys. Condens. Matter*, **18** (15), L163.

149 O'Dell, L.A., Savin, S.L.P., Chadwick, A.V., and Smith, M.E. (2005) Structural studies of silica- and alumina-pinned nanocrystalline SnO_2. *Nanotechnology*, **16** (9), 1836.

150 Schumacher, B. (2005) Nanogranulare SnO_2-Schichten für Gassensor-Mikroarrays. Master Thesis, Albert-Ludwigs-Universität, Freiburg, Germany.

151 Fuchs, M. (2009) Synthese und Charakterisierung von SnO_2/SiO_2-Kern/Hülle Nanopartikeln für gasanalytische Anwendungen. Master Thesis, Regensburg, Regensburg, Germany.

152 Li, N., Martin, C.R., and Scrosati, B. (2001) Nanomaterial-based Li-ion battery electrodes. *J. Power Sources*, **97–98**, 240–243.

153 Wang, Y., Lee, J.Y., and Chen, B.H. (2004) Microemulsion syntheses of Sn and SnO_2-graphite nanocomposite anodes for Li-ion batteries. *J. Electrochem. Soc.*, **151** (4), A563–A570.

154 Zhao, Y., Zhou, Q., Liu, L., Xu, J., Yan, M., and Jiang, Z. (2006) A novel and facile route of ink-jet printing to thin film SnO_2 anode for rechargeable lithium ion batteries. *Electrochim. Acta*, **51** (13), 2639–2645.

155 Liang, Y., Fan, J., Xia, X., and Jia, Z. (2007) Synthesis and characterisation of SnO_2 nano-single crystals as anode materials for lithium-ion batteries. *Mater. Lett.*, **61** (22), 4370–4373.

156 Gao, M., Chen, X., Pan, H., Xiang, L., Wu, F., and Liu, Y. (2010) Ultrafine SnO_2 dispersed carbon matrix composites derived by a sol-gel method as anode materials for lithium ion batteries. *Electrochim. Acta*, **55** (28), 9067–9074.

157 Ohzuku, T., Iwakoshi, Y., and Sawai, K. (1993) Formation of lithium-graphite intercalation compounds in nonaqueous electrolytes and their application as a negative electrode for a lithium ion (shuttlecock) cell. *J. Electrochem. Soc.*, **140** (9), 2490–2498.

158 Bruce, P.G., Scrosati, B., and Tarascon, J.M. (2008) Nanomaterials for rechargeable lithium batteries. *Angew. Chem. Int. Ed.*, **47** (16), 2930–2946.

159 Centi, G., and Perathoner, S. (2009) The role of nanostructure in improving the performance of electrodes for energy storage and conversion. *Eur. J. Inorg. Chem.*, **2009** (26), 3851–3878.

160 Chen, Y.-C., Chen, J.-M., Huang, Y.-H., Lee, Y.-R., and Shih, H.C. (2007) Size effect of tin oxide nanoparticles on high capacity lithium battery anode materials. *Surf. Coating. Technol.*, **202** (4–7), 1313–1318.

12
Oxidation, Purification and Functionalization of Carbon Nanotubes under Microwave Irradiation

Davide Garella and Giancarlo Cravotto

This chapter attempts to give a comprehensive overview of the use of microwaves for the purification, oxidation and functionalization of CNTs. This simple, efficient and scalable technique makes the preparation of pure CNTs both simpler and faster. Dielectric heating generates particular conditions and effects which are not obtainable under more common heating methods. This promising technique should promote easier access to new technological materials for biomedical applications, composite materials and nanoelectronics.

12.1
Introduction

The huge international research effort into nanostructured materials for advanced technological applications is expected to furnish important innovations of high industrial relevance in the short to medium term. Carbon nanotubes (single wall and multi-wall) have been widely investigated since their discovery in the early 1990s [1]. The quantum aspects of single wall carbon nanotubes (SWCNTs) remain largely unfamiliar to the scientific world; however, the more general desirable properties of CNT are already being exploited commercially. Although their mainback is their lack of solubility in physiological solutions, CNTs have recently found interesting applications in medicinal chemistry and, despite the huge number of studies describing such unique materials, it would appear that the greatest part of their substantial potential is still to be realised.

The limit that plays a fundamental role in the diffusion of the carbon nanotubes is certainly their poor reactivity. Their carbon atom sp^2 hybridization is similar to graphite, and confers upon these nanostructures a unique strength which, unfortunately, also constitutes a considerable limit on modification and functionalization.

Non-conventional techniques, and in particular microwave (MW) irradiation, may play a pivotal role in enhancing the purification and the structural

---

Microwaves in Nanoparticle Synthesis, First Edition. Edited by Satoshi Horikoshi and Nick Serpone.
© 2013 Wiley-VCH Verlag GmbH & Co. KGaA. Published 2013 by Wiley-VCH Verlag GmbH & Co. KGaA.

modification of the CNTs [2]. Kappe et al. proved that specific MW effects are inconclusive and purely thermal effects were responsible for results obtained under dielectric heating [3]. The interaction of MW with CNTs is influenced by several factors which are all attributed to specific superheating phenomenon rather than bulk solvent temperature. The most important limitation during irradiation with MW is temperature detection. Although the average temperature is usually detected directly inside the oven (using infrared or a thermocouple), there is no way to localize and check hot spot conditions.

The interaction between MW and carbonaceous materials in general, and in particular CNTs, generates strong substrate energy adsorption that in turn generates a considerable increase in temperature. This behavior is the result of additional phenomena.

The main interaction between MW irradiation and molecules with a dipolar moment is dipolar polarization; rapid molecular rotation, induced by the electric field oscillation, causes homogeneous solution heating and mixing during MW irradiation. [Eq. (12.1)] [4].

$$\frac{dT}{dt} = \frac{c\varepsilon'' f E_{rms}^2}{\rho C_p} \tag{12.1}$$

where c = constant, ρ = density, C_p = heat capacity, f = frequency, E^2_{rms} = r.m.s. field intensity, and ε'' = dielectric loss.

In addition to dipolar polarization, the presence of electrical conductors causes conduction effects that contribute to system heating. In this mechanism the electrons are induced to move by the electric field causing sample heating due to electrical resistance. This phenomenon is known as joule heating.

The presence of solvent ions and metal catalyst impurities, which are used in CNT preparation, are the main cause of localized hot spots (very high temperatures) when the CNTs are irradiated with MW. For this reason Harutyunyan et al. attributed superior purification ability to MW versus conventional conductive heating [5]. Localized hot spots found around metal nanoparticles may damage the carbon surface, giving acids access to the metal [6]. However, the studies of Imholt et al. showed minor temperature variations in the presence of impurities when a sample was subjected to MW irradiation [7]. It has been shown theoretically that a perfect CNT is a ballistic conductor: its resistance is quantized and thus independent of length, due to its unique 1D structure [8, 9]. This means that energy is not dissipated in the CNT, because of electron movement, and so extremely high stable current densities can exist. Because of this behavior, the current induced during MW irradiation is not converted to heat in the CNT and must be dissipated directly into any impurities, generating extremely high and localized temperatures. CNTs have unavoidable imperfections, allowing joule heating to occur causing localized superheating. Although joule heating may be the main contributor to system heating, quantifying the amount of joule heating in a CNT sample is difficult as it is heavily dependent upon the nature of the CNTs and the presence of contaminants. CNTs may also undergo more specific MW interactions where a longitudinal vibration force, under vacuum, is superior to joule heating. Under this condition the CNT samples are heated very quickly, but the presence of any impurities, such as amorphous carbon, metallic particles or absorbed gases, dampens the vibrational mode and reduces the localized temperature attained under MW radiation [10].

Therefore, it cannot be said with certainty which MW effects act on CNTs. It is likely that a combination of all the effects causes the increase in temperature and the formation of hot spots. We herein report the latest findings for the purification/oxidation and functionalization of CNTs under MW irradiation.

12.2
Oxidation and Purification

The purification and the chemical grafting of nanomaterials, in particular of CNTs, is a controversial and uncertain topic and, as such, is often debated in the scientific community. A universal purification protocol cannot be defined as reaction conditions are heavily influenced by typology, structure and CNT purity; the high curvature strain of SWCNTs requires milder conditions to prevent damage, while harsh procedures are used for multi-walled carbon nanotubes (MWCNTs) as they are more resistant to chemical attack [11]. Several CNT purification/oxidation protocols are reported in the literature and these include liquid and gas phase oxidation processes, chemical functionalization, filtration and chromatographic techniques. We will focus our survey on the methods carried out under MW irradiation (2.45 GHz). Besides the clear advantages that are provided by the technique's efficiency and rapidity, MW-promoted protocols are also suitable for scaling-up.

An accurate study was recently carried out by MacKenzie *et al.* [12] which assessed the best purification conditions using the following five parameters as the basis for comparison: residual weight, maximum oxidation peak, product homogeneity, Raman IG/ID ratio and functionalization. In this work three acids were considered: HNO_3 offers poor impurity removal, damaged CNTs, and promotes the introduction of carboxyl and C–O functional groups; HCl and H_2SO_4 behave similarly. Whereas HCl does not introduce functional groups and affords less homogeneous products, H_2SO_4 is preferred despite the sulfur-based side-functionalization it causes. The overall process can be reduced to a function of three variables: temperature, time and the stoichiometric excess of acid to impurities. Temperature, heating ramp rate (12–60 °C min^{-1}) and treatment time strongly affect CNT purity. A compromise that provides a high degree of CNT oxidation, but moderate damage, requires reagent concentration, bulk temperature and irradiation time to be well balanced. A common ratio of acid to impurity was found to be 4.5 times stoichiometry.

Further purification improvements may be achieved with higher temperatures or the addition of chelating agents. Mitra *et al.* [13] have shown that, so far, only ethylenediaminetetraacetic acid (EDTA) in combination with acetic acid removed the iron metal impurities in the CNTs. Other chelating agents, such as

N-(2-acetamido)iminodiacetic acid (ADA), *N,N*- bis(carboxymethyl)glycine (NTA) and diethylenetriaminepentaacetic acid (DTPA) were ineffective.

An interesting and very powerful procedure for CNT oxidation under acidic conditions was performed by the same authors. The experiment was carried out under MW irradiation (450 W), using a mixture of 70% nitric acid and 97% sulfuric acid (1:1) for 3 min. The presence of –COOH and –SO$_3$H groups was confirmed by the IR analysis, while elemental analysis estimated that one in three carbon atoms had been carboxylated, while one in ten carbon atoms had been sulfonated. Moreover, SEM images of SWCNT thin films, deposited from an aqueous suspension, revealed that the CNTs were in a parallel alignment which is probably due to hydrophobic forces and CNT–CNT interactions. This behavior may allow the synthesis of nanoscale device architectures and aligned nanocomposite structures to be carried out.

Cravotto et al. [14] recently described an efficient MWCNT purification/oxidation procedure (Mitsui CNTs: defect-free, with a characteristic amorphous layer) under acidic conditions (3:1 H$_2$SO$_4$/HNO$_3$ aqueous solution). The authors compared several non-conventional techniques: MW, 20 and 300 kHz high-intensity ultrasound (US), and simultaneous MW/US irradiation. The best results were achieved with MW and with 300-kHz US. The selective rupture of CNT tips was observed under MW or combined MW/US irradiation as CNT tips act as antennas and cause local superheating. Typical crystallinity was preserved and the amorphous external layer was almost completely removed. The Raman spectrum of the samples (Figure 12.1) shows an increased ID/IG band ratio, which confirms the partial formation of defects.

Grennberg et al. described an efficient acid-free MW-assisted CNT purification protocol in organic solvents [15]. In a typical experiment the CNT sample (20 mg of SWCNTs or MWCNTs) was suspended in CH$_2$Cl$_2$ (5 ml) in a MW closed vessel. The mixture was then rapidly heated to 100–150 °C for 5 min. Interestingly, in most experiments some iron particles were found and, being up to several millimeters in size, were easily removed by filtration, centrifugation, and even with a pair of tweezers.

Outstanding CNT purity, moderate damage and lower waste levels were obtained by repeating the entire procedure at least 2–5 times.

Other procedures entail the pre-treating of CNTs under air-flow in a MW oven. Harutyunyan et al., for example, irradiated a SWCNT sample under MW with the aim of eliminating the amorphous carbon and open cracks in the CNT structure. Subsequently, an acid treatment (HCl 4 M) at low temperature removed the residual metal catalyst impurities [5]. Under dielectric heating, the increase in the temperature was enough to either burst the different carbon structures covering the catalyst particles (thermal expansion), or to drive the preferential oxidation of the carbon coating over that of the catalyst particles. In either case, the carbon coating protecting the metal impurity was weakened, or removed by means of a milder acid (e.g., HCl). The best experimental conditions were found to be 150 W MW power, a dry-air flow (500 °C). More damage was observed at higher power. The authors also compared different heating times at constant power: 10, 40, and

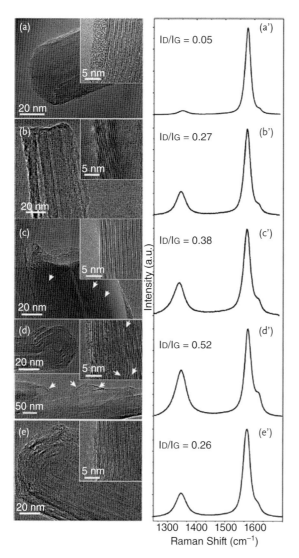

Figure 12.1 TEM images (a–e) and Raman spectra (a'–e') of MWCNTs treated using different techniques: (a/a') pristine MWCNTs; (b/b') MW; c/c' sequential MW/US 20 KHz; d/d' US 20 KHz; e/e' US 300 KHz, respectively. The arrows shown in the inset of (d) show the resulting stacking faults (inset top part) and graphene sheet fragments. Ref. [14]. Reproduced by permission of The Royal Society of Chemistry (RSC) on behalf of the Centre National de la Recherche Scientifique (CNRS) and the RSC.

60 min, followed by refluxing with 4 M HCl for 6 h. Less than 0.2 wt% metal impurities were detected in the samples. With 10 min MW treatment the sample still contained about 1.5 wt% metal residues. Longer MW treatments caused high SWCNT loss and also led to additional wall damage. This procedure caused the

amorphous carbon to be removed and the multi-shell carbon covering the metallic particles weakened, however, this oxidation step cannot be achieved without a certain loss of material.

In conclusion, it can be said that purification strategies vary according to the features of the raw CNTs and the specific requirements and target applications of the purified CNTs. Despite significant advances, a simple, effective CNT purification method is still to be finalized. MWCNTs are relatively easy and effective to purify, whereas SWCNTs (because of their length, diameter, transport properties, and even chirality) require a much more elaborate purification and so low yields and defects are often observed.

CNT applications are strictly linked to the efficiency of the purification procedure that can be greatly improved under MW irradiation.

12.3
Functionalization

CNTs are one of the most interesting and highly studied materials in existence, although their diffusion has been limited by their lack of solubility. Almost all applications require pristine CNTs to be heavily chemically modified to enhance their solubilization or dispensability, and consequently facilitate their manipulation. All the operations that involve the structural modification of CNT inevitably imply structural damage, therefore it is important that functionalization preserves structural integrity. CNT functionalization with the assistance of biological molecules remarkably improves solubility in aqueous or organic environments, and thus facilitates the development of novel biotechnological, biomedicinal, and bioengineering applications.

The carbon atoms that constitute CNTs are sp^2 hybridized and form a hexagonal tubular network which is closed at the extremities by hemispherical end caps. The functionalization approach to materials with new properties [16–18] may be carried out via covalent bond formation [19–23] or noncovalent interactions [24–26]. Chemical grafting mainly occurs at CNT sidewalls and tips. Typically CNTs are oxidized under acidic conditions and heating. As well as being purified, new oxygenated functional groups (–OH, –CO–, –COOH) are generated on the surface favoring aqueous dispersion and further functionalization (i.e., esterification, amidation etc.). Often the reactions on these materials require several hours, or even days, to reach a good degree of chemical grafting. Non-conventional techniques can increase reactivity, and MW irradiation, in particular, has proven to be a powerful tool in the rapid functionalization of CNTs.

Della Negra and coworkers grafted CNT surfaces with polyethylene glycol (PEG) under MW irradiation with the aim of increasing solubilizing properties [27]. The procedure involved a first step where the SWCNTs were treated under acidic conditions to produce carboxylic groups which were subsequently converted into acyl chlorides with thionyl chloride. These SWCNT derivatives were mixed with PEG-amine in the presence of a few drops of chloroform and ground in a mortar. The mixture was spread to form a film and dried under a nitrogen flow. The sample was put in a glass vessel connected to a nitrogen line, inside the MW oven, and irradiated for an overall time of 76 min. To prevent superheating, several irradiation cycles at low power (2 min of irradiation at 30/40 W and 30 s stop) were carried out. The black powder obtained showed higher solubility in organic solvents ($CHCl_3$, THF and CH_3CN), less solubility in polar solvents (H_2O, acetone, CH_3OH), and was totally insoluble in carbon disulfide, ethyl ether, and toluene.

An amidation reaction, under conventional heating, usually proceeds through three stages: carboxylation, acyl chlorination and amidation, however, one-pot conversion was feasible when the reaction was performed under MW irradiation and thus reaction time was cut down from 3–5 days to 15–20 min. After the SWCNTs were initially treated with HNO_3, the oxidized SWCNTs were reacted with 2,6-dinitroaniline in DMF (MW irradiation at 670 W and a pressure of 125 psi (15–20 min) [28].

The radical addition of 4-methoxyphenylhydrazine hydrochloride carried out under MW irradiation gave the same results as the reaction carried out under conventional heating, but with a dramatic reduction in the reaction time [29].

Cycloaddition reactions are one of the most powerful methods for the modification of CNT surfaces, a strategy that has previously been investigated for fullerene derivatization. Although both SWCNTs and fullerenes behave similarly, the former take much longer to react and so the application of MW to promote cycloaddition reactions on CNTs has become widely successful.

Prato *et al.* published a report on MW-assisted cycloaddition to pristine CNTs in solvent-free conditions [30]. Azomethine ylides are an important class of 1,3-dipoles in cycloaddition reactions and are obtained from aziridines. Once it was clear that azomethine ylides could be successfully added to fullerenes and CNTs, SWCNTs were suspended in dichloromethane (DCM) solution in the presence of aziridine. After 5-min sonication the mixture was concentrated under vacuum and irradiated, solventless, in a monomodal MW oven for 1 h. The sample was finally filtered over a Millipore membrane (Fluoropore, 0.2 μm), washed with DCM, and fully characterized (Raman, UV–vis–NIR spectroscopies, TGA, and TEM). The product was slightly soluble in DMF and dichloromethane (0.13 and 0.11 mg ml^{-1}) and the characterization confirmed that the cycloaddition of aziridine on the SWCNTs had been carried out. When the reaction was performed under conventional heating the reaction time was around 5 days. The author supposed that, in the absence of solvent, MW effects were magnified because the radiation is completely absorbed by the reagents. This condition makes the procedure environmentally friendly because the use of solvents is limited and the work-up procedures are simplified.

Further proof of the covalent sidewall functionalization of CNTs can be found in a work by Martin and coworkers [31]. Product analysis confirmed the loss of functional groups, and subsequently the sample exhibited no solubility in any solvent.

From theoretical calculations it would appear that the 1,3-dipolar cycloaddition between nitrile oxide and SWCNTs is not as favorable as the same reaction performed with azomethine ylides [32]. Langa *et al.* described an interesting procedure that opened new possibilities for the preparation of SWCNT derivatives [33]. The method was based on the typical fullerene functionalization via a 1,3-dipolar cycloaddition. 4-Pyridyl isoxazolino-SWCNT (Py-SWCNT) was prepared by reaction of the nitrile oxide (generated *in situ*) and a solution of pentyl ester-SWCNT in the presence of triethylamine (TEA) and *o*-dichlorobenzene. The mixture was irradiated with MW (150 W) for 45 min. At the end of the reaction the solvent was removed and the solid was washed several times with $CHCl_3$/water, $CHCl_3$/ethyl ether, and $CHCl_3$/pentane, yielding 4-pyridyl isoxazolino-SWCNT (Py-SWCNT). The product forms complexes with zinc porphyrin (ZnPor) in a way which is parallel to that of pyridyl-derivatized C60.

The same author described the first example of Diels–Alder (DA) cycloaddition to the SWCNT sidewall [34]. Thermodynamically, the products could undergo a retro-DA reaction, but stable cycloadducts were prepared using different *o*-quinodimethanes. The reaction was performed by irradiating the esterified nanotubes with MWs in pentanol (150 W, 50 min) in the presence of *o*-quinodimethane generated *in situ* from the corresponding 4,5-benzo-1,2-oxathiin-2-oxide. The solubility of the resulting SWCNTs was much improved.

Cravotto's group investigated several solvent-free cycloaddition reactions under MW irradiation (SynthWAVE–Milestone). An example is the 1,3-dipolar cycloaddition of carbonyl ylides, generated from a series of oxiranes, to SWCNTs (Figure 12.2). The procedure is extremely fast and repeatable. The reaction of SWCNTs with benzylidenmalononitrile epoxide afforded a gem-dicyano derivative which is well suited to further transformation to esters or amides via the Pinner reaction [35].

Bonifazi *et al.* described the simple functionalization of double-walled carbon nanotubes (DWCNTs) with Br_2 under MW conditions [36]. The Br atoms were linked covalently to DWCNTs in a weight percentage of about 5–8% (Br-DWCNTs). In a typical experiment, the DWCNTs (Nanocyl-2100) were mixed in an aqueous solution of Br_2 ($35\,g\,l^{-1}$, corresponding to the solubility of Br_2 in H_2O at $20\,°C$) in a glass tube. After a fast sonication to increase sample homogeneity, the mixture was irradiated with MW at $180\,°C$ for 9 min. When the orange color disappeared the reaction was over. The Br-DWCNTs were filtered and dried under vacuum. The same procedure was repeated several times with the aim of increasing Br content. The last purification was carried out by washing the sample with a solution of NaS_2O_3 (~1 M), rinsing with few milliliters of MeOH, and drying under vacuum (yield Br-DWCNTs about 65% w/w). The halogenated CNTs may represent a promising class of intermediates because they are suitable substrates for metal-catalyzed cross-couplings and useful precursors to organometallic species.

Recently Maghrebi and coworkers [37] developed a one-pot procedure for MWCNT functionalization in the presence of diamines (EDA: ethylenediamine, HMDA: 1,6-hexamethylenediamine and DAB: 1,4-diaminobenzene). The typical procedure involved pristine CNT pre-sonication in the presence of diamine and $NaNO_2$ (30 min at $50\,°C$), until a homogeneous suspension was obtained. The suggested reaction mechanism consists of a radical reaction between the semi-stable

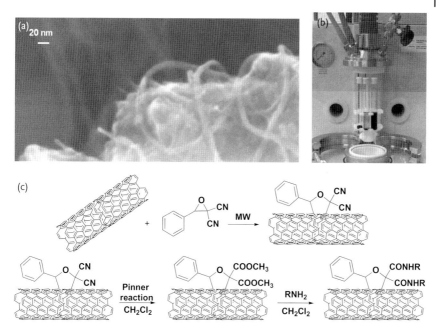

Figure 12.2 SWCNT grafting via solvent-free MW-assisted epoxides cycloaddition and further functionalization via Pinner reaction (c). (a) SEM image of SWCNTs after 1,3-cycloaddition and (b) the MW reactor (SynthWAVE, Milestone, Italy).

diazonium ions formed *in situ* and the CNTs. The suspension was heated in MW (500 W) at 90 °C for 15 min. EDA was the most reactive of the diamines, while HMDA and DAB required higher temperatures (120 °C). Several techniques were employed for CNT characterization and both functionalization and some structural damage were observed. TEM images showed some bridge moieties that could contribute to the formation of CNTs agglomerates. Using Xu's work as a base [38], Cheng *et al.* introduced the electrophilic addition of 3-chloropropene to the SWCNTs via a MW-assisted reaction [39]. In the first step of the reaction, 3-chloropropene formed carbocation species that underwent electrophilic addition in the presence of $AlCl_3$. The same authors described a procedure that involves the use of alcohols in the presence of a Lewis acid as a catalyst and traces of HCl [40]. The protonated alcohols were able to interact with SWCNTs through electrophilic addition (MW irradiation: 450 W, 60–80 °C, 30 min). Several techniques were employed to confirm the presence of the alkyl and hydroxy groups on the SWCNT surface.

Sugar moieties can be covalently bonded to SWCNTs via cycloaddition reactions. The method proposed by Leinonen *et al.* starts with sugar modification which furnishes acetyl bromides then azido derivatives (nucleophilic substitution with sodium azide) [41]. The sugars containing the $-N_3$ group were dissolved in 1,1,2,2-tetrachloroethane in the presence of SWCNTs and irradiated with MW at

150 °C (20 min). The sugars decomposed to form the corresponding nitrenes, which are particularly reactive and able to graft the SWCNTs, dramatically enhancing water solubility (Table 12.1).

A peculiar non-perturbative technique for CNT functionalization is MW plasma treatment [42–48]. This technique generates metastable ions and radicals that react with CNTs to form covalent bonds while preserving the peculiar structure. The selection of active radicals is subjected to mass diffraction, just as in common mass spectroscopy. The plasma device for MWCNT functionalization consists of a cylindrical stainless vacuum chamber equipped with a 2.45 GHz MW generator (Figure 12.3).

An inlet for plasma gases is set up, and a rotary and turbo-molecular pump system is connected to the chamber. The MW irradiation is guided into the chamber through antennas.

The sample is placed in a glass beaker usually kept at a 20 cm distance from the plasma showering quartz plate. Before the CNT treatment is carried out, Ar pretreatment is used to enhance surface activation, so a mixture of gaseous reagent/Ar is introduced through mass flow meters. This technique enables the activation of the less reactive substrates into reactive radicals, and thus allows the surface

Table 12.1 Solubility in water of the N-linked sugar-functionalized SWCNTs.

Sugar moiety	Solubility (mg ml^{-1})	Dispersed SWCNTs (%)	Solubility original sugar (g ml^{-1})
Glucopyranosyl	0.6	16.3	0.9
Galactopyranosyl	1.3	35.2	0.47
Mannopyranosyl	0.6	16.3	2.5

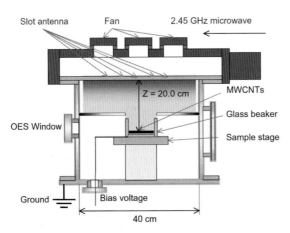

Figure 12.3 Experimental set-up. Reproduced with permission from Ref. [48]. Copyright 2010 by Elsevier.

functionalization of even the most inert materials without creating structural damage.

12.4
Conclusion

In conclusion, we have given a summary of the state-of-the-art CNT purification and functionalization methods under MW irradiation. CNT modification is one of the most interesting and challenging topics in materials chemistry today and can have wide ranging implications for several CNT applications in many different fields, such as nanotechnology, nanoelectronics and composite materials.

References

1. Iijima, S. (1991) Helical microtubules of graphitic carbon. *Nature*, **354** (6348), 56–58.
2. Kappe, C.O. (2004) Controlled microwave heating in modern organic synthesis. *Angew. Chem. Int. Ed.*, **43** (46), 6250–6284.
3. Herrero, M.A., Kremsner, J.M., and Kappe, C.O. (2008) Nonthermal microwave effects revisited: on the importance of internal temperature monitoring and agitation in microwave chemistry. *J. Org. Chem.*, **73** (1), 36–47.
4. Mingos, D.M.P., and Baghurst, D.R. (1991) Applications of microwave dielectric heating effects to synthetic problems in chemistry. *Chem. Soc. Rev.*, **20**, 1–47.
5. Harutyunyan, A.R., Pradhan, B.K., Chang, J.P., Chen, G.G., and Eklund, P.C. (2002) Purification of single-wall carbon nanotubes by selective microwave heating of catalyst particles. *J. Phys. Chem. B*, **106** (34), 8671–8675.
6. Wadhawan, A., Garrett, D., and Perez, J.M. (2003) Nanoparticle-assisted microwave absorption by single-wall carbon nanotubes. *Appl. Phys. Lett.*, **83** (13), 2683–2685.
7. Imholt, T.J., Dyke, C.A., Hasslacher, B., Perez, J.M., Price, D.W., Roberts, J.A., Scott, J.B., Wadhawan, A., Ye, Z., and Tour, J.M. (2003) Nanotubes in microwave fields: light emission, intense heat, outgassing, and reconstruction. *Chem. Mater.*, **15** (21), 3969–3970.
8. Dresselhaus, M.S., Dresselhaus, G., and Avouris, P. (eds) (2001) Carbon nanotubes. Synthesis, structure, properties, and applications, in *Topics in Applied Physics*, Springer-Verlag, Berlin, Heidelberg.
9. Frank, S., Poncharal, P., Wang, Z.L., and de Heer, W.A. (1998) Carbon nanotube quantum resistors. *Science*, **280** (5370), 1744–1746.
10. Ye, Z. (2005) Mechanism and the effect of MW-carbon nanotube interaction. University of North Texas. Dissertation.
11. Monthioux, M., Smith, B.W., Burteaux, B., Claye, A., Fischer, J.E., and Luzzi, D.E. (2001) Sensitivity of single-wall carbon nanotubes to chemical processing: an electron microscopy investigation. *Carbon*, **39** (8), 1251–1272.
12. MacKenzie, K.J., Dunens, O.M., Hanus, M.J., and Harris, A.T. (2011) Optimisation of microwave-assisted acid digestion for the purification of supported carbon nanotubes. *Carbon*, **49** (13), 4179–4190.
13. Chen, Y., Iqbal, Z., and Mitra, S. (2007) Microwave-induced controlled purification of single-walled carbon nanotubes without sidewall functionalization. *Adv. Funct. Mater.*, **17** (18), 3946–3951.

14 Cravotto, G., Garella, D., Calcio Gaudino, E., Turci, F., Bertarione, S., Agostini, G., Cesano, F., and Scarano, D. (2011) Rapid purification/oxidation of multi-walled carbon nanotubes under 300 KHz-ultrasound and microwave irradiation. *New J. Chem.*, **35** (4), 915–919.

15 Chajara, K., Andersson, C.-H., Lu, J., Widenkvistc, E., and Grennberg, H. (2010) The reagent-free, microwave-assisted purification of carbon nanotubes. *New J. Chem.*, **34**, 2275–2280.

16 Fujiwara, A., Matsuoka, Y., Suematsu, H., Ogawa, N., Miyano, K., Kataura, H., Maniwa, Y., Suzuki, S., and Achiba, Y. (2001) Hybridized microcrystals composed of metal fine particles and π-conjugated organic microcrystals. *Jpn J. Appl. Phys.*, **40**, L1229–L1231.

17 Freitag, M., Martin, Y., Misewich, J.A., Martel, R., and Avouris, P.H. (2003) Photoconductivity of single carbon nanotubes. *Nano Lett.*, **3**, 1067–1071.

18 Cao, L., Chen, H., Wang, M., Sun, J., Zhang, X., and Kong, F.J. (2002) Photoconductivity study of modified carbon nanotube/oxotitanium phthalocyanine composites. *Phys. Chem. B*, **106**, 8971–8975.

19 Georgakilas, V., Kordatos, K., Prato, M., Guldi, D.M., Holzinger, M., and Hirsch, A. (2002) Organic functionalization of carbon nanotubes. *J. Am. Chem. Soc.*, **124** (5), 760–761.

20 Dyke, C.A., and Tour, J.M. (2003) Solvent-free functionalization of carbon nanotubes. *J. Am. Chem. Soc.*, **125** (5), 1156–1157.

21 Holzinger, M., Abraham, J., Whelan, P., Graupner, R., Ley, L., Hennrich, F., Kappes, M., and Hirsch, A. (2003) Functionalization of single-walled carbon nanotubes with (R-)oxycarbonyl nitrenes. *J. Am. Chem. Soc.*, **125** (28), 8566–8580.

22 Murakami, H., Nomura, T., and Nakashima, N. (2003) Noncovalent porphyrin-functionalized single-walled carbon nanotubes in solution and the formation of porphyrin-nanotube nanocomposites. *Chem. Phys. Lett.*, **378**, 481–485.

23 Guldi, D.M., Marcaccio, M., Paolucci, D., Paolucci, F., Tagmatarchis, N., Tasis, D., Vazquez, E., and Prato, M. (2003) Single-wall carbon nanotube-ferrocene nanohybrids: observing intramolecular electron transfer in functionalized SWNTs. *Angew. Chem. Int. Ed.*, **42** (35), 4206–4209.

24 Star, A., Steuerman, D.W., Heath, J.R., and Stoddart, J.F. (2002) Starched carbon nanotubes. *Angew. Chem. Int. Ed.*, **41**, 2508–2512.

25 Fukushima, T., Kosaka, A., Ishimura, Y., Yamamoto, T., Takigawa, T., Ishii, N., and Aida, T. (2003) Molecular ordering of organic molten salts triggered by single-walled carbon nanotubes. *Science*, **300**, 2072–2074.

26 Sun, J., Gao, L., and Iwasa, M. (2004) Noncovalent attachment of oxide nanoparticles onto carbon nanotubes using water-in-oil microemulsions. *Chem. Commun.*, **7**, 832–833.

27 Della Negra, F., Meneghetti, M., and Menna, E. (2003) Microwave-assisted synthesis of a soluble single wall carbon nanotube derivative. *Fullerene Sci. Technol.*, **11**, 25–34.

28 Wang, Y., Iqbal, Z., and Mitra, S. (2005) Microwave-induced rapid chemical functionalization of single-walled carbon nanotubes. *Carbon*, **43**, 1015–1020.

29 Liu, J., Rodríguez, M., Zubiri, I., Vigolo, B., Dossot, M., Fort, Y., Ehrhardt, J.J., and MacRae, E. (2007) Efficient microwave-assisted radical functionalization of single-wall carbon nanotubes. *Carbon*, **45**, 885–891.

30 Brunetti, F.G., Herrero, M.A., Muñoz, J.M., Giordani, S., Díaz-Ortiz, A., Filippone, S., Ruaro, G., Meneghetti, M., Prato, M., and Vázquez, E. (2007) Reversible microwave-assisted cycloaddition of aziridines to carbon nanotubes. *J. Am. Chem. Soc.*, **129** (47), 14580–14581.

31 Martín, N., Altable, M., Filippone, S., Martín-Domenech, A., Echegoyen, L., and Cardona, C.M. (2006) Retro-cycloaddition reaction of pyrrolidinofullerenes. *Angew. Chem. Int. Ed.*, **45**, 110–114.

32 Lu, X., Tian, F., Wang, N., and Zhang, Q. (2003) Theoretical exploration of the 1,3-dipolar cycloadditions onto the sidewalls of (n,n) armchair single-wall carbon nanotubes. *J. Am. Chem. Soc.*, **125** (34), 10459–10486.

33 Alvaro, M., Atienzar, P., de la Cruz, P., Delgado, J.L., Troiani, V., Garcia, H., Langa, F., Palkar, A., and Echegoyen, L. (2006) Synthesis, photochemistry, and electrochemistry of single-wall carbon nanotubes with pendent pyridyl groups and of their metal complexes with zinc porphyrin. Comparison with pyridyl-bearing fullerenes. *J. Am. Chem. Soc.*, **128** (20), 6626–6635.

34 Delgado, J.L., de la Cruz, P., Langa, F., Urbina, A., Casado, J., and López-Navarrete, J.T. (2004) Microwave-assisted sidewall functionalization of single-wall carbon nanotubes by Diels-Alder cycloaddition. *Chem. Commun.*, 1734–1735.

35 Tagliapietra, S., Cravotto, G., Calcio Gaudino, E., Visentin, S., and Mussi, V. (2012) Functionalization of single-walled Carbon Nanotubes via the 1,3-Cycloaddition of Carbonyl ylides under Microwave Irradiation. *Synlett*. doi: 10.1055/s-0031-1290681

36 Colomer, J.-F., Marega, R., Traboulsi, H., Meneghetti, M., Van Tendeloo, G., and Bonifazi, D. (2009) Microwave-assisted bromination of double-walled carbon nanotubes. *Chem. Mater.*, **21**, 4747–4749.

37 Amiri, A., Maghrebi, M., Baniadam, M., and Zeinali Heris, S. (2011) One-pot, efficient functionalization of multi-walled carbon nanotubes with diamines by microwave method. *Appl. Surf. Sci.*, **257** (23), 10261–10266.

38 Xu, Y., Wang, X., Tian, R., Li, S., Wan, L., Li, M., You, H., Li, Q., and Wang, S. (2008) Microwave-induced electrophilic addition of single-walled carbon nanotubes with alkylhalides. *Appl. Surf. Sci.*, **254** (8), 2431–2435.

39 Tian, R., Wang, X., Xu, Y., Li, S., Wan, L., Li, M., and Cheng, J. (2009) Microwave-assisted functionalization of single-walled carbon nanotubes with 3-chloropropene. *J. Nanopart. Res.*, **11** (5), 1201–1208.

40 Tian, R., Wang, X., Li, M., Hu, H., Chen, R., Liu, F., Zheng, H., and Wan, L. (2008) An efficient route to functionalize singe-walled carbon nanotubes using alcohols. *Appl. Surf. Sci.*, **255** (5), 3294–3299.

41 Leinonen, H., Pettersson, M., and Lajunen, M. (2011) Water-soluble carbon nanotubes through sugar azide functionalization. *Carbon*, **49** (4), 1299–1304.

42 Bogaertsa, A., Neytsa, E., Gijbelsa, R., and van der Mullenb, J. (2002) Gas discharge plasmas and their applications. *Spectrochim. Acta B*, **57**, 609–658.

43 Khare, B., Wilhite, P., Tran, B., Teixeira, E., Fresquez, K., Mvondo, D.N., Bauschlicher, C., and Meyyappan, M.J. (2005) Functionalization of carbon nanotubes via nitrogen glow discharge. *J. Phys. Chem. B*, **109** (49), 23466–23472.

44 Chen, C., Liang, B., Ogino, A., Wang, X., and Nagatsu, M. (2009) Oxygen functionalization of multiwall carbon nanotubes by microwave-excited surface-wave plasma treatment. *J. Phys. Chem. C*, **113** (8), 7659–7665.

45 Wu, Z., Xu, Y., Zhang, X., Shen, G., and Yu, R. (2007) Microwave plasma treated carbon nanotubes and their electrochemical biosensing application. *Talanta*, **72** (4), 1336–1341.

46 Ruelle, B., Felten, A., Ghijsen, J., Drube, W., Johnson, R.L., Liang, D., Erni, R., Van Tendeloo, G., Peeterbroeck, S., Dubois, P., Godfroid, T., Hecq, M., and Bittencourt, C. (2009) Functionalization of MWCNTs with atomic nitrogen. *Micron*, **40**, 85–88.

47 Kalita, G., Adhikari, S., Aryal, H.R., Ghimre, D.C., Afre, R., Soga, T., Sharon, M., and Umeno, M. (2009) Functionalization of multi-walled carbon nanotubes (MWCNTs) with nitrogen plasma for photovoltaic device application. *Curr. Appl. Phys.*, **9** (2), 346–351.

48 Chen, C., Liang, B., Lu, D., Ogino, A., Wang, X., and Nagatsu, M. (2010) Amino group introduction onto multiwall carbon nanotubes by NH_3/Ar plasma treatment. *Carbon*, **48** (4), 939–948.

Index

Page numbers in italics refer to figures and tables.

1-dodecanethiol   13–14
1,3-dipolar cycloaddition reactions   317
2-[4-(2-hydroxyethyl)-1-piperazinyl] ethanesulfonic acid (HEPES)   157
2,7-dihydroxynaphthalene (2,7-DHN)   154, 155
4-methoxyphenylhydrazine hydrochloride   317

A

acorn structures   150
aerosol technology   281
aggregation ratios   14
alkanediols   190
alloy nanoparticles   17, 165, 168, 169
alternating electric fields   41
alumina (Al_2O_3)   43
alumina (Al_2O_3) nanoparticles   72
aluminum phosphate ($AlPO_4$)   253
amidation reactions   317
amino acids   30
antibacterial technology   154, 155
aqueous sol–gel process   186, 212
aspect-ratio tuning   116–118
atomic species   39

B

bacterial infections   28, 29
barium titanate ($BaTiO_3$) layers   195, 196
barium titanate ($BaTiO_3$) nanoparticles   100
batch synthesis   249
benzyl alcohol   113, 191–197, 199–201
bimetallic nanoparticles (see also alloy nanoparticles, core shell nanoparticles)   119, 164, 172
biofilm formation   152

biomedical nanoparticles   214
biotin–streptavidin interaction/linkage   219, 222
bismuth dioxide (BiO_2) nanoparticles   100
bismuth(III) selenide (Bi_2Se_3) nanoparticles   101
bisphosphonates   218, 220
bottom-up (build-up) method   8, 9, 107, 108
bovine serum albumin (BSA)   152
breakdown/build-up method (see top-down/bottom-up method)
Burkholderia cepacia   29
butyl mercaptan   201

C

cadmium hydroxide ($Cd(OH)_2$) nanoparticles   100
cadmium selenide (CdSe) nanocrystals   129
cadmium selenide (CdSe) nanoparticles   100
cadmium sulfide (CdS) nanocrystallites   127
cadmium sulfide (CdS) nanoparticles   58–61, 100, 101
cadmium telluride (CdTe) nanocrystals   129
cadmium telluride (CdTe) nanoparticles   100
cadmium zinc selenide (CdZnSe) nanoparticles   102
cadmium-telluride–cadmium-sulfide (CdTe–CdS) core–shell nanocrystals   127
cancer cells   22, 23
carbodiimide chemistry   220
carbon nanotubes (CNTs)   173, 174, 311–321
carbon-coated nanomaterials   167, 168
carbon-supported metal catalysts   171

catalytic properties, nanoparticles   19–21
centrifugal separation   15
ceramic–ceramic core–shell
    nanoparticles   280
ceramics   31
cerium(IV) oxide (CeO_2) nanoparticles   100
characterization, nanoparticles   278, 279
chelating agents   313
chemical vapor deposition (CVD)
    method   9
chemically converted graphene (CCG)
    sheets   173
chemisorption   217
chromium (Cr) nanoparticles   99, 158, 160
citric acid   18
click chemistry   219, 221, 222
cobalt ferrite ($CoFe_2O_4$) nanocrystals   68
cobalt ferrite ($CoFe_2O_4$) nanoparticles   *101*
cobalt (Co) nanoparticles
– applications   159, 164
– synthesis   99, 158, 160, 164
cobalt–nickel (Co–Ni) alloys   175
colloidal nanocrystal clusters
    (CNCs)   111–113
colloidal semiconductor nanocrystals   125, 126
communications   25
conduction heating   43, 45
continuous flow methods   90–96, 249–268
conventional heating methods   31, 49, 62, 63, 64, 67, 68, 71, 86–90, 108, 147
cooling chamber   57
cooling conditions   56, 84–87
copper (Cu) alloy nanoparticles   169
copper (Cu) nanoparticles
– application to electronics   22, 158
– synthesis   98, 158, 159
copper–copper(I)-oxide (Cu–Cu_2O)
    core–shell nanoparticles   165
copper–silver (Cu–Ag) core–shell
    nanoparticles   165
copper(I) oxide (Cu_2O) nanoparticles   99
copper(II) oxide (CuO) nanofluids   68
copper(II) oxide (CuO) nanoparticles   99
copper(II) sulfide (CuS) nanoparticles   99
copper(II) sulfide (CuS) nanotubes   70
coprecipitation method   211
core-alloy-shell nanoparticles   167
core–shell nanoparticles   17–18, 150, *151*, 164–168, 280, 281
covalent linkages   219
cryptomelane-type manganese oxide
    (K-OMS-2)   257, 258, *259–261*, 262, 263

cycloaddition reactions   317–319
cytotoxicity   153

D

D-glucose   154
diamond films   69
dielectric constants   42, 43, 44, 248, *249*
dielectric heating   41, 43, 45, 247–249
dielectric loss factor   42, 43, 44, 47, 48, 248
dielectric materials   39, 40
Diels–Alder (DA) cycloaddition
    reaction   318
digestive ripening method   15
dimethyl formamide   201
dimethyl sulfoxide (DMSO)   43, 258, 263–265
dipolar polarization   248, 312
dipole interactions   109
direct functionalization   217–219
direct micelle microemulsion method   212, 224, 225
dispersing agents   13
domestic microwave systems   109, 110, 250
dopants/doping   192–195
double-walled carbon nanotubes
    (DWCNTs)   318
drug targeting   210
dry grinding   8
dynamical theory of the grating   10

E

electrical properties, nanoparticles   19
electrodeless lamps   32, 33
electromagnetic waves   39, *40*
electron spin resonance (ESR)
    spectroscopy   26
electronic applications, nanoparticles   22
electrostatic repulsion   217, 218
elongated nanoparticles   21
energy transmission   25
ethanol   155, 197
ethylene glycol   19, 43, 155, 200
ethylenediaminetetraacetic acid
    (EDTA)   313
europium (Eu) nanoparticles   235
europium(III) oxide (Eu_2O_3)
    nanoparticles   235

F

Faraday, Michael   3
ferromagnetic nanoparticles   294–296
Fischer–Tropsch reactions   159
flower-like mixed-phase titania   196
fluorophores   232

Index | 327

food defrosting/heating   76
forced hydrolysis   211, 212
frequency selection   43, 46, 47, 76–81
– continuous flow methods   250
– plasma methods   283, 284
fuel cells   174
functionalization (*see also* surface functionalization)   316–321

G

gas-phase reactions   272
gas-sensing nanoparticles   297, 298
gel electrophoresis   15
gel filtration   15
gold (Au) clusters   21
gold (Au) films   20
gold (Au) hydrosols   18
gold (Au) nanoparticles
– applications
– – chemistry   21, 171
– – electronics   22
– – medicine   22, 152
– – paints   20
– color   1, 11, 12
– functionalization   230, 231
– history   3
– synthesis   75, 77–79, 86, 97, 98, 152–154, 162
– – continuous flow methods   267
– – plasma methods   96
gold (Au) nanorods
– applications   22, 23, 160
– synthesis   15–17
gold (Au) nanospheres   12
gold (Au) nanotubes   18
gold (Au) nanowires   161
gold/CNT hybrid nanostructures   64, 65
gold–gold-palladium (Au–Au-Pd) core–alloy-shell nanoparticles   168
gold–palladium (Au–Pd) alloys   168
gold–palladium (Au–Pd) core–shell nanoparticles   166, 167
gold–platinum (Au–Pt) core–shell nanoparticles   166
gold–silver (Au–Ag) alloy nanoparticles   169
gold–silver (Au–Ag) core–shell nanoparticles   17, 119, 166
graphene   132, 172, 173
graphene nanosheet–gold (GNS-Au) nanocomposites   69
graphene nanosheets   65, 133, 134
graphite oxide (GO)   133, 134, 173
green chemistry   84, 145, 146

H

hermatite (α-Fe_2O_3) nanostructures   114–118, 224–256
hexadecyltrimethylammonium bromide (CTAB)   16, 154, 162
high quality nanoparticles   272
hot spots   56, 118
hot surfaces   118
human serum albumin (HSA)   152
hybrid core–shell nanoparticles   281
hydrophobic nanoparticles   215
hydrothermal method   84, 85, 212
hydroxide ions   43
hydroxyapatite (HAP)   64
hyperthermia   210

I

icosahedrons   21
immunoassays   208, 209
in situ nozzle with conventional heating (INC) method   67
in situ nozzle with microwave heating (INM) method   65–67, 251, 252, *253*, 255, 266
indium trioxide (InO_3)   69
infrared radiation   42
ink-jet method   22
International Telecommunication Union (ITU)   47
ionic conduction   109, 248
ionic liquids   198, 199
ionic surfactants   158
iridium (Ir) nanoparticles   99, 158, 160
iron (Fe) nanoparticles   99
iron–palladium (Fe–Pd) alloy nanoparticles   169
iron–platinum (Fe–Pt) alloy nanoparticles   169
iron(III) oxide (Fe_2O_3) nanocrystals
– applications   207
– synthesis   198
iron(III) oxide (Fe_2O_3) nanocubes   229
iron(III) oxide (Fe_2O_3) nanoparticles (*see also* superparamagnetic iron oxide nanoparticles)
– functionalization   221, 231–234
– synthesis (conventional heating)   *211*, 224
– synthesis (microwave heating)   99, *100*, *211*, 224–227
– – plasma methods   295
iron(III) oxide (Fe_2O_3) nanorods   226

iron(III) oxide (Fe$_2$O$_3$) nanospheres   228
iron(III) oxide (Fe$_2$O$_3$) nanowires   226
ISM (industrial, scientific and medical)   47, 48, 76

J
Johnson Space Center   28

K
Karlsruhe micro-nose (KAMINA)   297, 298
Karlsruhe microwave plasma process (KMPP)   280

L
L-arginine   154
L-lysine   154
lead chalcogenides (PbS, PbSe, PbTe)   129–131
lead hydroxide bromide (Pb(OH)Br) nanowires
– synthesis (conventional heating)   62, 63
– synthesis (ultrasound/microwave irradiation)   61–63
lead selenide (PbSe) nanoparticles   *101*, 130, 131
lead sulfide (PbS) nanoparticles   *101*, 129
lead titanate (PbTiO$_3$) nanoparticles   100
lead titanate (PbTiO$_3$) nanopowders   68
ligands   215, 216, 219
liquid phase methods   9
liquid plasma methods (*see* plasma synthesis methods)
lithium ion (Li-ion) batteries   200, 287, 299–300
lithography   107
localized surface plasmon resonance (LSPR)   11, 12
loss angles/tangent δ   43, *249*
Lycurgus cup   2, 3

M
maghemite (γ-Fe$_2$O$_3$) nanocrystals   231–233
maghemite (γ-Fe$_2$O$_3$) nanoparticles   210, 224, 229
maghemite (γ-Fe$_2$O$_3$) nanorods   226
magnetic field   34, 35, 36, 45, *53*
magnetic heating   45, 81–83
magnetic nanoparticles   150, 159, 163
magnetic properties, nanoparticles   18–20
magnetic resonance imaging (MRI)   209
magnetite (Fe$_3$O$_4$) nanoparticles   99, 192
manganese ferrite (MnFe$_2$O$_4$) nanoparticles   *101*

manganese (Mn) nanoparticles
– applications   159
– synthesis   99, 158, 160
manganese zinc ferrite (MnZnFe$_2$O$_4$) nanoparticles   *102*
manganese(IV) oxide (MnO$_2$) nanoparticles   262, 263, *264, 265*
mass production, nanoparticles   178, 179
mechanical properties, nanoparticles   19
medical applications, nanoparticles   22, 23
metal catalysts   20, 21
metal nanoparticles   1, 152, 163
metal–metal-oxide core–shell nanoparticles   164, 165
metal-decorated graphene   173
metal-oxide nanoparticles   197
metal-oxide nanotube arrays   170, 171
metal-oxide-supported nanoparticles   170, 171
methanol   155
micelle synthesis (*see* direct micelle microemulsion method)
microbial contamination   152
microemulsion   212
microfluidic reactors   92, 93
microwave chemistry   25, 26, 28, 108, 109
– equipment   33–36, 109, 110
microwave discharge electrodeless lamps (MDELs)   32, 33
microwave drying   68
microwave extraction   32
microwave heating   25
– advantages   55, 76, 108, 150, 185
– applications   26–29
– disadvantages   56
– efficiency   36, 42, 43, 50
– limitations   42
– performance optimization   52, 53
– types of heating   45
microwave plasma chemical vapor deposition (MPCVD)   69, 70
microwave plasmas   274–278
microwave power, plasma systems   291, 292
microwave radiation
– applications   25, 26, *27*
– history   25
microwave-assisted coupling   232
microwave-assisted hydrothermal (MW-HT) method   223–227
microwave-assisted hydrothermal reduction/carbonization (MAHRC)   121
microwave-assisted organic synthesis (MAOS)   222

microwave-assisted solvothermal method (MW-ST)   70, 71, 223, 227–229
microwave-transparent materials   56
mixed metal oxides   266
molecular assemblies   40
molecular sieve nanomaterials   253–259
molecular species   39, 40
molybdenum (Mo) nanoparticles   99, 158, 160
molybdenum(VI) selenide (MoSe$_3$) nanoparticles   101
monodisperse colloidal nanoparticles   90
monomers   186
multilayer ceramics capacitors (MLCC)   178, *179*
multimode microwave apparatus   33, 34, 35
multiwall carbon nanotubes (MWCNTs)   174–176, 314, 318

N

nanoalloys   168
nanomaterials   2
nanomedicine   208
nanometallic inks   95
nanoparticles
– applications
– – chemistry   20, 21
– – electronics   22
– – medicine   22, 23
– – paints   20
– color   11, 12, 20
– definition   1, *2*
– history   1–7
– properties   1, 19, 20
nanotechnology   1
National Nanotechnology Initiative (NNI)   1
nickel ferrite (NiFe$_2$O$_4$) nanoparticles   *101*
nickel ferrite (NiFe$_2$O$_4$) spinel nanoparticles   65, *66*, 67
nickel (Ni) nanocrystals   163
nickel (Ni) nanoparticles
– applications   159, 161, 162
– synthesis   98, 147, 159, 178
nickel (Ni) nanorods   162
nickel/carbon supported nanoparticles   171
nickel/silver (Ni/Ag) composite nanoparticles   83
nickel–cobalt (Ni–Co) core–shell nanoparticles   167
nickel–nickel-oxide (Ni–NiO) core–shell nanoparticles   165
nickel-coated carbon nanofibers   64

noble metals   157
non-aqueous sol–gel process   186–189, 213
non-covalent interactions   222
non-equilibrium local heating   150, *151*
non-thermal microwave effects   191
nozzle microwave (NMW) method   251, *253*, 266
nucleation   278
nucleation–aggregation–dissolution–recrystallization   118
nucleation–aggregation–recrystallization   118

O

olefin hydrogenation   157
oleic acid   216, *217*
one-dimensional nanostructures   160
optical applications, nanoparticles   13
optical properties, nanoparticles   1, 19, 20
organic solvents   190, 192, 199
organic synthesis   29, 30, 75, 76
organometallic precursors   285, 286
osmium (Os) nanoparticles   99, 158, 160
oxidation, CNTs   314

P

palladium (Pd) nanoparticles
– applications   157, 171, 177
– color   1
– morphology   163
– synthesis   75, *98*, 157, 158, 162
palladium–copper (Pd–Cu) alloy nanoparticles   176, 177
palladium–gold (Pd–Au) core–shell nanoparticles   167
particle mass spectrometry (PMS)   278
penetration depths   45, 46, 79, 249, 250
permittivity   42
phosphorus   18
photocatalysis   170
physical adsorption   217
physical vapor deposition (PVD) method   9
plasma jet reactors   252, 253, *254*, 265, 266
plasma synthesis methods   96, 271, 272, 279–300
plasma treatments   320, 321
plasmas   96, 272–278
platinum (Pt) nanoparticles
– applications   156, 161, 171
– color   1

– synthesis (microwave heating) 75, *98*, 156–158, 162
– – continuous flow methods 267, 268
– – plasma methods 96
– – rapid deposition method 170
– synthesis (ultrasound/microwave irradiation) 58
platinum/carbon catalysts 68
platinum/carbon-nanotube (Pt/CNT) supported nanoparticles 174
platinum/cerium(IV)-oxide (Pt/CeO$_2$) supported nanoparticles 170
platinum/graphene hybrids 134, *135*
platinum/H-ZSM5 catalysts 71, 72
platinum/multiwall-carbon-nanotube (Pt/MWCNT) supported nanoparticles 175
platinum/titanium-dioxide (Pt/TiO$_2$) supported nanoparticles 170, 171
platinum–cobalt (Pt–Co) alloy nanoparticles 172
platinum–gold (Pt–Au) core–shell nanoparticles 167
platinum–palladium (Pt–Pd) core–shell nanoparticles 18
platinum–ruthenium (Pt–Ru) alloy nanoparticles 172, 176, 177
poly(ethylene oxide)-poly(propylene oxide)-poly(ethylene oxide) (PEO-PPO-PEO) 162
poly(N-vinylpyrrolidone) (PVP) 153–157
polyacrylamide–metal nanocomposites 155
polyalcohols 189
polyethylene glycol (PEG) 316, 317
polymer–metal nanocomposites 156
polymers
– synthesis 30, 31
– use as surfactants 153, 155–157
polyol method 19, 82, 119, 153, 156, 158, 164, 189–191, 199, 200, 213, 214
– continuous flow systems 267
polypeptides 30
polystyrene micro/nano-particles 229
polyvinyl alcohol (PVA) 153
precipitation separation 14, 15
precursors
– concentration 287, 288
– selection 284–287
purification, CNTs 313, 314
Purple of Cassius 3

R
Raman spectra 147–149
rapid deposition method 170
red grape pomace 153, 156, 157

reducing agents 18, 19
residence times, plasma synthesis methods 288–293, 296
reverse micelles microemulsion method 213
rhenium (Re) nanoparticles 99, 158
Rhodamine 6G 147
Rhodamine B 233, 234
rhodium (Rh) nanoparticles
– applications 157
– synthesis 99, 158, 160
room temperature ionic liquids (RTILs) 122, 123
ruthenium (Ru) nanoparticles
– applications 157, 171
– synthesis 99, 158, 160
ruthenium/carbon-nanotube (Ru/CNT) supported nanoparticles 174

S
scanning electron microscopy (SEM) 279
scientific microwave systems 110
sedimentation method 9, 10
seeded growth processes 196
selective heating 50–52, 196
selenium dandelions 124
selenium/carbon (Se/C) colloids 123
selenium–carbon (Se–C) composites 125
sequential ultrasound/microwave irradiation 56–58, 63–72
shape-controlled synthesis 15–17, 75, 116
silane (SiH$_4$) 277
silica nanoparticles 229, 259, 261
silica shells 217, 234, 235
silica-coated gold nanoparticles 230
silicon dioxide (SiO$_2$) nanoparticles 5
silver (Ag) alloy nanoparticles 169
silver (Ag) nanoclusters 121
silver (Ag) nanoparticles
– applications 154, 163
– color 1, 12
– history 13
– synthesis (conventional heating) 86–90
– synthesis (microwave heating) 75, 79–90, 97, 147, *148*, 154–156, 163
– – continuous flow methods 92, 93, 156, 267
– – plasma methods 96
silver nanorods
– applications 160
– synthesis 160, 161
silver (Ag) nanowires 18
silver/carbon (Ag/C) nanocables 121, 122

silver–copper (Ag–Cu) core–shell nanoparticles 150, *151*, 165
silver–gold (Ag–Au) core–shell nanoparticles 166
silver–nickel (Ag–Ni) core–shell nanoparticles 165
silver–silicon-dioxide (Ag–SiO$_2$) core–shell nanoparticles 94, 171, 262
simultaneous (US/MW) irradiation 56, *57*, 58–63
single-mode microwave apparatus 33, 35, 36
single-wall carbon nanotubes (SWCNTs) 311, 314–318
size-controlled synthesis 12–15
sol–gel processes 186, 189
sonochemistry 56
specific microwave effect 126, 127
stained glass 3, 4
structure-controlled synthesis 17, 18
sugar moieties 319, 320
sulfonated multi-walled carbon nanotubes (SF-MWCNTs) 63, 175
superheating 49, 50, *51*
superparamagnetic iron oxide nanoparticles (SPIONs) 208, 209
superparamagnetic nanoplatforms 232, 233
supported nanoparticles 196
surface functionalization 207, 208, 214–222, 229–235
surface plasmon resonance (SPR) 10–12, 15, 21
surface-enhanced Raman scattering (SERS) 147, *148–150*, 154, 160, 163
surfactants 188, 189, 191
Suzuki–Miyaura cross coupling reaction 29

T
tails interdigitation 216, 117
tellurium (Te) 122
tellurium (Te) nanoparticles *98*, 122, 123
tellurium (Te) nanowires 123
temperature, plasma synthesis methods 288, 289
temperature distributions, microwave heating 47–49
tetradecylammonium bromide (TOAB) 161
tetraethylorthosilicate (TEOS) 235, 259
tetramethylammonium hydroxide (TMAOH) 235
thallium(III) oxide (Tl$_2$O$_3$) nanoparticles *100*
thermal plasmas 273

thermal properties, nanoparticles 19
thermolysis 214
thermophoretic deposition 281
tin (Sn) nanoparticles 299
tin dioxide (SnO$_2$) 297
tin dioxide (SnO$_2$) nanoparticles *100*, 299, 300
tin-dioxide–silicon-dioxide core–shell nanoparticles 297, 298
titanium dioxide (TiO$_2$) nanoparticles
– history 5
– synthesis *99*, 197, 198
top-down (breakdown) method 8, 9, 107
transmission electron microscopy (TEM) 278
transparent conducting oxides (TCOs) 193–195
transverse electrical (TE) mode 279
tungsten (W) nanoparticles *99*, 158, 160
tungsten trioxide (WO$_3$) nanoparticles *101*, 197
twinned shell structure 17

U
ultrasound (US) irradiation 56
ultrasound/microwave (US/MW) irradiation 55, 56

V
vulcanization 30, *31*

W
water
– dielectric constant 43
– molecular structure 42
– use as a solvent 84
water-soluble nanoparticles 214, *216*
waveguides 40–41
wet grinding 8–9
wet-chemical methods 170, 194, 272
wood drying 76

X
X-ray diffraction (XRD) 278

Y
yttria-stabilized zirconia (YSZ) nanocrystals 65

Z
zeolites 254, 255–7
zinc aluminate (ZnAl$_2$O$_4$) spinel nanoparticles 65
zinc oxide (ZnO) 110

zinc oxide (ZnO) nanocrystals 110–113
zinc oxide (ZnO) nanoparticles/
  nanostructures
  – applications 111
  – doping 193
  – inhomogeneous deposition 195
  – synthesis (microwave heating) *101*, 113,
    114, 189, 190, 197, 201
  – synthesis (ultrasound/microwave
    irradiation) 58

zinc phosphate ($Zn_3(PO_4)_2$)
  nanocrystals 63, 64
zinc selenide (ZnSe) nanoparticles *101*
zinc selenide (ZnSe) quantum dots 131,
  132
zinc sulfide (ZnS) nanoparticles *101*, 131
zinc tunstate ($ZnWO_4$) nanoparticles *101*
zirconium oxide (ZrO_2) nanoparticles *101*,
  259, *261*
ZSM-5 molecular sieve 254